PREFRONTAL CORTEX:
From Synaptic Plasticity to Cognition

PREFRONTAL CORTEX:
From Synaptic Plasticity to Cognition

edited by

Satoru Otani
Universite de Paris VI, Paris, France

KLUWER ACADEMIC PUBLISHERS
Boston / Dordrecht / London

Distributors for North, Central and South America:
Kluwer Academic Publishers
101 Philip Drive
Assinippi Park
Norwell, Massachusetts 02061 USA
Telephone (781) 871-6600
Fax (781) 681-9045
E-Mail: kluwer@wkap.com

Distributors for all other countries:
Kluwer Academic Publishers Group
Post Office Box 322
3300 AH Dordrecht, THE NETHERLANDS
Telephone 31 786 576 000
Fax 31 786 576 254
E-Mail: services@wkap.nl

Electronic Services <http://www.wkap.nl>

Library of Congress Cataloging-in-Publication Data

A C.I.P. Catalogue record for this book is available
from the Library of Congress.

Prefrontal Cortex: From Synaptic Plasticity to Cognition
ISBN 1-4020-7766-1

Printed on acid-free paper.

Printed in the United States of America.

Table of Contents

Chapters

Contributors

Claudio BABILONI
IRCCS, Brescia, Italy
Dipartimento di Fisiologia Umana e Farmacologia, Università La Sapienza, Rome, Italy

Joseph E. BATES
Department of Psychology, Yale University, New Haven, CT, USA

Stefano CAPPA
Centro di Neuroscienze Cognitive, Università Salute-Vita S. Raffaele, Milan, Italy

Jan CIOE
Okanagan University College, Lethbridge, AB, Canada

Roshan COOLS
Department of Experimental Psychology, University of Cambridge, Cambridge, UK

Jan P. C. de BRUIN
Netherlands Institute for Brain Research, Amsterdam, The Netherlands

Matthijs G. P. FEENSTRA
Netherlands Institute for Brain Research, Amsterdam, The Netherlands

Shintaro FUNAHASHI
Department of Cognitive and Behavioral Sciences, Graduate School of Human and Environment Studies, Kyoto University, Kyoto, Japan

René GARCIA
Neurobiologie Comportementale, Université de Nice-Sophia Antipolis, Nice, France

Yukiori GOTO
Department of Neuroscience, University of Pittsburg, Pittsburg, PA, USA

Hirac GURDEN
Neurobiologie de l'Apprentissage et de la Mémoire, Université Paris XI, Orsay, France

Cyril HERRY
Neurosciences Cognitives, Université de Bordeaux I, Talence, France

Maïté HOTTE
Neurobiologie de l'Apprentissage et de la Mémoire, Université Paris XI, Orsay, France

Thérèse M. JAY
Physiopathologie des Maladies Psychiatriques, INSERM EMI 0117, Paris, France

Andreas KEIL
Department of Psychology, University of Konstanz, Konstanz, Germany

Bryan KOLB
University of Lethbridge, Lethbridge, AB, Canada

Bogdan KOLOMIETS
Neurobiologie des Processus Adaptatifs, Université Paris VI, Paris, France

Barbara L. LEWIS
Center for Neuropharmacology and Neuroscience, Albany Medical College, Albany, NY, USA

Christy MARSHUETZ
Department of Psychology, Yale University, New Haven, CT, USA

Carlo MINIUSSI
IRCCS, Brescia, Italy

Kazushige MIZOGUCHI
Pharmacology Department, Central Research Laboratories, Tsumura and Company, Ibaraki, Japan

Patricio O'DONNELL
Center for Neuropharmacology and Neuroscience, Albany Medical College, Albany, NY, USA

Satoru OTANI
Neurobiologie des Processus Adaptatifs, Université Paris VI, Paris, France

Angela C. ROBERTS
Department of Anatomy, University of Cambridge, Cambridge, UK

Cyril ROCHER
Neurobiologie de l'Apprentissage et de la Mémoire, Université Paris XI, Orsay, France

Simone ROSSI
Dipartimento di Neuroscienze, Sezione Neurologia, Università di Siena, Siena, Italy

Paolo Maria ROSSINI
IRCCS, Brescia, Italy
Neurologia, Università Campus Biomedico, Rome, Italy
AFaR-Dipartimento Neuroscienze, Rome, Italy

Jeremy K. SEAMANS
Department of Physiology, MUSC, Charleston, SC, USA

Michael SPEDDING
Neurobiologie de l'Apprentissage et de la Mémoire, Université Paris XI, Orsay, France

Kuei-Yuan TSENG
Center for Neuropharmacology and Neuroscience, Albany Medical College, Albany, NY, USA

Preface

This volume, *Prefrontal Cortex: from Synaptic Plasticity to Cognition*, is an interdisciplinary approach to characterize the function of the anterior portion of the frontal lobe in rodents and human and non-human primates. The specific topics discussed in the chapters of this volume are purposefully diverse: they range from membrane properties of prefrontal neurons to cognitive psychology. Nevertheless, this volume must not be regarded as a mere collection of writings with the different sub-themes. As you will see, chapters often vigorously encompass domains of the prefrontal field in effort to provide a big picture. That is actually what we attempted to do in this volume.

On one hand, we have accumulated knowledge on the properties of neurons and synapses in the prefrontal cortex as well as the actions of critical neuromodulators such as dopamine. On the other hand, behavioral and cognitive neurosciences have begun to reveal the fascinating role of the prefrontal cortex in such mental processes as working memory, attention switching and rule following, and long-term memory. Needless to say, our ultimate goal as neurobiologists is to know what relationship there is between these cellular and cognitive processes. This volume is meant to serve as a comprehensive introduction towards that goal. Readers will be informed, for example, of how plasticity of prefrontal neurons is regulated, how it is involved in certain cognitive processes in rodents, and how the rodent models can apply to the primates. Equally, the prefrontal cortex-dependent cognitive processes in human and non-human primates are themselves analyzed in detail, which will invite the readers to refer to the underlying cellular processes.

The prefrontal cortex is a most important brain region to study with a multidisciplinary attitude. It is regarded by many as the highest-order executive controller, which determines an appropriate coupling between a sensory input and a motor output to meet environmental demands. It is obvious that our cognitive ability heavily relies on the function of the prefrontal cortex. By analyzing the behavior of prefrontal neurons and synapses as well as modulatory inputs, and by relating them to the high-order cognitive processes, we may be able to pave the way for understanding mechanistic properties of our cognition. In the near future, we hope that our knowledge will be placed in a broader context of the neuroscience, and more details on the interactions between prefrontal cortex and the anatomically remote brain areas such as the thalamus, hippocampus, amygdala, and striatum will be analyzed.

When this volume was in the final stage of the editorial process, in the beginning of August, we were struck by the news that the leading prefrontal scientist Patricia Goldman-Rakic tragically died from a car accident. Her contribution to the field, particularly the cellular basis of working memory, was enormous. Although the detailed account on her contribution is beyond the scope of this Preface (see the tribute by Shintaro Funahashi in the following pages), we would like to dedicate this volume to the achievement of Patricia Goldman-Rakic.

Satoru Otani
University of Paris VI

September 2003, Paris

In Memoriam
Professor Patricia S. Goldman-Rakic (1937-2003)

My mentor and friend, Patricia S. Goldman-Rakic, Professor of Neuroscience at Yale University School of Medicine, died on July 31, 2003. She was a world-renowned neuroscientist specializing in the study of the functions of the prefrontal cortex, the most important cortical structure for understanding human beings. Since 1979, she had been a professor at Yale University School of Medicine, where Professor John Fulton and Dr. Carlyle Jacobsen first started experimental studies on the prefrontal cortex using primates, and found in 1930s that the bilateral lesion of the prefrontal cortex impairs delayed-response performances. Until the early 1960s, Yale University School of Medicine was a world center for the prefrontal research. When Pat was invited to the Fulton Lecture held at Yale School of Medicine in the mid 80s, she told the audience about the legacy of Prof. John Fulton and expressed her hope that she would once again make the Section of Neuroanatomy (now Department of Neurobiology) a world center for prefrontal research. As she had hoped, many today certainly regard the Department as a world center.

Among her many contributions to neuroscience and translational research, the particularly important one is the introduction of the concept "working memory" to understand prefrontal cortical functions. Although her concept of working memory was somewhat different from the model of working memory proposed by Baddeley and others, her idea triggered a number of imaging studies in 1990s and the current flourish of prefrontal researches in humans as well as in animals. She also focused on translational research, especially the neurobiological basis of schizophrenia, and achieved significant advances in the understanding of the cause of this disease. Thus, she achieved great contributions to both basic and translational researches of the prefrontal cortex. Her death is a great loss for the neuroscience world.

Personally, I was a member of her research group from November 1983 until August 1990. When I joined her at Yale, she already had the biggest prefrontal research group in the world. She energetically organized a variety of research projects including anatomical, psychological, pharmacological, developmental, and physiological studies. These projects were all directed toward the understanding of prefrontal cortex function. In 1983, I was the only neurophysiologist in her group, but two other neurophysiologists (Fraser Wilson and Jeff Moran) joined soon afterwards, and the group continued to grow.

Pat was always kind to me, encouraged me all the time, and was very patient. She taught me a lot of things, from basic animal handling and surgical skills, to how to choose a "sexy" title for posters. Her surgical skill was excellent. I followed her surgical skill when we made lesions in the prefrontal cortex. She visited my laboratory to watch neurophysiological recordings occasionally, but she always participated in surgery, especially when we determined the position of the chamber for single-neuron recording. She was patient. She spent a lot of time with me at her office or in my laboratory to discuss on the results, even though our discussion was interrupted often by telephone calls. She quietly waited for our first publication for almost 6 years. I still remember her cheerful face when our first paper was published in Journal of Neurophysiology in 1989. When I obtained a faculty position at Kyoto University, in August 1990, I left Yale. However, I think that the period at Yale when I worked with Pat was the happiest time in my life.

I sometimes disappointed her, but she never disappointed me. Last year (2002) she was a recipient of Ralph W. Gerald Prize in Neuroscience from Society for Neuroscience. She sent me an invitation card for the reception of the prize and asked me to attend it. But she did not tell me who the prize recipient was. I skipped the reception, and the next day, I met her and found out that she and Pasko were the recipients! She said that she was disappointed when she did not see me at the reception. I had to apologize to her for my fault. However, during the rest of the meeting, I saw Pat everyday, had time to discuss with her the prefrontal research of ourselves and others, and even had a chance to have coffee together at a restaurant. She enthusiastically explained to me her recent results regarding schizophrenia and the many difficulties in conducting schizophrenia research. I invited Pat to visit Japan for the symposium on prefrontal functions that was to be held at Japan Neuroscience Society Meeting in July 2003. She agreed to do so. However, because she was required to attend other meetings in Europe, she could not come to Japan. A few days after the symposium finished, I received the unexpected message from a member of Pat's group that she had had a car accident and was in serious condition in Yale New Haven Hospital. We all hoped that Pat would soon recover. I did not expect that saying "see you soon" to her in front of the elevator in Double Tree Castle Hotel with Graham Williams on November 6 2002, would be the last chance to talk to her in my life. I regret her early death.

Shintaro Funahashi
Kyoto University

October 2003, Kyoto

Chapter 1

ORGANIZATION AND PLASTICITY OF THE PREFRONTAL CORTEX OF THE RAT

Bryan Kolb[1] and Jan Cioe[2]
[1]University of Lethbridge and [2]Okanagan University College, Lethbridge, AB, Canada

Keywords: Class-common behavior, executive control, neuromodulation, experience-dependent plasticity, hormones, cortical injury.

Abstract: The rat is probably the most-studied species both in behavioral neuroscience in general as well as in studies of brain plasticity. A discussion of the organization and plasticity of the prefrontal cortex (PFC) of rodents is therefore germane to the general topic of the current volume. Nonetheless, controversy remains over the question of whether the frontal regions of the rodent can legitimately be viewed as relevant models of prefrontal cortical organization in primates (e.g. Preuss, 1995). One problem with the rat is that the behavioral repertoire of rodents would appear to be considerably simpler than that of primates. To the extent that the prefrontal regions of primates are involved in the complex executive functions, it is therefore critical to determine if rodents even have such behavioral processes. A second problem with the rat as a prefrontal model is that the gross organization and cytoarchitecture of the frontal cortical regions of rodents and primates show some marked differences. For example, whereas layer IV of the PFC of primates is distinctly granular in appearance, layer IV is virtually absent in the rat frontal cortex. Consider too, that the volume of the cerebral cortex of a rat is about a hundred times smaller than that of the cerebral cortex of a rhesus monkey, and about a thousand times smaller than that of a human being (Uylings and Van Eden, 1990). A thorough discussion of issues relating to homology and brain organization are beyond the scope of this chapter, but before examining the organization and plasticity of the PFC of rats, it will be necessary to at least superficially consider the question of rodent-primate comparisons. We then review the functional organization of the PFC of the rat before considering the nature of frontal cortical plasticity in rats. As might be anticipated, we shall argue here that the rat is an excellent model for studying frontal functions and plasticity in humans and other primates.

1. INTRODUCTION

One of the major obstacles in comparing the behavior of different species of mammals is that each species has a unique behavioral repertoire that permits the animal to survive in its particular environmental niche. There is, therefore, the danger that neocortical organization is uniquely patterned in different species in a way that reflects the unique behavioral adaptation of those different species. One way to address this problem is to recognize that although the details of behavior may differ somewhat, mammals share many behavioral traits and capacities (e.g. Warren and Kolb, 1978; Kolb and Whishaw, 1983a). For example, all mammals must detect and interpret sensory stimuli, relate this information to past experience, and act appropriately. Similarly, all mammals appear to be capable of learning complex tasks under various schedules of reinforcement (e.g. Warren, 1977). The details and complexity of these behaviors clearly vary, but the general capacities are common to all mammals. Warren and Kolb (1978) proposed that behaviors and behavioral capacities demonstrable in all mammals could be designated as class-common behaviors. In contrast, behaviors that are unique to a species and that have presumably been selected to promote survival in a particular niche are designated as species-typical behaviors. This distinction is important because it has implications for the organization of the cerebral cortex. We note that just because mammals have class-common behaviors does not prove that they have not independently evolved solutions to the class common problems. There is little evidence in support of this notion, however. Neurophysiological, anatomical, and lesion studies reveal a similar topography in the motor, somatosensory, visual, and auditory cortices of the mammals, a topography that provides the basis for class-common neural organization of fundamental capacities of mammals.

Kaas (1987) has argued, for example, that all mammalian species have similar regions devoted to the analysis of basic sensory information (e.g. areas V1, A1, S1), the control of movement (M1), and a frontal region involved in the integration of sensory and motor information. We can extend Kaas's idea by suggesting that these regions have class-common functions. To be sure, there are large species differences in the details of the class-common behaviors. Monkeys (and humans) have chromatic vision compared to the largely achromatic vision of cats or rats. Nevertheless, in all mammalian species studied, removal of visual cortex severely disrupts object recognition. Indeed, although the visual cortex of the rat has often been portrayed as primitive in organization, the visual acuity of rats is surprisingly good and the tuning characteristics of visual neurons is strikingly similar to that of larger-brained mammals. Similarly, rats and cats have a large somatosensory representation of the whiskers whereas monkeys and humans have no such representation, but in all species the somatosensory cortex functions to represent skin-related receptors for tactile sensations. Thus, both the visual and tactile recognition of objects are class-common functions, even though the details of this recognition may vary in a species-typical manner. A similar argument can be made for motor functions. Intracortical stimulation studies have shown that all mammals have a motor map (e.g. Woolsey, 1958) in which the relative motor facility of different body regions is reflected by the size of the motor representation. Curiously, although there are clear interspecies differences in the capacity to use the forelimbs for object manipulation (Iwaniuk and Whishaw, 2000), it has become apparent from the work of Whishaw and his colleagues that the capacity for independent digit manipulation, and the cerebral organization of this control, is strikingly similar between rodents and primates (Whishaw et al., 1992a).

2. WHITHER THE PREFRONTAL CORTEX OF THE RAT?

It is less obvious just what the class-common functions of the frontal cortex might be, but we would anticipate that if the frontal cortex of mammals developed because all mammals face common functional problems, then we should be able to identify class-common functions of the frontal cortex. One place to begin searching for class-common frontal functions is to consider what animals use sensory inputs for. The most obvious function is to guide behavior on line, such as in the visuomotor control of movements in space or the identification of food items using visual, tactile, and olfactory information. But the sensory world has far more information available than the brain can handle at one time so there must be some system to select information as well as to focus and maintain attention. Similarly, although behavior can be directed to sensory stimuli on-line, it can also be related to information that is stored or expected. Stored information may be in a type of scratch-pad memory system, which is often referred to as working memory and implies a short-term erasable storage of information, or by a type of long-term memory system in which information is stored for an extended time. In both instances, the stored information is used to select and generate behavior that is appropriate for the particular context. Behaviors that are generated may be novel and directly related to the sensory events, or they may be preprogrammed behavioral chains that are innate but still must be selected with respect either to ongoing sensory information or to internal states. Thus, there must be some type of master (sometimes referred to as executive) control system that selects behavior. It is our contention that the class-common function of the prefrontal cortex (PFC) is to select and generate behavior patterns. In addition, it is proposed that this system has a working memory subsystem but that it uses a long-term memory store that is largely a function of the medial temporal regions. Although this general view of prefrontal functioning is hardly novel (see reviews by Kolb, 1984; Goldman-Rakic, 1987; Fuster, 1997; Passingham, 1993), it is the idea that a prefrontal system with such functions will be found in all mammals that is the key concept in the current discussion.

2.1 Anatomical Organization

It has been traditional to define the organization of cortical regions by their connectivity with the thalamus. Rose and Woolsey (1948) first noted that all mammalian species had a dorsal medial thalamic nucleus (MD) that uniquely projected to regions of the frontal lobe, and they concluded that the MD-projection field could be considered PFC. In 1972, Leonard (1972) first demonstrated that there were two distinct regions of the frontal cortex of the rat that received projections from discrete portions of MD, a medial prefrontal region (mPFC) and an orbital region (OFC). Later behavioral work led to the conclusion that these regions were functionally dissociable and possibly homologous to the dorsolateral and orbital regions of primates (Kolb, 1984).

One difficulty with this simple story is that with the advent of more sophisticated anatomical tracing techniques, it has become clear that thalamic nuclei are more promiscuous than was previously believed. Thus, it is now known that MD projects beyond the frontal lobe and that other thalamic nuclei also project into the frontal lobe (e.g. Uylings et al., 2003). This turn of events led to questions about the utility of single anatomical criterion for establishing valid cross species comparisons. It is now generally agreed that cross species comparisons can be made by examining the pattern of specific thalamic, cortico-cortical, and corticosubcortical connections, the functional (i.e. electrophysiological and behavioral) properties of subregions, and the presence and specific distribution of different neuroactive substances and neurotransmitter receptors. However, on the basis of these criteria, there are strong grounds for accepting the rat as a good model of prefrontal function in primates (see Uylings et al., 2003 for a detailed review).

The frontal cortex of the rat now can be subdivided into a number of subregions as illustrated in Figure 1. These regions can be grossly grouped into a mPFC region and an OFC region on the basis of thalamo-cortical and cortico-cortical connections. Within the mPFC cortex, it is likely that there

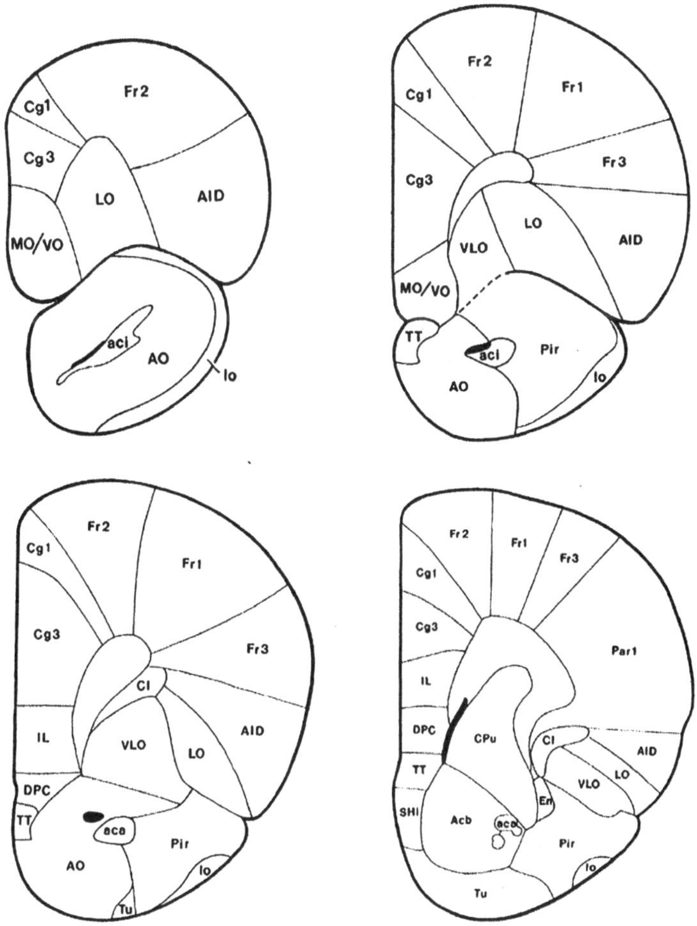

Figure 1: Schematic representation of coronal sections through the PFC of the rat. The medial prefrontal regions (mPFC) include the shoulder cortex (Fr2), dorsal anterior cingulate cortex (Cg 1, 2), prelimbic cortex (Cg 3), infralimbic cortex (IL), and medial orbital cortex (MO). The orbital frontal cortex (OFC) includes the ventral and lateral orbital regions (VO, VLO, LO), and agranular insular cortex (AI). Adopted from Zilles (1985) with permission.

are at least five distinct functional regions: anterior cingulate cortex (Zilles' areas Cgl, Cg2), the prelimbic cortex (Zilles' area Cg 3), infralimbic/prelimbic cortex, the shoulder cortex (Zilles' Fr2), and the medial orbital areas. Similarly, the OFC likely be dissociated into the lateral orbital regions (Zilles' areas LO and VLO) and the insular regions (Zilles' AId and AIv; Zilles, 1985). Although the direct relationship between these subregions and subregions of the primate frontal lobe are unlikely to be easy to determine, we do know that the general pattern of frontal to basal ganglia, hippocampal formation, amygdala, and brainstem projections are strikingly similar in rodents and primates (Groenewegen, 1988). Similarly, there are clear parallels between the pattern of monaminergic and cholinergic projections in rodents and primates as well as general parallels in the effects of lesions in the two orders (see below).

2.2 Cholinergic and Monoaminergic Gating Systems

The PFC of rats and primates plays a role in gating the inputs of the cholinergic and monoaminergic systems to the rest of the cerebral mantle (e.g. Ragozzino, 2000). Thus, although the entire neocortex receives inputs from cholinergic, noradrenergic, and serotinergic systems, only the PFC sends reciprocal connections to the basal forebrain, locus coeruleus, and the dorsal and median raphe (e.g. Uylings and Van Eden, 1990; Arnsten, 1997; Everitt and Robbins, 1997). This feedback system is presumed to modulate these inputs and thus drugs, such as antidepressants, that affect these systems likely have a significant impact upon frontal lobe functioning.

The prefrontal and entorhinal regions of the rat are the primary recipients of dopaminergic inputs from the ventral tegmental area (VTA), and again, the prefrontal regions send reciprocal connections back to the VTA (Kalsbeek et al., 1990). The dopaminergic projections have been the subject of intense study in recent years because of the putative different roles of the different dopamine receptors (e.g. Dl, D2, D5) in behavioral modulation (e.g. Robbins, 2002). It is generally assumed that behavioral syndromes such as schizophrenia and attention deficit disorders are related to abnormalities in one or more of the dopamine receptor subtypes in the PFC (also see Chapters 2, 3, and 7 in this volume).

The cholinergic and monoaminergic inputs are presumed to modulate whatever functions are ongoing in the prefrontal areas. In recent years, there has been an attempt to demonstrate how these inputs contribute to working memory and attention, in particular (e.g. Sagawachi and Goldman-Rakic, 1994; Ragozzino, 2000). In addition, various lines of work suggest that there are dynamic changes in dopamine release in the mPFC when there are changes in the environmental demands on animals, especially under conditions of stress, fear, or other affective stimuli (e.g. Rosenkranz and Grace, 2001; Pezze et al., 2003). It appears that the cholinergic and dopaminergic modulations may have selective effects on different subregions of the mPFC, although the details are still sketchy (see review by Ragozzino, 2000).

2.3 Effects of Lesions to the mPFC

It was demonstrated in the early 1970s that lesions to the mPFC and OFC in rats produced very different behavioral syndromes, and that these behavioral changes were strikingly similar to those observed in primates with lesions to the dorsolateral and OFC regions, respectively (Table 1; for reviews see Kolb, 1984, 1990). For example, damage to the mPFC area produces severe deficits in acquisition and retention of working memory tasks such as delayed response (Kolb et al., 1974), delayed alternation (Wikmark et al., 1973), different types of delayed nonmatching-to-sample tasks (e.g. Dunnett, 1990; Otto and Eichenbaum, 1992; Kolb et al., 1994a), and related tasks (e.g. Kesner and Holbrook, 1987). More recently, deficits have been shown in various types of attentional tasks (e.g. Muir et al., 1996) and in a task requiring a shift of attention from one set of cues to another (Birrel and Brown, 2000). Medial frontal lesions also produce disruptions to the production of various motor and species-typical behaviors that require the ordering of motor sequences, such as in nest building, food hoarding, or latch opening (e.g. Shipley and Kolb, 1977; Kolb and Whishaw, 1983b).

Although these types of experiments were viewed by many as convincing evidence of parallel (and perhaps homologous) functions in rodents and primates, Preuss (1995) remained unconvinced. Indeed, he has argued that given the significant anatomical differences and the failure to find prolonged or long-lasting deficits after mPFC lesions in rodents that are equivalent to those observed in primates with dorsolateral lesions, the research on the mPFC of the rat has little to offer those interested in understanding frontal lobe functioning in primates. Preuss was most certainly wrong on his conclusion that rats with mPFC lesions do not have significant memory deficits (e.g. Kolb et al., 1974, 1994a), but the fact that most studies of mPFC function had made lesions including all of the medial subregions did provide grist for his skepticism. Accordingly, in the past decade, there has been considerable interest in dissociating the different subregions of the rat's mPFC. It has now become clear that the dorsal anterior cingulate region and prelimbic/infralimbic region can be functionally dissociated. In general, it appears that the prelimbic region is involved in attentional and response selection functions as well as visual working memory (e.g. Granon and Poucet, 2000), whereas the more dorsal regions (anterior cingulate) are

Table 1: Summary of the effects of mPFC and OFC lesions in rats

Behavioral Impairment	Basic References
mPFC	
Visual Working memory	Kolb et al., 1974
Strategy formation	Kolb et al., 1994a
Spatial reversal	Divac, 1971
Habituation	Kolb, 1974a,c
Skilled reaching	Whishaw et al., 1992b
Motor sequencing	Kolb and Whishaw, 1983b
Attention	Muir et al., 1996
Attention set shift	Birrel and Brown, 2000
Food hoarding	Kolb, 1974b
Fear Extinction	Quirk et al., 2000
Conditioned emotional responses	Frysztak and Neafsey, 1994
OFC	
Hyperactivity	Kolb, 1974a
Social behaviour	Kolb, 1974d; de Bruin et al., 1990
Incentive association	Gallagher et al., 1999
Odor working memory	Otto and Eichenbaum, 1992
Configural odor learning	Whishaw et al., 1992c
Odor Reversal learning	Saddoris et al., 2001
Feeding	Kolb and Nonneman, 1975

involved with generating rules associated with temporal ordering and motor sequencing of behavior (see reviews by Gisquet-Verrier et al., 2000; Kesner, 2000). Indeed, on the basis of such behavioral studies, Kesner (2000) has gone so far as to suggest that the anterior cingulate region is homologous to Brodmann's areas 6/46 whereas the prelimbic/infralimbic regions are homologus with Broadmann's areas 45 and 47. Additionally, although less is known about its precise role in behavior, it appears that the infralimbic region plays a special role in autonomic control, and especially in the modulation of fear-related behaviors (e.g. Quirk et al., 2000; Morgan et al., 2003).

Kesner's hypothesis will be a difficult one to unequivocally demonstrate to skeptics like Preuss, but it is not necessary for the current argument, which is simply that the mPFC regions have class-common functions that are similar to those of the dorsolateral and possibly medial regions in the monkey frontal lobe. We suggest that these class-common functions include functions that are often referred to as executive functions in primates. These functions would include working memory, the selection of information (often referred to as attention), and the shifting of attention from one stimulus attribute to another (e.g. Brown and Bowman, 2002). Tests purported to measure such functions in rats and primates show deficits following mPFC or dorsolateral frontal lesions in rats and primates, respectively.

2.4 Effects of Lesions to the OFC

There is much more parsimony in reviews comparing the effects of OFC lesions in rodents and primates (e.g. Schoenbaum and Setlow, 2002). The OFC receives significant olfactory and taste input, and although OFC lesions do not produce deficits in olfactory or taste discriminations, they do produce deficits in tasks requiring working memory for odor or taste information (e.g. Otto and Eichenbaum, 1992; DeCoteau et al., 1997; Ragozzino and Kesner, 1999). Furthermore, lesions to the OFC disrupt the learning of cross-modal associations that involve odor or taste cues (e.g. Whishaw et al., 1992c). More recently, studies by Schoenbaum and his colleagues (e.g. Gallagher et al., 1999; Schoenbaum and Setlow, 2002) have emphasized a role of the OFC in the encoding of the acquired incentive value of cues. For example, both rats and primates can show intact performance on discriminations that require responding to neutral cues (such as a light) that predicts reward, while at the same time showing marked deficits when the incentive value of the stimulus is reduced. Such deficits can be seen during extinction when the incentive value of a stimulus is reduced to zero, yet animals continue to respond to the cue as though reward is expected (e.g. Gallagher et al., 1999; Baxter et al., 2000).

The role of the OFC in stimulus-reward associations is further seen in studies measuring the tuning characteristics of neurons in the OFC of both rats and monkeys (see review by Schoenbaum and Setlow, 2002). Finally, damage to the OFC produces deficits in social and play behavior in rats (e.g. Kolb, 1974; de Bruin, 1990).

The overall pattern of deficits related to OFC lesions leads to a general conclusion that there is a class-common function related to making higher order use of olfactory and taste information. This can be seen easily in behaviors that require the association of such information with events in the world, whether they are learned associations such as neutral cues and reward or natural stimuli (such as conspecific odors) or rewards that may be more abstract (such as social bonding). Although odors obviously play a reduced role in the control of social behaviors in humans, the neural networks underlying many social functions remain related to the OFC.

In summary, we argue that all mammals have a PFC and that damage to this region produces a parallel set of deficits in different species. Although the details of anatomical organization are clearly different across different taxa, and certainly between rodents and primates, there are relatively discrete regions across both orders that are involved in higher order cognitive functions (e.g. working memory, directed attention) as well as social and affective behavior and motor programming. As we look for models of prefrontal plasticity, it thus appears that the rat is an excellent model for understanding prefrontal function and plasticity in primates. We now turn our attention to the nature of prefrontal plasticity in rodents.

3. PLASTICITY AND THE PREFRONTAL CORTEX OF THE RAT

In thinking about the relationship between brain and behavior, there is a tendency to focus on constancy, rather than on change, and on similarities, rather than on differences. Thus, as we try to find parallels between the organizations of the frontal regions in different species of mammals, we focus on the constancies and similarities in the organization and function across species. But another way to examine brain-behavior relations is to focus on variability and change in organization. The recognition of the importance of change and variability in brain function has led to the study of the role of environmental events in shaping brain structure and function.

In principle, there are three ways that experience could alter the brain: either by modifying the ontogenetic unfolding of brain structure, by modifying existing brain circuitry, or by creating novel circuitry. It is reasonable to suppose that the environment influences the frontal cortex in all three ways, although it is likely that a particular type of change will vary with the developmental stage of the animal. The goal of this section is to examine the plastic changes in the PFC of rats that occur (or do not occur) in response to a variety of experiential factors (Table 2). Few studies have compared the effect of experience on specific subregions of the frontal cortex as most studies have focused on the motor cortex and the anterior cingulate and insular regions. The emphasis here, therefore, will be on these regions. Further, as we begin to examine the data showing plasticity in the frontal cortex, we will see that the studies to date have led to more questions than answers.

The underlying assumption of studies of brain and behavioral plasticity is that if behavior changes, there must be a change in the neural networks in the nervous system that produce the behavior. Similarly, we assume that if neural networks are changed by experience, there must be a corresponding change in behavior. The challenge for those interested in frontal lobe plasticity, however, is to determine what types of behavioral changes are likely to reflect changes in frontal circuitry. In view of the frontal lobe's central role in the control of behavior, and especially in behaviors often referred to as

Table 2: Summary of the effects of factors on frontal cortical plasticity

Factor	Result	Basic Reference
Complex housing	Increased spine density	Kolb et al., 2003b
Psychoactive drugs	Increased dendritic length and increased spine density	Robinson and Kolb, 1999
Cocaine self administration	Pathological structure of prefrontal neurons	Robinson et al., 2001
Nerve growth factor	Increased dendritic length and increased spine density	Kolb et al., 1997
Gonadal hormones	Altered neuronal structure during development	Kolb and Stewart, 1991
	Removal of estrogen stimulates dendritic growth	Forgie and Kolb, 2003
Hippocampal injury	Infant hippocampal lesions reduce synapse number	Gorny et al., 2001

executive functions, it would seem reasonable to predict that frontal lobe plasticity would be relatively easy to demonstrate in response to a variety of experiences. We shall see, however, that the PFC is less responsive to sensory and motor experience than we might have expected.

As we begin the examination of frontal plasticity and behavior, we are faced with the basic question of how can we measure changes in circuits? Because circuits are composed of individual neurons, each of which connects with a subset of other neurons at synapses to form the circuits, the logical place to look for plastic changes is at the junction of neurons, which is at the synapse. The examination of synapses is a daunting task, however, because there are so many in even a relatively restricted region of brain. It is clearly impractical to use electron microscopic (EM) techniques to examine synaptic change directly because of the sheer number of synapses that would have to be examined. One way to approach the task is to assume that changes in synaptic organization can be inferred from grosser, light microscopic, studies of dendritic space. Previous EM studies have shown that there is a good correlation between changes in synapse number observed in EM studies and estimates of synaptic space from Golgi-stain studies (e.g. Siervaag and Greenough, 1988). That is, it appears that if dendrites grow longer or spines density increases, so does the number of synapses counted in EM studies (Fig. 2). Stated differently, because the dendrites of a cell function as the scaffolding for synapses, if we can measure total dendritic length, then we can begin to guesstimate how many synapses are on a cell. Further, if we know the density of synapses on the dendrites, we could estimate synapse number because we know that about 95% of a cell's synapses are on its dendrites. One way to estimate synaptic density is to measure the density of dendritic spines (see Fig. 2). Spines are the location of up to 95% of all excitatory synapses so by knowing spine density we can use simple arithmetic to estimate the number of excitatory synapses. Of course, we could completely miss the synaptic changes if what actually happens is a change in some characteristic of existing synapses (such as size), rather than a change in the number of synapses. Unfortunately, this question takes us back to the need for an EM analysis, but as first step EM remains impractical. What the Golgi procedure allows us to do, however, is to take a faster look at how a wide variety of factors might influence neural circuitry in the frontal regions. Nonetheless, there must still eventually be EM studies to look at the ultrastructure of the synapses.

3.1 Effects of Sensory and Motor Experience

There is an extensive literature showing that the structure of cortical neurons is influenced by various types of sensory and motor experience (for a review, see Kolb and Whishaw, 1998). For example, if laboratory animals ranging from rats to cats and monkeys are placed in complex environments versus living in standard lab cages, there are large changes in dendritic length and synapse number throughout the primary visual and somatosensory cortex (e.g. Greenough et al., 1985; Beaulieu and Colonnier, 1987). Similarly, if rats are trained on neuropsychological learning tasks such as a visual maze or a skilled motor learning task, then there are changes in cells in occipital cortex and motor cortex respectively (Greenough and Chang, 1988). These changes are specific, however, as visual training does not influence motor cortex neurons and *visa versa*.

Curiously, examination of the prelimbic region (Zilles' Cg3) of the mPFC and nearby parietal region (Par 1) in animals that were placed in complex environments for 4 months in adulthood showed an unexpected result: whereas the parietal cortex showed a large (10%) increase in dendritic length in response to this experience, there was no obvious change in the dendritic length of the neurons in Cg3 (Kolb et al., 2003b). This contrasting effect was especially surprising given that we have found this experience to increase dendritic length throughout the sensory and motor cortices, striatum, and nucleus accumbens. There is clearly something different about the effect of

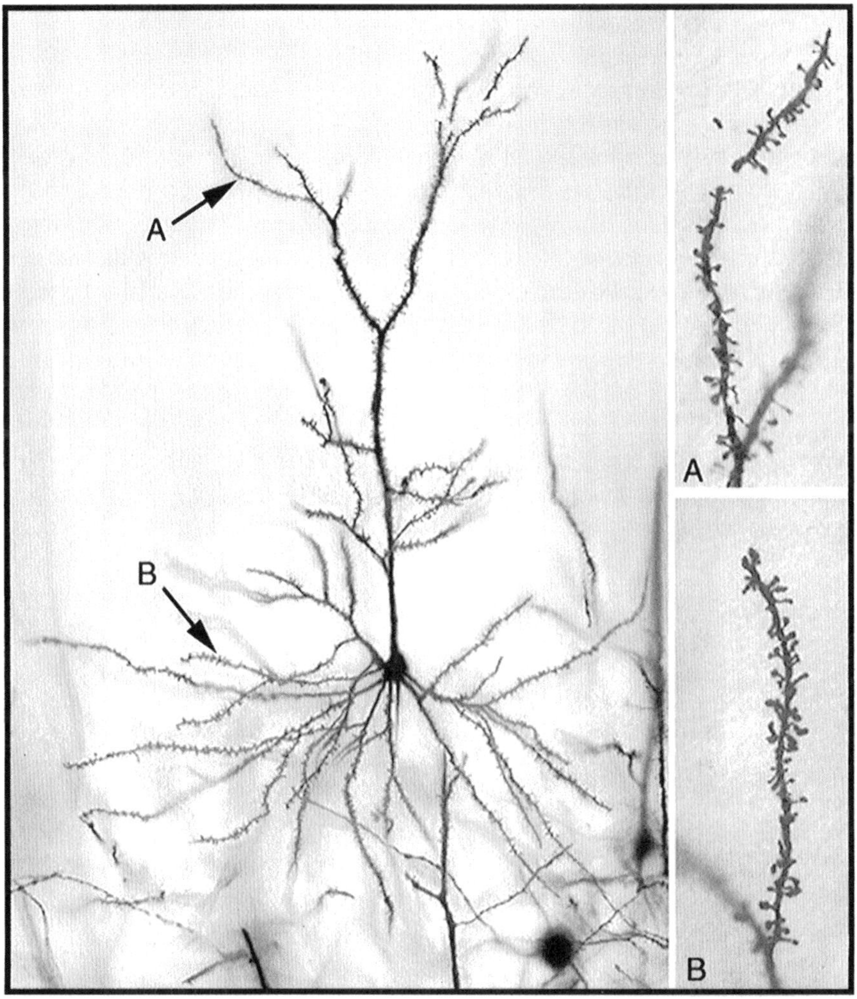

Figure 2: A Golgi-stained neuron in the mPFC of the rat.

experience on the neurons in the PFC versus other regions in the forebrain. We next examined the effect of the experience on spine density, expecting that there would be no change in the Cg3 cells, but again we were mistaken: the cells showed an increase in spine density that was as large as we had seen in other cortical regions. These changes in spine density were intriguing for at least two reasons. First, this was the first time that we had observed changes in spine density in the absence of a change in dendritic length.

Second, we had previously shown that changes in spine density in response to experience are age-dependent. That is, whereas animals placed in complex environments in adulthood or senescence show significant increases in spine density in parietal and occipital cortex, animals placed in similar environments as juveniles show a significant decrease in spine density (Kolb et al., 2003a). When we looked at spine density in Cg3 of mPFC in juvenile rats, we were surprised to find that there was an increase in spine density, a result that was opposite to what we had found in sensory cortex (Fig. 3).

The failure to find parallel effects of experience on prefrontal and other cortical pyramidal cells leads to the question of whether training animals in neuropsychological tasks, which are known to be sensitive to prefrontal injuries, would produce changes similar to those observed in motor or occipital cortex of animals that have been trained in motor or visual tasks respectively. We are unaware of any systematic study of this possibility, but in an unpublished study of rats trained in a radial arm maze, a task that is sensitive to medial frontal cortex lesions in rats, we found no evidence of

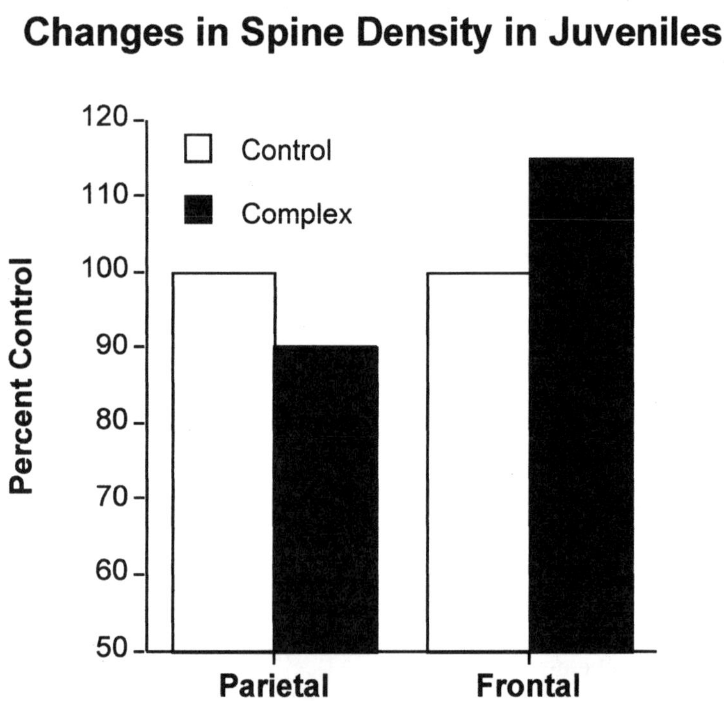

Figure 3: The effects of complex housing on spine density in the parietal and prefrontal cortex of juvenile rats. There is a decrease in spine density in the parietal cortex that contrasts with an increase in spine density in the mPFC.

changes in dendritic length in mPFC (B. Kolb and G. Winocur, unpublished data). This study needs replication and extension to other neuropsychological tests, but it does suggest that once again the PFC responds differently to experience than other cortical regions. Furthermore, although the firing properties of cells in the OFC have been shown to change with the development of olfactory memories (Ramus and Eichenbaum, 2000; Schoenbaum et al., 2000; Alvarez and Eichenbaum, 2002), we are unaware of any morphological studies showing synaptic changes, and this is clearly an obvious topic for study.

The simplest conclusion from the complex housing and learning results is that placing animals in complex environments for several months or training animals to criterion in neuropsychological tests does not engage prefrontal neurons the same way that it engages sensory or motor cortical neurons. It is quite possible that the PFC is engaged only until behavioral strategies are developed, after which time it is no longer necessary, an idea that was proposed first by Hebb (1949). In contrast, sensory and motor areas are engaged as long as animals are displaying particular behaviors. The question, however, is whether there is evidence that the prefrontal cells ever change their synaptic organization or if they are simply engaging a relatively unplastic system to generate behavioral strategies. Hebb proposed that during development the PFC was especially important because it was during this time that the frontal lobe was developing schemas to solve problems that would be encountered later in life (Hebb, 1949). If this hypothesis is correct, it may be that the prefrontal cells are particularly responsive to experience during development but less so in adulthood.

Other forms of environmental stimulation do appear to produce changes in prefrontal neurons, however. For example, animals given chronic injections of corticosterone, which presumably mimics the effects of stress, show a change in organization of dendritic morphology in mPFC (Wellman, 2001). Similarly, animals given daily injections of saline, which again was presumed to be stressful, showed increased spine density in mPFC neurons (Seib and Wellman, 2003).

3.2 Effects of Drugs and Natural Reinforcers

Many people commonly take stimulant drugs like nicotine, amphetamine, cocaine, or depressant drugs like morphine or alcohol, all of which affect behavior and are thus said to be psychoactive. The long-term consequences of abusing psychoactive drugs are now well-documented, and it has been hypothesized that some of the behavioral symptoms observed in drug addicts or alcoholics are related to abnormalities in the functioning of the prefrontal regions (Robbins and Everitt, 2002). One experimental demonstration of drug-induced changes in the brain is known as drug-induced behavioral sensitization, often referred to just as behavioral sensitization. Behavioral sensitization is the progressive increase in the behavioral actions of a drug that occurs after repeated administration of a constant dose of the drug. Behavioral sensitization occurs with most psychomotor stimulant drugs (e.g. amphetamine, nicotine) and sometimes to morphine. For example, when a rat is given a small dose of amphetamine, it may show a small increase in motor activity. When the rat is given the same dose on subsequent occasions, the increase in activity is progressively larger, thus showing behavioral sensitization. This drug-induced behavioral change persists for weeks or months so that if the drug is given in the same dose as before, the behavioral sensitization is still present. In a sense, the brain has some memory of the effects of the drug.

The parallel between the drug actions and memory led to the question of whether there might be permanent changes in the neurons of the brain that could account for the persistence of the behavior (e.g. Robinson and Kolb, 1999). Indeed, there are. Figure 4 compares the effects of amphetamine and saline treatments on the structure of neurons in Cg3 of the PFC. It can be seen that neurons in the amphetamine-treated brains have greater dendritic material as well as more densely organized spines; the latter being the location of a large percentage of the synapses on these cells. These plastic changes were not found throughout the brain, however, but rather they were localized to regions such as the PFC and nucleus accumbens, both of which are implicated in the rewarding properties of these drugs.

In contrast to the increased synaptic density in the Cg3 neurons exposed to stimulants, there was a decrease in dendritic length and spine density in the insular cortex. This result was completely unexpected and shows that different subregions of the rat PFC may respond dramatically differently to the same stimulation. Further studies showed a similar asymmetry in the medial/orbital regions in response to morphine. In this case, there was a decrease in dendritic length and spine density in the anterior cingulate neurons but an increase in the insular neurons (Robinson et al., 2002). Thus, not only were the effects of stimulants and depressants on the prefrontal neurons qualitatively different, but in both cases there were qualitatively different effects of the drugs on different prefrontal subfields. The contrasting effects of the psychoactive drugs on the two subfields of the rat PFC are intriguing and are reminiscent of the differences seen in metabolic levels of the dorsolateral and orbital regions of human depressed patients (Drevets et al., 1999). These patients show an increase in activity in the orbital regions and a decrease in the dorsolateral region. The parallel between drug effects and depression is intriguing and suggests that plastic changes in the two subfields may act in a reciprocal manner.

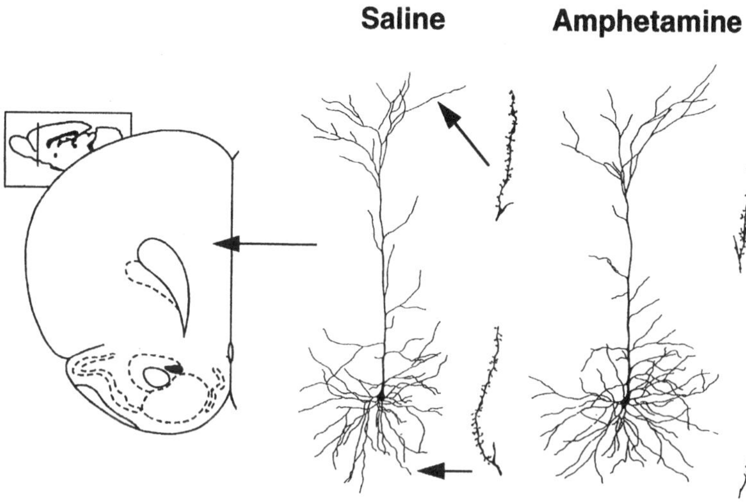

Figure 4: The contrasting effects of saline and amphetamine injections on the structure of cells in the mPFC of rats. Amphetamine increases dendritic length and spine density. Adopted from Robinson and Kolb (1999) with permission.

The effects of psychoactive drugs on cells in the PFC are presumed to be due, at least in part, to actions of the drugs on dopaminergic cells in the brainstem that project to the prefrontal regions. But, not only do drugs affect dopaminergic afferents to the prefrontal neurons but so do naturally-occurring rewards such as sex (Fiorino and Kolb, 2003) and social interaction (Hamilton and Kolb, 2003). For example, analysis of prefrontal neurons of male rats paired daily for two weeks with receptive females confirmed that sex produces changes in prefrontal neurons that are strikingly similar to those observed in rats treated with psychomotor stimulants. In contrast to the drugs, however, similar changes were not seen in nucleus accumbens, a result that may explain why people are addicted to drugs but not normally to natural rewards like sex.

Many questions remain. Why, for example, do rewarding events change synaptic organization? Drugs produce changes in a variety of trophic factors and immediate early genes, but there is as yet no direct evidence of how such changes may alter synaptic organization. Similarly, are there age-related changes in the effects of rewarding events? Given that the reward value of many events, including drugs, appears to wane with age, it would not be surprising to find age-related differences in reward-induced synaptic reorganization. Finally, how do reward-induced synaptic changes interact with other experience-dependent changes? For instance, if neurons in the

PFC are changed by drugs, how do they now respond to experience (e.g. Kolb et al., 2003c)?

3.3 Effects of Gonadal Hormones

One consistent finding in studies of the effects of prefrontal lesions in rats is that there are sex-related differences in the effects of injury to both the mPFC and OFC (e.g. Kolb and Cioe, 1996), a result that is similar to observations in both humans (e.g. Kimura, 1999) and rhesus monkeys (Clark and Goldman-Rakic, 1989). This leads to the possibility that there might be sex-related differences in the structure of cells in the mPFC and/or OFC. There are. Males show more extensive dendritic fields in the mPFC whereas females showed more extensive dendritic fields in the OFC (Kolb and Stewart, 1991; Markham and Juraska, 2002). These differences are hormone-dependent, as neonatal castration or ovariectomy eliminates the differences in adulthood (Kolb and Stewart, 1991).

There are no comparable studies of cell structure in humans, but a recent MRI (magnetic resonance imaging) study looking at the volume of different cortical areas is intriguing. Goldstein et al. (2001) showed that there are complementary sex-related differences in the relative volume of the dorsolateral and medial versus the orbital cortex: Females have a larger volume of dorsolateral and medial regions whereas males have a larger volume of orbital cortex. This result would seem to be the opposite to the rodent results, but this may not be the case. The rat studies measured the synaptic space of individual neurons, whereas in the human study, the measure was the total volume, relative to the rest of the cortical area. There are at least two reasons why these results may be compatible. First, differences in volume could reflect differences in non-neuronal elements, such as glial or vascular differences. Second, it is quite possible that areas of high numbers of synapses per neuron may have fewer overall numbers of neurons. Thus, a difference in neuron number may be compensated by differences in synapses per neuron. Following this logic, if there are fewer neurons in the female orbital cortex and the male medial cortex, then there may be a compensatory increase in synapse numbers per neuron. This idea is speculative but easily testable.

The presence of sexually dimorphic cell structure in different regions of the frontal cortex of rats presumably reflects differences in the distribution of hormone receptors during cortical development and thus reflects a hormone-dependent organizational effect on synaptic organization. But what about activational differences in adulthood? That is, might circulating gonadal hormones affect synaptic organization in adulthood in a manner parallel to the well-known effects of circulating levels of estrogen in the hippocampus

(Wooley et al., 1990)? To test this possibility, we removed the ovaries of adult rats, waited 3 months, and then examined the structure of cortical neurons. The results were surprising: Ovariectomy resulted in an extensive increase of both dendritic length and spine density of pyramidal cells in both the medial frontal and parietal cortex (Stewart and Kolb, 1994; Forgie and Kolb, 2003). Furthermore, this dendritic growth could be blocked by the administration of estrogen. We have not yet analyzed the effects of hormonal manipulations in adulthood on the orbital cortex, but by now we should not be surprised if the effects were different than those observed in the mPFC.

In sum, the hormone studies have shown that the synaptic organization of neurons in both the mPFC and the OFC is altered by gonadal hormones both during development and in adulthood. Furthermore, it appears that like psychoactive drugs, hormones differentially affect the medial and orbital subregions.

3.4 Effects of Growth Factors

Neurotrophic factors are proteins that are manufactured in the brain and act to influence development and maintenance of neurons. Nerve growth factor (NGF) was the first neurotrophic factor to be described, and it is still the best characterized. Intraventricular infusions of NGF stimulate dendritic growth and increased spine density in cortical pyramidal cells, including those in the medial frontal region (Kolb et al., 1997). Indeed, the effect of NGF on neurons in Cg 3 is even larger than the effect of psychomotor stimulants. It is not known whether other neurotrophic factors might also influence cortical organization, but it does seem likely, and there is reason to think that at least some actions might be specific to the PFC. For example, Flores and Stewart (2000) have found that rats given sensitizing doses of amphetamine show an increase in basic fibroblast growth factor (bFGF) expression in medial frontal cortex but not in more posterior cortex. They note that bFGF may thus participate in the development of structural changes brought about by amphetamine. Importantly, although the structural changes in neurons are long lasting, and possibly permanent, the changes in bFGF are not maintained. This makes sense if the bFGF activity is involved in stimulating the dendritic changes, because once changed, the bFGF would no longer be needed. It is not known whether administration of bFGF might selectively change the structure of neurons in the frontal lobe, but it seems likely. The interesting question is how this might be manifested behaviorally.

3.5 Frontal Lobe Plasticity in the Injured Brain

When the brain is injured, there are likely to be compensatory changes in the remaining neural networks that will reflect either the reorganization of existing circuits or the creation of new circuits, most likely reflecting the formation of new connections among remaining neurons. Because the PFC is connected with most of the posterior cortical regions, as well as the motor regions of the frontal lobe, it seems reasonable to predict that damage to other cerebral regions will influence the synaptic organization of the prefrontal regions. Curiously, there are few studies bearing on this issue. Nonetheless, we can predict at least two different outcomes. First, it is reasonable to expect that if regions of the brain that have extensive connections with the frontal cortex are damaged, then the absence of such connections could produce dendritic atrophy, and there is at least one study showing this. Thus, when rats are given strokes that involve the motor and somatosensory regions, there is an atrophy of the neurons in Cg3 (Kolb et al., 1997), presumably reflecting the loss of afferents from the damaged regions. Second, we could predict that if there were some form of adaptation to the injury, either because of endogenous compensatory responses or in response to some type of exogenous treatment, then there might be changes in prefrontal cortical organization. Indeed, if animals are given intraventricular injections of nerve growth factor following stroke of the sensorimotor cortex, there is a partial restitution of both motor and cognitive functions, and this is correlated with an expansion of dendritic fields and increased spine density in Cg3. This result implies that the reorganization of the neural networks involving Cg3 neurons is somehow related to the functional recovery. Of course, the presence of such changes need not be causal. For example, it is quite possible that the changes in behavior produce the changes in the prefrontal circuitry. Whatever the cause, the point remains that behavioral compensation is correlated with changes in prefrontal circuits.

But what happens if there is cerebral injury during development? Recall that Hebb (1949) emphasized that the development of the PFC is especially important to problem solving in adulthood. We could predict that early injury to regions with intimate connections with the frontal lobe could disrupt normal frontal lobe development. This idea has not been well studied, with the exception of Weinberger and his colleagues who have studied the effects of neonatal injury to the ventral hippocampus (e.g. Raedler et al., 1998). In adulthood, these animals show various symptoms characteristic of rats with prefrontal injuries, such as hyperactivity and deficits in social behavior and working memory (e.g. Sams-Dodd et al., 1997). These functional deficits are ameliorated by antipsychotic drugs and are associated with a decrease in the metabolites of dopamine in the medial frontal region, which has led the

Table 3: Summary of the effects of frontal cortical injury at different ages

Age at Injury	Result	Basic Reference
E18	Cortex regrows with odd structure Functional recovery	Kolb et al., 1998a
P1-P6	Small brain, dendritic atrophy Dismal functional outcome	Kolb and Gibb, 1991
P7-P12	Dendrite and spine growth Cortical regrowth Functional recovery	Kolb and Gibb, 1991 Kolb et al., 1998b
P120	Dendritic atrophy, then growth Partial return of function	Kolb, 1995

Abbreviations: E18, embryonic day 18; Px, postnatal day x

authors to propose that schizophrenia might result from developmental abnormalities in the hippocampal formation. Given that psychomotor stimulants enhance dopaminergic-mediated activity in the frontal lobe and produce an expansion of dendritic fields in medial frontal cortex in rats, it is reasonable to predict that decreased prefrontal dopaminergic activity after infant hippocampal lesions might decrease dendritic arborization and spine density in PFC. This is indeed the case, as there is a reduction in dendritic arborization and a drop in spine density in prefrontal, but not parietal cortex, in rats with early hippocampal lesions (Gorny et al., 2001). This is an exciting result, because it suggests that injury elsewhere in the brain may alter connectivity in the frontal lobe and that, in turn, may alter behavior.

3.6 Cerebral Plasticity after Prefrontal Lesions

There is an extensive literature examining the effect of prefrontal injury during development on the structure and function of the remaining brain (e.g. Kolb, 1995). The details of these studies are beyond the scope of this chapter, so we will review this topic only briefly and with emphasis upon synaptic plasticity (for more thorough review, see Kolb and Gibb, 2001).

As a general rule, damage to the medial or orbital subfields of the rat PFC between about 7-12 days of age produce markedly attenuated behavioral effects relative to injuries in the first few days of life or after about 15 days of life. Indeed, on some behavioral measures sensitive to prefrontal injuries in adulthood, animals with prefrontal lesions at 10 days of age perform as well in adulthood as sham-operated littermates (see Table 3). Animals with similar injuries during the first week of life do not show this functional capacity and are often severely impaired even relative to adults with similar injuries. The obvious explanation for this age-dependent effect of early injury is that there are plastic changes after injury on day 10 that are not seen after similar injury either before or after.

One of the most obvious, and consistent, changes in the brain after early frontal injury is that brain size in adulthood is directly related to the postnatal age at injury: the earlier the injury, the smaller the brain and the thinner the cortical mantle. Thus, rats with perinatal lesions have very small brains whereas those with lesions at day 10 have larger brains. Curiously, however, the day 10 brains still are markedly smaller than the brains of rats with lesions later in life, such as day 25, even though the behavioral outcome is far better (Kolb and Whishaw, 1981; Kolb et al., 1996). Therefore, it must be the *organization of the brain* rather than its *size* that predicts recovery in the day 10 animal. Changes in organization can be inferred from an analysis of dendritic organization, cortical connectivity, and evidence of neurogenesis.

Dendritic analyses of cortical neurons of rats with perinatal lesions consistently show a general stunting of dendritic arborization and a drop in spine density across the cortical mantle (e.g. Kolb and Gibb, 1991, 1993; Kolb et al., 1994b). In contrast, rats with cortical lesions around 10 days of age show an increase in dendritic arbor and an increase in spine density relative to normal control littermates. Thus, animals with the best functional outcome show the largest dendritic fields whereas animals with the worst functional outcome have the smallest dendritic arbor relative to control animals. The development of the functional recovery and dendritic hypertrophy in the day 10 operates is especially intriguing. Kolb and Gibb (1993) compared the spatial navigation behavior of rats with day 1 or 10 medial frontal lesions when the animals were either 22 or 56 days old. When tested as weanlings, both brain-injured groups were equally impaired, and subsequent dendritic analysis revealed that both groups had dendritic atrophy, relative to littermate sham controls, in pyramidal cells across the remaining cortex. In contrast, when animals were tested as adolescents, the day 10, but not the day 1, animals showed almost complete recovery of function, and this was associated with dendritic hypertrophy across the remaining cortical mantle. It certainly appears that reorganization of the neural circuitry in the remaining cortex was supporting the functional recovery.

In the course of studies of the effect of restricted lesions of the medial frontal cortex or olfactory bulb, we discovered that, in contrast to lesions elsewhere in the cerebrum, midline telencephalic lesions on postnatal day 7-12 led to spontaneous regeneration of the lost regions, or at least partial

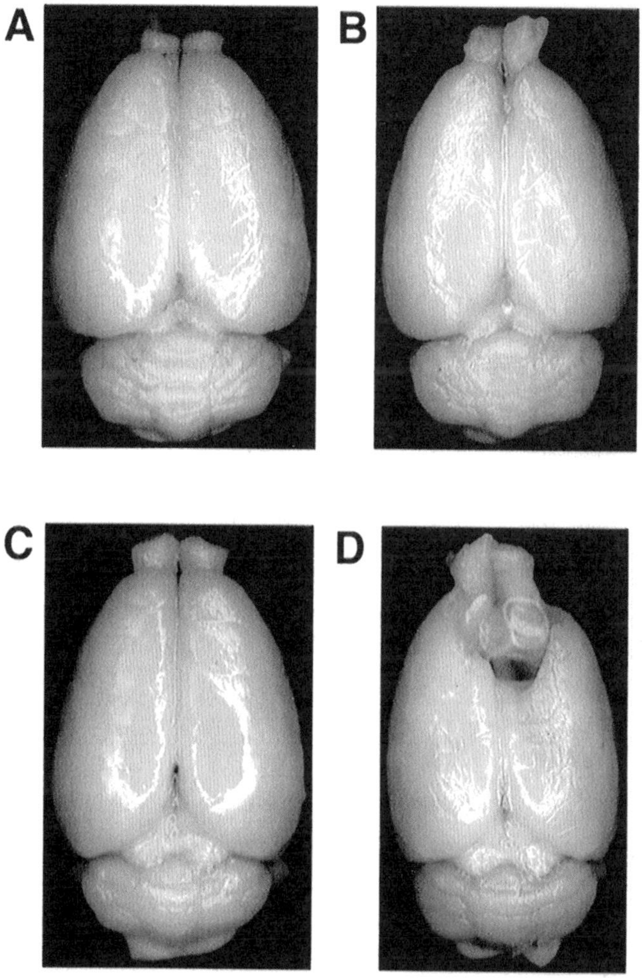

Figure 5: Regenerated brains with and without BrdU. A. Control brain. B. Day 10 mPFC lesion brain showing the spontaneous regeneration of the lost tissue. C. BrdU-treated (on E13) control brain. D. BrdU-treated day 10 mPFC lesion brain. Note that the BrdU blocked the spontaneous regeneration in the lesion brain.

regeneration of the lost regions (Fig. 5). Similar injuries either before or after this temporal window did not produce such a result. Analysis of the medial frontal region showed that the area contained newly-generated neurons that formed at least some of the normal connections of this region (Kolb et al., 1998b). Furthermore, animals with this regrown cortex appeared virtually normal on many, although not all, behavioral measures (e.g. Kolb et al., 1996). Additional studies showed that if we blocked regeneration of the tissue with prenatal injections of the mitotic marker bromodeoxyuridine (BrdU), the lost frontal tissue failed to regrow and there was no recovery of function (Kolb et al., 2003c), a result that implies that the regrown tissue was supporting recovery. Parallel studies in which we removed the regrown tissue found complementary results: removal of the tissue eliminated the functional recovery (Dallison and Kolb, 2003). Thus, in the absence of the regrown tissue, either because we blocked the growth or because we removed the tissue, function was lost.

4. CONCLUSIONS

We began by asking whether the PFC of the rat can be seen as a useful model for studying the organization and plasticity of the frontal lobe of primates. Although there are clear differences in the gross anatomical organization of the mPFC and OFC of rats and primates, there is a convergence of behavioral evidence showing that the functions of these areas are remarkably similar across primates and rats. It is argued that this is so because mammals have a set of behavioral demands that are similar across the entire mammalian order, which has led to the evolution of class-common solutions. It is presumed that those extinct mammalian ancestors that gave rise to at least some of the modern mammalian taxa, but certainly to rodents and to primates, also faced similar class-common problems and that they developed a primitive prefrontal area to solve these problems. One characteristic of most brain areas is that they change with experience, the property of plasticity, but not all brain regions change in response to all experiences. The prefrontal regions are interesting in this regard because although they are highly plastic relative to adjacent sensorimotor regions in response to hormonal and drug manipulations, they are less influenced by sensory and motor experience than the adjacent sensorimotor regions. This difference is somewhat surprising but is presumed to provide some insight into the functions of the PFC of mammals.

REFERENCES

Alvarez P, Eichenbaum H (2002) Representations of odors in the rat orbitofrontal cortex change during and after learning. Behav Neurosci 116:421-433.

Arnsten AF (1997) Catecholamine regulation of the prefrontal cortex. J Psychopharmacol 11:151-162.

Baxter MG, Parker A, Lindner CCC, Izquierdo AD, Murray EA (2000). Control of response selection by reinforcer value requires interaction of amygdala and orbitofrontal cortex. J Neurosci 20:4311-4319.

Beaulieu C, Colonnier M (1987) Effect of the richness of the environment on the cat visual cortex. J Comp Neurol 266:478-494.

Birrel JM, Brown VJ (2000) Medial frontal cortex mediates perceputal attentional set shifting in the rat. J Neurosci 20:4320-4324.

Brown VJ, Bowman EM (2002) Rodent models of prefrontal cortical function. Trends Neurosci 25:340-343.

de Bruin JPC (1990) Social behaviour and the prefrontal cortex. In: Progress in Brain Research, vol 85 (Uylings H.B.M., van Eden C., de Bruin JPC, Corner MA, and Feenstra MGP, eds), pp 485-500, Elsevier, Amsterdam.

Campbell CBG, Hodos, W (1970) The concept of homology and the evolution of the nervous system. Brain Behav Evol 3:353-367.

Clark AS, Goldman-Rakic PS (1989) Gonadal hormones influence the emergence of cortical function in nonhuman primates. Behav Neurosci 103:1287-1295.

Dallison A, Kolb B (2003) Recovery from infant frontal cortical lesions in rats can be reversed by cortical lesions in adulthood. Behav Brain Res (in press).

DeCoteau WE, Kesner RP, Williams JM (1997) Short-term memory for food reward magnitude: The role of the prefrontal cortex. Behav Brain Res 88:239-249.

Divac I (1971) Frontal lobe system and spatial reversal in the rat. Neuropsychologia 9:171-183.

Drevets WC, Gadde KM, Krishnan KR (1999) Neuroimaging studies of mood disorders. In: The Neruobiology of Mental Illness (Charney DS, Nestler EJ, and Bunney BS, Eds) pp 246-257, Oxford University Press, New York.

Dunnett SB (1990) Role of the prefrontal cortex and striatal output systems in short-term memory deficits associated with ageing, basal forebrain lesions, and cholinergic-rich grafts. Can J Psychol 44:210-232.

Everitt BJ, Robbins TW (1997) Central cholinergtic systems and cognition. Annu Rev Psychol 48:649-684.

Fiorino D, Kolb B (2003) Sexual experience alters dendritic organization in prefrontal cortex and nucleus accumbens. Soc Neurosci Abstr (in press).

Flores C, Stewart J (2000) Changes in astrocytic basic fibroblast growth factor expression during and after prolonged exposure to escalating doses of amphetamine. Neuroscience 98:287-293.

Forgie ML, Kolb B (2003) The influence of circulating ovarian steroids on neural plasticity of the prefrontal cortex in the female rat. (submitted).

Frysztak RJ, Neafsey EJ (1994) The effect of medial frontal cortex lesions on cardiovascular conditioned emotional responses in the rat. Brain Res 643:181-193.

Fuster JM (1997) The Prefrontal Cortex: Anatomy, Physiology, and Neuropsychology of the Frontal Lobe (3rd Ed) Raven: New York.

Gallagher M, McMahan RW, Schoenbaum G (1999) Orbitofrontal cortex and representations of incentive value in associative learning. J Neurosci 19:6610-6614.

Gisquet-Verrier P, Winocur G, Delatour B (2000) Functional dissociation between dorsal and ventral regions of the medial prefrontal cortex in rats. Psychobiology 28:248-260.

Goldman-Rakic PS (1987) Circuitry of primate prefrontal cortex and regulation of behavior by representational memory. In: Handbook of Physiology: Section 1. The nervous system: vol 5. Higher functions of the brain (Plum and Mountcastle V, eds), pp 373-417, American Physiological Society, Bethesda, MD.

Goldstein JM, Seidman LJ, Horton NJ, Makris N, Kennedy DN, Caviness VS, Faraone SV, Tsuang MT (2001) Normal sexual dimorphism of the adult human brain assessed by in vivo magnetic resonance imaging. Cereb Cortex 11:490-497.

Gorny G, Kolb B, Weinberger DR, Sheridan R, Lipska BK (2001) Synaptic pathology in the prefrontal cortex and nucleus accumbens of rats with neonatal damage of the ventral hippocampus. Soc Neurosci Abstr 27:876.16.

Granon S, Poucet B (2000) Involvement of the rat prefrontal cortex in cognitive functions: A central role for the prelimbic area. Psychobiology 28:229-237.

Greenough WT, Chang FF (1988) Plasticity of synapse structure and pattern in the cerebral cortex. In: Cerebral Cortex vol 7 (Peters A and Jones EG, eds), pp 391-440, Plenum Press, New York.

Greenough WT, Larsen JR, Withers GS (1985) Effects of unilateral and bilatearl training in a reaching task on dendritic branching of neurons in the rat motor-sensory forelimb cortex. Behav Neural Biol 44:301-314.

Groenewegen HJ (1988) Organization of the the afferent connections of the mediodorsal thalamic nucleus in the rat, related to the mediodorsal-prefrontal topography. Neuroscience 24:379-431.

Groenewegen HJ, Uylings HBM (2000) The prefrontal cortex and the integration of sensory, limbic, and autonomic information. In: Progress in Brain Research vol 126, (Uylings HBM, van Eden CG, de Bruin JPC, Feenstra MGP, and Pennartz CMA, eds), pp 3-28, Elsevier, Amsterdam.

Hamilton D, Kolb B (2003) Social interaction with novel partners alters dendritic organization in the orbital prefrontal cortex of rats. Soc Neurosci Abst (in press).

Hebb DO (1949) The Organization of Behavior: A Neuropsychological Theory. John Wiley and Sons, New York.

Iwaniuk A, Whishaw IQ (2000) On the origin of skilled forelimb movements. Trends Neurosci 23:372-376.

Kaas JH (1987) The organization of neocortex in mammals: Implications for theories of brain function. Annu Rev Psychol 38:129-151.

Kalsbeek A, deBruin JPC, Matthijs GP, Feenstra GP, Uylings HBM (1990) Age-dependent effects of lesioning the mesocortical dopamine system upon prefrontal cortex morphometry and PFC-related disorders. In: Progress in Brain Research, vol 85 (Uylings H.B.M., van Eden C., de Bruin JPC, Corner MA, and Feenstra MGP, eds), pp 257-283, Elsevier, Amsterdam.

Kesner RP (2000) Subregional analysis of mnemonic functions of the prefrontal cortex in the rat. Psychobiology 28:219-228.

Kesner RP, Holbrook T (1987) Dissociation of item and order spatial memory in rats following medial prefrontal cortex lesions. Neuropsychologia 25:653-664.

Kimura D (1999) Sex and Cognition. MIT Press, Cambridge MA.

Kolb B (1974a) Dissociation of the effects of lesions of the orbital or medial aspect of the prefrontal cortex of the rat with respect to activity. Behav Biol 10:329-343.

Kolb B (1974b) Prefrontal lesions alter eating and hoarding behavior in rats. Physiol Behav 12:507-511.

Kolb B (1974c) Some tests of response habituation in rats with prefrontal lesions. Can J Psychol 28:260-267.

Kolb B (1974d) The social behavior of rats with chronic prefrontal lesions. J Comp Physiol Psychol 87:466-474.

Kolb B (1984) Functions of the frontal cortex of the rat: A comparative review. Brain Res Rev 8:65-98.

Kolb B (1990) Animal models for human PFC-related disorders. In: Progress in Brain Research, vol 85 (Uylings H.B.M., van Eden C., de Bruin JPC, Corner MA, and Feenstra MGP, eds), pp 491-509, Elsevier, Amsterdam.

Kolb B (1995) Brain Plasticity and Behavior. Lawrence Erlbaum, Mahwah NJ.

Kolb B, Cioe J (1996) Sex-related differences in cortical function after medial frontal lesions in rats. Behav Neurosci 110:1271-1281.

Kolb B, Gibb R (1991) Sparing of function after neonatal frontal lesions correlates with increased cortical dendritic branching: A possible mechanism for the Kennard effect. Behav Brain Res 43:51-56.

Kolb B, Gibb R (1993) Possible anatomical basis of recovery of spatial learning after neonatal prefrontal lesions in rats. Behav Neurosci 107:799-811.

Kolb B, Gibb R (2001) Early brain injury, plasticity and behavior. In: Handbook of Developmental Cognitive Neuroscience (Nelson CA and Luciana M, eds) MIT Press, Cambridge MA.

Kolb B, Nonneman AJ (1975) Prefrontal cortex and the regulation of food intake in the rat. J Comp Physiol Psychol 88:806-815.

Kolb B, Stewart J (1991) Sex-related differences in dendritic branching of cells in the prefrontal cortex of rats. J Neuroendocrinol 3:95-99.

Kolb B, Whishaw IQ (1981) Neonatal frontal lesions in the rat: sparing of learned but not species-typical behavior in the presence of reduced brain weight and cortical thickness. J Comp Physiol Psychol 95:863-879.

Kolb B, Whishaw IQ (1983a) Generalizing in neuropsychology: problems and principles underlying cross-species comparisons. In: Behavioral Contributions to Brain Research (Robinson TE, ed) Oxford University Press, New York.

Kolb B, Whishaw IQ (1983b) Dissociation of the contributions of the prefrontal, motor and parietal cortex to the control of movement in the rat. Can J Psychol 37:211-232.

Kolb B, Whishaw IQ (1998) Brain plasticity and behavior. Annu Rev Psychol 49:43-64.

Kolb B, Nonneman AJ, Singh R (1974) Double dissociation of spatial impairment and perseveration following selective prefrontal lesions in the rat. J Comp Physiol Psychol 87:772-780.

Kolb B, Buhrmann K, MacDonald R, Sutherland RJ (1994a) Dissociation of the medial prefrontal, posterior parietal, and posterior temporal cortex for spatial navigation and recognition memory in the rat. Cereb Cortex 4:15-34.

Kolb B, Gibb R, van der Kooy D (1994b) Neonatal frontal cortical lesions in rats alter cortical structure and connectivity. Brain Res 645:85-97.

Kolb B, Petrie B, Cioe J (1996) Recovery from early cortical damage in rats. VII. Comparison of the behavioural and anatomical effects of medial prefrontal lesions at different ages of neural maturation. Behav Brain Res 79:1-13.

Kolb B, Cote S, Ribeiro-da-Silva, Cuello AC (1997) NGF stimulates recovery of function and dendritic growth after unilateral motor cortex lesions in rats. Neuroscience 76:1139-1151.

Kolb B, Cioe J, Muirhead D (1998a) Cerebral morphology and functional sparing after prenatal frontal cortex lesions in rats. Behav Brain Res 91:143-155.

Kolb B, Gibb R, Gorny G, Whishaw IQ (1998b) Possible brain regrowth after cortical lesions in rats. Behav Brain Res 91:127-141.

Kolb B, Gibb R, Gonzalez C (2001) Cortical injury and neuroplasticity during brain development. In: Toward a Theory of Neuroplasticity (Shaw CA and McEachern JC, eds) Elsevier, New York.

Kolb B, Gibb R, Gorny G (2003a) Experience-dependent changes in dendritic arbor and spine density in neocortex vary with age and sex. Neurobiol Learn Mem 79:1-10.

Kolb B, Gorny G, Sonderpalm A, Robinson TE (2003b) Environmental complexity has different effects on the structure of neurons in the prefrontal cortex versus the parietal cortex or nucleus accumbens. Synapse 48:149-153.

Kolb B, Pedersen B, Gibb R (2003c) Recovery from frontal cortex lesions in infancy is blocked by embryonic pretreatment with bromodeoxyuridine. (submitted).

Leonard CM (1972) The connections of the dorsomedial nuclei. Brain Behav Evol 6:524-541.

Markham JA, Juraska JM (2002) Aging and sex influence the anatomy of the rat anterior cingulate cortex. Neurobiol Aging 23:579-588.

Morgan MA, Schulkin J, LeDoux JE (2003) Ventral medial prefrontal cortex and emotional perseveration: the memory for prior extinction training. Behav Brain Res (in press).

Muir JL, Everitt BJ, Robbins TW (1996) The cerebral cortex of the rat and visual attentional function: Dissociable effects of mediofrontal, cingulate, anterior dorsolateral and parietal cortex lesions on a five-choice serial reaction time task. Cereb Cortex 6:470-481.

Otto T, Eichenbaum H (1992) Complementary roles of the orbital prefrontal cortex and the perirhinal-entorhinal cortices in an odor-guided delayed-nonmatching-to-sample task. Behav Neurosci 106:762-775.

Passingham R (1993) The Frontal Lobes and Voluntary Action. Oxford Unversity Press: New York.

Pezze MA, Bast T, Feldon J (2003) Significance of dopamine transmission in the rat medial prefrontal cortex for conditioned fear. Cereb Cortex 13:371-380.

Preuss TM (1995) Do rats have a prefrontal cortex? The Rose-Woolsey-Akert program reconsidered. J Cog Neurosci 7:1-24.

Quirk GJ, Russo GK, Barron JL, Lebron K (2000) The role of ventromedial prefrontal cortex in the recovery of extinguished fear. J Neurosci 20:6225-6231.

Raedler TJ, Knable MB, Weinberger DR (1998) Schizophrenia as a developmental disorder of the cerebral cortex. Curr Opin Neurobiol 8:157-161.

Ramus SJ, Eichenbaum H (2000) Neural correlates of olfactory recognition memory in the rat orbitofrontal cortex. J Neurosci 20:8199-8208.

Ragozzino ME (2000) The contribution of cholinergic and dopaminergic afferents in the rat prefrontal cortex to learning, memory, and attention. Psychobiology 28:238-247.

Ragozzino ME, Kesner RP (1999) The role of the agranular insular cortex in working memory for food reward value and allocentric space in rats. Behav Brain Res 98:103-112.

Ramus SJ, Eichenbaum H (2000) Neural correlates of olfactory recognition memory in the rat orbitofrontal cortex. J Neurosci 20:8199-8208.

Robbins TW (2002) The 5-choice serial reaction time task: behavioural pharmacology and functional neurochemistry. Psychopharmacol 163:362-380.

Robbins TW, Everitt BJ (2002) Limbic-striatal memory systems and drug addiction. Neurobiol Learn Mem, 78:625-636.

Robinson TE, Kolb B (1999) Alterations in the morphology of dendrites and dendritic spines in the nucleus accumbens and prefrontal cortex following repeated treatment with amphetamine or cocaine. Eur J Neurosci 11:1598-1604.

Robinson TE, Mitton E, Gorny G, Kolb B (2001) Self administration of cocaine modifies neuronal morphology in nucleus accumbens and prefrontal cortex. Synapse 39:257-266.

Robinson TE, Gorny G, Savage V, Kolb B (2002) Widespread but regionally-specific effects of self-administered versus experimenter-administered morphine on dendritic spines in the nucleus accumbens, hipocampus, sensory cortex, and prefrontal cortex of the rat. Synapse 46:271-279.

Rose JE, Woolsey CN (1948) The orbitofrontal cortex and its connections with the mediodorsal nucleus in rabbit, sheep, and cat. Research Publications of the Association for Nervous and Mental Disease 27:210-232.

Rosenkranz JM, Grace AA (2001) Dopamine attenuates prefrontal cortical suppression of sensory inputs to the basolateral amygdala of rats. J Neurosci 21:4090-4103.

Saddoris MP, Setlow B, Nugent S, Schoenbaum G (2001) A reexamination of the role of orbitofrontal cortex and basolateral amygdala in acquisition and reversal of odor-guided go, no go discrimination task. Soc Neurosci Abstr 27:189.5.

Sams-Dodd F, Lipska BK, Weinberger DR (1997) Neonatal lesions of the rat ventral hippocampus result in hyperlocomotion and deficits in social behaviour in adulthood. Psychopharmacol 132:303-310.

Sawaguchi T, Goldman-Rakic PS (1994) The role of D1-dopamine receptor in working memory: local injections of dopamine antagonists into the prefrontal cortex of rhesus monkeys performing an oculomotor delayed-response task. J Neurophsiol 71:515-528.

Schoenbaum G, Chiba AA, Gallagher M (2000) Changes in functional connectivity in orbitofrontal cortex and basolateral amygdala during learning and reversal training. J Neurosci 20:5179-5189.

Schoenbaum G, Setlow B (2002) Integrating orbitofrontal cortex into prefrotnal theory: common processing themes across species and subdivisions. Learn Mem 8:134-147.

Seib LM, Wellman CL (2003) Daily injections alter spine density in rat medial prefrontal cortex. Neurosci Lett 337:29-32.

Shipley JE, Kolb B (1977) Neural correlates of species typical behavior in the Syrian Golden hamster. J Comp Physiol Psychol 91:1056-1073.

Sirevaag AM, Greenough WT (1988) A multivariate statistical summary of synaptic plasticity measures in rats exposed to complex, social and individual environments. Brain Res 441:386-392.

Stewart J, Kolb B (1994) Dendritic branching in cortical pyramidal cells in response to ovariectomy in adult female rats: suppression by neonatal exposure to testosterone. Brain Res 654:149-154.

Uylings HBM, Van Eden CG (1990) Qualitative and quantitative comparison of the prefrontal cortex in rat and in primates. In: Progress in Brain Research, vol 85 (Uylings H.B.M., van Eden C., de Bruin JPC, Corner MA, and Feenstra MGP, eds), pp 31-62, Elsevier, Amsterdam.

Uylings HBM, Groenewegen HJ, Kolb B (2003) Do rats have a prefrontal cortex? Behav Brain Res (in press).

Warren JM (1977) Functional lateralization of the brain. Ann NY Acad Sci 299:273-280.

Warren JM, Kolb B (1978) Generalizations in neuropsychology. In: Brain Damage, Behavior and the Concept of Recovery of Function (Finger S, ed), Plenum Press, New York.

Wellman CL (2001) Dendritic reorganization in pyramidal neurons in medial prefrontal cortex aftger chronic corticosterone administration. J Neurobiol 49:245-253.

Whishaw IQ, Pellis SM, Gorny BP (1992a) Skilled reaching in rats and humans: evidence of parallel development or homology. Behav Brain Res 47:59-70.

Whishaw IQ, Pellis SM, Gorny BP (1992b) Medial frontal cortex lesions impair the aiming component of rat reaching. Behav Brain Res 50:93-104.

Whishaw IQ, Tomie J, Kolb B (1992c) Ventrolateral frontal cortex lesions in rats impair the acquisition and retention of a tactile-olfactory configural task. Behav Neurosci 106:597-603.

Wikmark RGE, Divac I, Weiss R (1973) Delayed alternationin rats wtih lesions of the frontal lobes: implications for a comparative neurospcyhology of the prefrontal system. Brain Behav Evol 8:329-339.

Woolley CS, Gould E, Frankfurt M, McEwen BS (1990) Naturally occurring fluctuation in dendritic spine density on adult hippocampal pyramidal neurons. J Neurosci 10:4035-4039.

Woolsey CN (1958) Organization of somatic sensory and motor areas of the cerebral cortex. In: Biological and Biochemical Bases of Behavior (Harlow HF and Woolsey CN, eds), University of Wisconsin Press, Madison.

Zilles K (1985) The Cortex of the Rat: A Stereotaxic Atlas. Springer-Verlag, Berlin.

Acknowledgements

The authors gratefully acknowledge the support of grants from NSERC and CIHR to BK and from OUC to JC.

Bryan Kolb's full corresponding address: Canadian Centre for Behavioural Neuroscience, University of Lethbridge, 4401 University Drive, Lethbridge, AB, Canada T1K 3M4.

tel: 403-329-2405; fax: 403-329-2775; e-mail: Kolb@uleth.ca

Chapter 2

WORKING MEMORY IN PREFRONTAL CORTEX AND ITS NEUROMODULATION

Jeremy K. Seamans
Department of Physiology, MUSC, 173 Ashley Avenue, Suite 403, Charleston, SC29425 USA. E-mail: seamans@musc.edu

Keywords: Short-term memory, delayed response, delay period activity, computational models, persistent activity.

Abstract: Working memory is conceptually different from short-term memory and likely relies on different neurobiological substrates. Working memory may be defined as the capacity to use mnemonic information to plan and organize forthcoming action. These processes rely on the prefrontal cortex (PFC), and neurons in this region appear to encode mnemonic information and forthcoming responses based on memory. The task related activity of PFC neurons and overall working memory performance is strongly regulated by dopamine. Dopamine might bias networks of PFC neurons to enter different processing modes, causing PFC networks to either process memory related information in a flexible manner (state 1) or to strongly maintain a single goal state in memory even in the presence of distracters (state 2). Dopamine levels in PFC fluctuate during different cognitive and emotional states, and such fluctuations could switch PFC networks between these two states. Dopamine may therefore dynamically regulate how PFC networks "work with memory" to guide future thought or action.

"Thinking is done by the cells of the brain behind the forehead... if the forehead cells do not know how to think, the mind cannot make use of memories. We say that such a person is a fool."

Overton (1897)

1. INTRODUCTION

Defining the neurobiology of working memory, Overton's statement made over a century ago was remarkably insightful in emphasizing that the cells behind the forehead (prefrontal cortex, PFC) are critical in the ability to make use of memories. This ability to make use of memories embodies the concept of working memory, which may be defined as the capacity to use mnemonic information to plan and organize forthcoming action. The term working memory has its origins in the work of cognitive and comparative psychologists such as Baddeley (1986; see also Baddeley and Hitch, 1974; Baddeley and DeSalla, 1996), Honig (1971), and Olton (Olton et al., 1979). Baddeley (1986) used the term working memory to replace the concept of passive short-term memory and to emphasize the on-line *manipulation* of information. According to Baddeley and Hitch (1974), working memory is composed of a central executive, which controls interconnected slave systems. One of these interconnected slave systems is a visuo-spatial sketchpad, which holds visuo-spatial information temporarily. The transient nature of information in the sketchpad separates working memory from other types of memory such as semantic or procedural memory which are long-lasting and which are thought to rely on 'passive' storage, whereby information is stored as changes in synaptic weights (e.g. Barnes, 1995).

Working memory appears to rely on the PFC. Goldman-Rakic (1991, 1995) and Fuster (1991) have argued that the activity of PFC neurons underlies the ability to hold transiently information that will be used to guide action (see below). Goldman-Rakic (1996) has stated that although damage to the PFC does not impair knowledge about the world or long-term memory, it does impair the ability to use such knowledge to guide behavior. Likewise, Fuster (1993) has stated, "frontal memory, above all, is memory for action". This type of memory for action embodies the concept of working memory as it emphasizes the executive control of memory used to guide action. However, there has been considerable confusion in the literature about exactly how working memory is defined experimentally and what separates it from short-term memory.

2. THE CONTRIBUTION OF THE PFC TO WORKING MEMORY, *NOT* SHORT-TERM MEMORY

In 1936, Jacobsen first demonstrated that lesions of the PFC of primates impair performance of the delayed-response working memory task and this finding has been replicated by numerous investigators (see Funahashi and Kubota, 1994 for review). However, there has been considerable difficulty in understanding the nature of this deficit. Working memory and short-term memory have been related theoretically, and therefore there has been a lasting tendency to view working memory processes mediated by the PFC simply as short-term memory processes. There is considerable evidence against the idea that the PFC subserves simply short-term memory processing.

First, short-term memory loss is generally not a result of selective PFC damage (Petrides, 1996). Patients with PFC damage show no deficits on traditional short-term memory tasks of recognition or recall, and such patients have a normal digit span and are unimpaired in the memory component of intelligence tests (Hebb, 1939, 1977; Stuss and Benson, 1986; Petrides, 1989; D'Esposito and Postle, 1999; Manes et al., 2002). Moreover, primates with PFC lesions perform normally on recognition memory tasks, delayed matching to sample tasks, and delayed object alternation tasks (Passingham, 1975; Bachevalier and Mishkin, 1986; Petrides, 1995, 2000a) that require short-term memory.

Consistent with the role of the PFC in working memory, PFC lesions affect the monitoring and manipulation of information in short-term memory. A classic demonstration of monitoring in memory is the self-ordered pointing task whereby different arrangements of stimuli are presented on each trial and the subject must choose a different stimulus until all are chosen (Petrides and Milner, 1982; Petrides, 1995). In this task, attention must be directed both to the stimulus under consideration as well as other stimuli in memory. Performance on this task is severely impaired by dorsolateral PFC lesions. Likewise the dorsolateral PFC is activated during the feedback portion of sorting tasks when current information must be related to earlier events (Monchi et al., 2001). PFC lesions also impair the ability to use memory to plan events in everyday life or plan responses on laboratory tasks (Shallice, 1982; Shallice and Burges, 1996; Robbins, 1996). PFC patients are particularly impaired on a modified version of the traditional Tower of London task that requires subjects to plan the moves from a starting state to a 'goal' configuration set by the experimenter (Robbins, 1996; Owen et al., 1990, 1995, 1996; Manes et al., 2002). In this way, the subject must plan moves internally by maintaining and comparing information about the initial, transition, and goal states in short-term memory. Thus, the deficit seen with patients with PFC damage is the result of an inability to monitor and manipulate information in memory, rather than the ability to actually hold the information in memory.

The distinction between the role of the PFC in *working*, as opposed to *short-term*, memory is made especially clear when one examines the effects of PFC lesions on tasks requiring response flexibility. On such tasks, PFC patients commit repeated errors that they are consciously aware of and that they can report, but cannot use to update behavior (Milner, 1963; Konow and Pribram, 1970). A classic example of this is observed in PFC patients with the Wisconsin Card Sorting task (Milner, 1963). This task requires subjects to formulate a card sorting strategy based on feedback from an experimenter. PFC-damaged patients are able to deduce, remember, and verbalize the correct sorting strategy to the experimenter, but are unable to alter their sorting strategy based on this knowledge. As a result, they perseverate in their initial response strategy, unable to shift to a strategy they know to be correct. Primates with lesions of the PFC also perseverate on their initial response strategy during performance of the analogous "A-not-B" task (Diamond and Goldman-Rakic, 1989). In the "A-not-B" task, primates must learn that one of two spatially distinct wells initially contains food while the other does not. After training, the well containing food is switched. Normal animals quickly go to the newly baited food well, while lesioned animals continue to revisit a previously rewarded spatial location, indicating that they had specific knowledge about the spatial location where food was presented previously, yet they could not use this knowledge to update their behavior. In contrast, primates with hippocampal damage perform normally at short delays (2-15s) but at 30s delays respond randomly on this task, not exhibiting the "AB error pattern" but rather alternating their responses between correct and incorrect food wells (Diamond et al., 1989). This indicates an anatomical dissociation between the retention of spatial-reward contingencies (at >15s intervals) and the ability to use this knowledge to guide behavior (working memory), with the former involving hippocampal regions and the latter involving the PFC.

According to Petrides (1994, 1995, 1996, 2000a), the PFC may act alone or in concert with other brain regions to guide working memory under different conditions. He has suggested that ventrolateral regions of the PFC are involved in the active organization of behavior based on the retrieval of information from posterior association corticies while dorsolateral regions are involved in holding information for monitoring and manipulation in accordance with willed actions. Based on this hypothesis, information may be retained within the PFC or in other brain regions but the critical function of the dorsolateral PFC relates to the ability to monitor, manipulate and use information to guide thought or action, i.e. working memory.

3. THE CELLULAR BASIS OF WORKING MEMORY

3.1 Functional Anatomy of the PFC

The PFC is a collection of distinct architectonic areas. It has traditionally been defined as the region rostral to motor and premotor areas as well as the prominent cortical projection area of the medial dorsal (MD) nucleus of the thalamus (Rose and Woolsey, 1948; Nauta, 1971; Groenewegen et al., 1990; Uylings and van Eden, 1990). The MD projects to the dorsolateral, ventrolateral and ventromedial PFC, and the medial and lateral PFC (Uylings and van Eden, 1990). In the primate, the mid-dorsolateral PFC has received the most attention as a locus for working memory processes, and encompasses the region within and above the principal sulcus (Brodmann's areas 46 and 9), anterior to area 8. Lesions to regions of the dorsolateral PFC that spare this principal sulcus mid-dorsolateral region, do not result in an impairment on the classic delayed response task (Goldman and Rosvold, 1970; Petrides, 2000a).

In the rat, the medial PFC is divided into several subregions, with the most dorsal region being the anterior cingulate, the middle region being the prelimbic (PL), and the most ventral region being the infralimbic cortex. According to Uylings and Van Eden (1990), the PL region of the rat is equivalent to area 32 or ventral medial PFC in the primate cortex. The rat lacks the anatomical equivalent of the mid-dorsolateral PFC (areas 46 and 9) in the primate. However, the PFC is thought to have evolved from both an archicortical and paleocortical moiety (Pandya and Yeterian, 1990). From the archicortical moiety arose proisocortical areas 24 (anterior cingulate), 25 (infralimbic), and 32 (prelimbic) which gave rise to dorsomedial and dorsolateral PFC regions in the primate (Panya and Yeterian, 1990). Thus, the prelimbic region may be viewed as a primitive version of the dorsolateral region of the primate PFC that is also anatomically related to the primate ventral medial PFC (Kolb, 1984).

The subiculum and nearby temporal corticies send projections to both the rat and primate PFC (Uylings and van Eden, 1990; Jay and Whitter, 1991; Condé et al., 1995). Likewise, the parietal cortex in the primate and the somatosensory cortex in the rat also project to the PFC (Goldman-Rakic, 1988; Condé et al., 1995; Mitchell and Cauller, 1997). Moreover, the PFC of both species also projects to the striatum (Sesack et al., 1989; Groenewegen et al. 1990; Uylings and van Eden 1990). The PFC is therefore situated to receive inputs from regions involved in the encoding and storage of spatial and object-related information (i.e. parietal and temporal cortices), while projecting to regions involved in response initiation (i.e. basal ganglia). Such an anatomical profile is required for a structure involved in using internal representations to guide action.

3.2 Cellular Analyses of Working Memory

The delayed response task has been used extensively to investigate the cellular bases of working memory processes (see Goldman-Rakic 1987, 1990, 1995 for reviews). In the classic delayed response task, monkeys observed an experimenter bait one of two covered food wells. An opaque screen was then lowered to block the monkey's view of the covered food wells. After a delay, the screen was raised and the monkey must choose the previously baited well to obtain the reward. More recently, an oculomotor delayed response task has been used to assess working memory. In this task, a monkey is placed in front of a video screen and must initially fixate on a center dot of light. During the *cue* phase, a light is flashed in one of 8 spatial locations on the screen that are equidistant from the center fixation light. The fixation and cue light are then extinguished for a few seconds in the *delay* period. During the *response* phase which follows the delay, the monkey is required to perform a saccade to the spatial location on the screen where the light was flashed. Since the cue light was extinguished, the saccade must be directed based on mnemonic information. In rats, a similar task has been used (Orlov et al., 1988; Bateuv et al., 1990), but a light was flashed above a food well to the right or left of the rat. After a delay of 5sec, the rat was allowed to visit the previously lighted food well. Approximately 54% of PFC neurons responded preferentially during the delay while the firing of 85% was correlated with the response (Orlov et al., 1988; Bateuv et al., 1990). These processes have been studied much more extensively in the primate dorsolateral PFC. Neurons in the primate PFC increase in activity during the cue, delay, and response phases of the original (Kubota and Niki, 1971; Fuster and Alexander, 1971; Fuster, 1973) and the oculomotor delayed response tasks (Funahashi et al., 1989). Most attention has been paid to the delay-active neurons in the PFC as the activity of these neurons may underlie the ability to retain information transiently (Goldman-Rakic, 1990, 1995). There are a number of findings that suggest that the activity of these neurons represents an active neural trace of previously encountered external stimuli.

First, delay-period activity is not observed on 'mock' trials, when the monkey does not observe a food well being baited (Fuster 1984, 1991). Second, delay-active neurons have 'memory fields' in that individual neurons fire during the delay period of the task, only if a cue was presented previously in a specific spatial location (Funahashi et al., 1989; Goldman-Rakic, 1990). Third, if the activity of these neurons decreases throughout the delay, the animal is highly likely to make an error (Niki and Watanabe, 1979; Funahashi and Kubota, 1994; Funahashi et al., 1989). Fourth, these neurons show sustained firing during the delay even if the animal is required to make a response in the opposite location from the initial cue, indicating that the activity is related to the memory of the previously presented stimuli and not the mechanics of the response itself (Funahashi et al., 1993). Finally, activity during the delay increases or decreases uniformly as the delay interval increases or decreases (Kojima and Goldman, 1982). These finding suggest that indeed neurons in the dorsolateral PFC seem to transiently and actively encode information about previously presented stimuli. While having this type of activity is a requirement for a working memory system, it does not imply that the short-term retention of information is the primary function of the PFC. Rather, information must be held transiently if it is to be manipulated and used to guide action.

The PFC is not unique in its ability to exhibit delay period activity. Delay-active neurons are also found in other areas of the brain such as the parietal and inferotemporal cortex and hippocampus (Watanabe and Niki, 1985; Koch and Fuster, 1989; Fuster, 1990), suggesting that copies of recently presented task-relevant stimuli are distributed. This may explain why PFC lesions alone do not impair short-term memory. However these brain areas interact during the performance of delayed tasks since PFC cooling disrupts delay-period activity in the inferotemporal cortex (Fuster et al., 1985), while cooling of the parietal cortex or inferotemporal cortex disrupt task related activity in PFC neurons (Fuster et al., 1985; Quintana et al., 1989; Chafee and Goldman-Rakic, 1998, 2000). Miller and Desimone (1994) and Miller et al. (1996) have pointed out key differences between activity in PFC and inferotemporal or parietal neurons. The activity of PFC neurons is less stimulus dependent but exhibits greater 'match-non-match' effects on delayed matching and nonmatching to sample tasks, again suggesting that PFC neurons are more involved in the manipulation of information in memory. In addition, PFC neurons exhibit progressive increases in activity during the delay period. The progressive increase in activity of PFC neurons during the delay has been termed "climbing activity" and is related to the probability of making a correct forthcoming response (Quintana and Fuster, 1992). The climbing activity in the PFC may be related to the prospective memory of the upcoming response. Response-correlated activity in PFC neurons is also observed on simple non-delayed tasks without a memory component, such as Go/No Go tasks (Watanabe, 1986a,b). Furthermore, on more complex conditional tasks, the activity of motor-set units can precede that of delay-active neurons in well trained animals. In such tasks, the color of a cue light instructs experienced monkeys where to direct their response following a delay. The activity of motor set neurons often begins to increase as soon as the light cue is presented, presumably because information about the direction of a forthcoming response is given completely by the color of the cue light (Fuster, 1991). Thus, on both working memory tasks and conditional memory tasks, the discharge of the motor-set units in the PFC may predict the direction of the impending motor response. Thus, there is a subclass of PFC neurons that encode impending actions based on memory.

If a response is guided by information that pertains to future actions not yet completed (i.e. to remember what needs to be done), it is said to be coded prospectively; if it is based on a comparison to stimuli/actions that have already been encountered (i.e. to remember what has already been done), then it is coded retrospectively (Cook et al., 1985). Clearly, motor set units are coding the prospective response, but many of the delay activity neurons encode memory prospectively as well. Rainer et al. (1999) used a type of conditional task that assessed prospective coding, the delayed paired associate task, and compared it to a simple delayed match to sample task. In the delayed paired associate task, three sets of sample and test stimuli were paired. Two sample stimuli and two test stimuli were similar in appearance. One sample was presented, and following a delay, a test stimulus was presented that may or may not have been previously paired with the sample stimulus. If the previously paired test stimulus appeared after a delay, the animal had to release a level to obtain reward. Reaction times were similar whether the test and sample stimuli were the same (delayed match to sample task) or for test stimuli predicted by a previously paired sample stimulus (delayed paired associate task). Moreover, errors occurred more frequently for similar looking test stimuli as opposed to similar looking sample stimuli. This suggested that the performance of the animals was dependent upon the anticipation of the forthcoming stimulus based on the memory of previous sample-test stimuli pairings, and is therefore indicative of a prospective code. Likewise, a number of neurons exhibited increased firing throughout the delay for a given test stimulus regardless of which sample stimuli preceded it. This increased activity for the forthcoming target occurred prior to the presentation of the test stimulus and was selective for certain forthcoming test stimuli, indicating that the neurons were encoding the anticipated test stimulus. These data provide evidence that neurons in the PFC are capable of encoding the prospective memory of a forthcoming stimulus.

There is also evidence for distinctly retrospective coding by delay active neurons in the PFC (e.g. Rainer et al., 1999; Fuster, 2000; Constantinidis et al., 2001). Yet as noted above, this activity is not unique to the PFC, and the integrity of the PFC is not necessary for short-term memory. Rather, this retrospective coding may only be necessary to hold information long enough so that it can be used to guide responding. Or as Fuster (1990, 1991, 1995) has proposed, mnemonic information encoded by delay-active cells may be communicated to response-active PFC neurons to ensure that a forthcoming response is directed to the correct location. In this way, the retrospective coding by PFC neurons may simply be required to maintain information in memory long enough to manipulate it and use it to guide the appropriate action.

If the delay period is very brief, the online maintenance and manipulation of information occur simultaneously and therefore cannot easily be dissociated. Yet even at short delays, Rainer et al. (1999) showed that activity of PFC neurons shifted from encoding the sample stimulus to anticipation of the test stimulus. If PFC neurons were primarily encoding the manipulation of information in memory, one would predict that at very delays, too long to maintain information actively, "delay-period" activity should begin to occur near the time of the response because it is at this point where information is manipulated and used to guide action. As a test of this hypothesis in rats, transient inactivation of the rat PFC by lidocaine impaired response phase performance on a delayed working memory radial arm maze task, only if given prior to the response phase and not prior to the sample phase or during the delay (Seamans et al. 1995). However, given the differences in the activity in rat and primate PFC neurons during working memory tasks (see Pratt and Mizumori, 2001), similar experiments with longer delays are required in experiments using primates.

The idea that the role of the PFC is in the manipulation of information in memory rather than its simple storage, removes a temporal component to working memory. Accordingly, dorsolateral PFC lesions do not produce delay-dependent deficits on working memory tasks (Petrides, 2000b). Yet some definitions of working memory emphasize the temporal nature of working memory. Working memory has been defined as memory for trial unique information, while reference memory was related to the memory of trial invariant stimuli (Olton et al., 1979). However, most tasks involving working memory and the prefrontal cortex require the implementation of trial invariant information such as the implementation of learned rules required to solve the task. While this type of information may be viewed as reference memory, it is related more to the abstract procedural rules rather than specific information such as the invariant location of food. A variety of lesion studies highlight the important role of the PFC in the application and use of abstract rules and the firing of delay-active neurons varies when different task rules are implemented (Milner, 1963; Passingham, 1993; Verin et al., 1993; Seamans et al., 1995; Wise et al., 1996; White and Wise, 1999; Wallis et al., 2001).

Collectively, it seems that the PFC provides much more than a short-term memory store. It maintains, monitors, and compares items in memory based on context dependent, abstract rules. In this way, the function of the PFC may be best described not as working memory but rather as "working with memory".

3.3 Cellular Working Memory and Behavioral Significance

A critical aspect in the ability to work effectively with memory is to determine which stimuli are appropriate in a given context. At the cellular level, a selection process must occur so that task relevant items are maintained and compared while the multitude of other stimuli potentially encoded by PFC afferents is ignored. The most effective way to accomplish this is to have the task-related activity of PFC neurons be related to the behavioral significance of the encoded stimuli.

Considerable evidence suggests that task-related activity of PFC neurons is regulated by the behavioral significance of the stimuli presented. Although PFC neurons respond to visual cues not associated with reward, such activity is significantly enhanced if stimuli are of particular behavioral significance (Bruce, 1988). In contrast, cells in other cortical association areas typically code only for specific stimuli, regardless of their significance (Miller et al., 1996). PFC neurons do not respond directly to the presence or absence of reward, but respond similarly to different stimuli with the same behavioral significance while responding differently to identical stimuli of varying behavioral significance (Watanabe, 1981, 1986a,b, 1990, 1996; Watanabe et al., 2002).

In non-delayed tasks, some PFC neurons respond simply to the presentation of a primary reward, and this activity is abolished if the rewarding value of the stimuli is decreased by adding quinine to the food (Inoue et al., 1985). Likewise, on delayed response tasks, the delay-period activity of PFC neurons is dependent strongly on the nature of the reward, as cues associated with palatable reward produce significantly greater activity in delay-active PFC neurons (Watanabe, 1996). Although PFC neurons fired more vigorously to stimuli predictive of reward relative to equivalent stimuli that were irrelevant to the monkey (Yajeya et al., 1988), the activity of delay-active neurons was more vigorous if food itself served as a cue relative to a stimulus previously paired with food.

Watanabe has investigated the issue of modulation of PFC firing by presentation of stimuli of behavioral significance in a series of insightful studies (Watanabe, 1981, 1986a,b, 1990, 1996; Watanabe et al., 2002). Watanabe (1990) tested the effect of associative significance on PFC unit activity using a novel associative learning task that varied the significance of stimuli, while keeping the mnemonic and response demands constant. On such tasks the animal must release a lever to begin a trial; however, reward is delivered only on trials where a discriminative cue had been presented several seconds earlier. As in the delayed response task, subsets of PFC units were active during the cue, delay, and response phases of the task. However, a majority of these task-related neurons showed increased activity only on trials where the discriminative cue was presented, regardless of its physical attributes. Using a similar approach, Watanabe et al. (2002) used the delay period of a modified delayed response task, in which the cue indicated whether and what type of reward would occur after a delay, rather than which response to make. The firing rate of dorsolateral PFC neurons was not only dependent on whether the cue indicated reward would be present, but also showed a quantitatively different delay-period activity depending on what type of reward (i.e. what food item) was expected to occur. Moreover, if the firing rate of the neuron was reward discriminate, the baseline firing rate often remained until the start of the next trial.

Collectively, these studies demonstrate that in PFC mnemonic information is modulated by reward and that PFC neurons encode not only the memory of a forthcoming response but also the memory of a forthcoming reward. Moreover, the "strength" of this memory (i.e. PFC unit activity) is directly related to the particular significance of the stimuli. Finally, the effect of reward exerts a tonic effect on PFC activity as the firing rate often remains across trials. Based on these data, one would conclude that there is a reward-related signal that is transferred to the PFC that alters the memory-related firing rate of PFC neurons.

3.4 Role of Dopamine in Determining Behavioral Significance

Reward related information in the brain appears to be coded specifically by the activity of midbrain dopamine (DA) neurons (Schultz, 1992a,b; Schultz et al., 1998). DA neurons respond in short phasic bursts to appetitive or novel stimuli (Romo and Shultz, 1990). However, the response of DA neurons to the same stimuli changes as the salience of the stimulus changes. For example, novel stimuli that evoke a vigorous response initially do not activate DA neurons when the animal is familiar with the stimuli. DA neurons also change their response to stimuli paired with reward (Romo and Schultz, 1990; Ljungberg et al., 1992). Initially, DA neurons respond immediately after the receipt of reward. With repeated pairings of the conditioned stimulus (CS) and the primary reward (unconditioned stimulus, US), the phasic activation of DA neurons shifts from the time of delivery of reward to the time of CS onset. Following this shift, DA neurons no longer respond to the primary reward. The shift in the activity of DA neurons is related to the shift in the monkeys' appetitive behavioral reaction from the US to the CS (Schultz et al., 1998). DA neurons therefore code for both the *a priori* and learned significance of stimuli.

Although the activity of both DA neurons and PFC neurons are related to the significance of stimuli, there are notable differences in their activation characteristics. First, DA neurons tend to respond homogeneously to a given stimulus (Schultz, 1992a,b). In contrast, the activity of PFC neurons exhibits considerable heterogeneity as the response of individual neurons varies depending on the attributes of the object, it's location, and when it is presented in time (Miler and Cohen, 2001; Freedman et al., 2002; Rainer and Miller, 2002). Second, the response of DA neurons to the same stimulus can be very different depending on its significance in a given context

(Watanabe, 1998; Watanabe et al., 2002). The response of PFC neurons is less dependent on the ascribed significance of the stimulus as the significance only serves to modify the responses of PFC neurons to other task- related variables (Watanabe, 1996). Given these properties, DA neurons appear to code specifically for the behavioral significance of a stimulus, while the activity of PFC neurons is only modified by behavioral significance.

Schultz (1992 a,b) postulated that behavioral significance of a stimulus might be signaled to the PFC via the release of DA from the terminals of DA neurons. He suggested that DA released in the PFC may focus the activity of PFC neurons such that this activity is restricted to the processing of the most prominent or behaviorally significant inputs. As such, the DA input functions much in the same way as does attention, altering brain circuits in response to stimuli of behavioral significance. Indeed, Redgrave et al. (1999) have suggested that the phasic burst of activity by DA neurons switches attentional and behavioral resources to behaviorally significant stimuli.

3.5 DA Modulation of Working Memory Processes Mediated by the PFC

DA strongly modulates both working memory performance and the task-dependent neuronal activity within the PFC. 6-OHDA lesions or microinjection of DA D1 receptor antagonist into the PFC disrupts performance on delayed-response tasks (Brozoski et al., 1979; Sawaguchi et al., 1990b, 1994; Seamans et al., 1998; Zahrt et al., 1997; Aujla and Benninger, 2001). Paradoxically, pharmacologically-induced high rates of DA turnover in the PFC also produce deficits in delayed-tasks (Murphy et al., 1996). Similarly, iontophoresis of either DA *or* a D1 antagonist at low ejection currents enhance delay period activity, relative to 'background', activity on a delayed response task (Sawaguchi and Matsumura, 1985; Sawaguchi et al. 1986, 1990a,b; Sawaguchi, 1987; Williams and Goldman-Rakic, 1995). Thus, the action of DA in the PFC is highly complex, and both increases and decreases in DA activity can enhance or attenuate performance on working memory tasks and task-related neural activity.

Another complex aspect of the DA dependent modulation of delayed responding is that DA neurons in the ventral tegmental area (VTA) do not show sustained activity throughout the delay period of a delayed response task (Shultz and Romo, 1990; Ljungberg et al., 1992). In order to reconcile these data, it has been argued that during delayed responding, DA release may be modulated at the terminal level in the PFC (Schultz, 1992a,b). Alternatively, DA released at the outset of the task may modulate the activity of PFC neurons for prolonged periods via second messenger signaling pathways coupled to the D1 receptor.

3.6 Electrophysiological Action of DA on PFC Neurons

Electrophysiological data indicate that DA exerts complex, long-lasting modifications in the properties of PFC neurons. The action of DA on pyramidal cells in the PFC has been studied traditionally using extracellular recording techniques on anaesthetized rats *in vivo*. Most studies have shown that DA exerts an inhibitory effect on pyramidal cell excitability (Bunney and Aghajanian, 1976; Mora et al., 1976; Mantz et al., 1988; Sesack and Bunney, 1989; Godbout et al., 1991; Pirot et al., 1992). However, in a critical experiment, Pirot et al. (1992) showed that this inhibitory effect was often abolished if the $GABA_A$ antagonist bicuculline was iontophoresed prior to DA, suggesting that the inhibitory action of DA was an indirect effect on GABAergic interneurons. This finding is consistent with the DA-mediated increased spontaneous IPSP frequency and evoked IPSC amplitude in PFC pyramidal neurons and the increased depolarization and firing of fast spiking interneurons (Penit-Soria et al., 1987; Zhou and Hablitz, 1999; Seamans et al., 2001b; Gorelova et al., 2002). Thus, DA may suppress the activity of PFC pyramidal neurons via interneurons.

In contrast, DA appears to enhance the effects of excitatory stimuli directly onto PFC pyramidal neurons. DA has been shown to enhance the intrinsic excitability of pyramidal neurons (Yang and Seamans, 1996; Gorelova and Yang, 2000; Henze et al., 2000; Gulledge and Jaffe, 2001), and the excitatory responses of PFC neurons to NMDA or acetylcholine (Yang and Mogenson, 1990; Cépeda et al., 1992; Zheng et al. 1999; Seamans et al., 2001a). Remarkably, these effects last for very prolonged periods of time, often tens of minutes after agonist offset. In this way, DA may exert a processing tone in the PFC that alters the way that PFC neurons respond to subsequent excitatory and inhibitory stimuli.

3.6.1 The Effect of DA in PFC: Enhancing Robustness

When viewed collectively, it is clear that DA has multiple, often contradictory effects on the activity of PFC pyramidal neurons. Durstewitz et al. (2000) and Durstewitz and Seamans (2002) have argued that these diverse actions mediated by DA converge on a single function: increasing robustness of working memory representations in PFC networks. Specifically, D1 enhancement of NMDA and persistent inward currents causes strongly activated assemblies of interconnected neurons to exhibit a significant boost in sustained activity levels. Since assemblies of neurons are thought to be regulated by interneuronal activity (Lewis et al., 1999), the increased activation of one assembly may quell activity in nearby competing assemblies. This effect is further enhanced by the D1-mediated increase in widespread (Seamans et al., 2001b) but not cell to cell unitary IPSCs (Gao et al., 2003). Collectively, this leads to the acceleration in the activation of one assembly at the expense of activity in other assemblies. If an assembly encodes items in working memory, this would imply that one item in working memory exerts greater control over working memory buffers. Thus, if DA encodes behavioral significance and transfers this to the PFC, then our models would predict that the effect of behavioral significance would be focusing working memory buffers on a limited set of stimuli for action.

3.6.2 The Role of D2 Receptors in PFC: Expanding Focus

This hypothesis outlined above is valid only for the case of strong D1 receptor stimuli, which would be a common occurrence, given the disproportionate densities of D1 relative to D2 receptors in the PFC (Vincent et al., 1993; Gaspar et al., 1995). In contrast, conditions favoring strong activation of D2 receptors would actually reduce pyramidal cell excitability (Gulledge and Jaffe, 1998, 2001), NMDA currents (Zheng et al., 1999), and $GABA_A$ currents (Seamans et al., 2001b) in pyramidal neurons. Strong D2 activation would therefore have the opposite effect from D1 activation, with assemblies showing spontaneous transitions to persistent activity states (Durstewitz et al., 2000) and multiple assemblies co-activated nearly simultaneously. Under this regime, many items may be encoded in working memory yet none particularly robustly. These ideas are shown graphically in Figure 1.

3.7 A Summarizing Hypothesis

Five main points were presented above; 1) Working memory within the PFC may be best represented as "working with memory" to incorporate the online monitoring and manipulation of mnemonic information, 2) Persistent delay-period activity in PFC underlies the ability to work with memories, 3) Persistent activity associated with working memory is affected by behaviorally significant stimuli, 4) DA neurons signal stimuli of behavioral significance, and 5) DA affects working memory performance and the cellular activity encoding working memory information in the PFC.

According to our hypothesis (Durstewitz et al., 2000; Durstewitz and Seamans, 2002), if a stimulus of significance is encountered, the increased DA release due to elevated VTA firing enhances the encoding of information in working memory without providing any specific information.

In other words, the DA signal alters the processing of information from other sources, but provides no information on its own. DA activation that results in a strong D2 tone (state 1) would allow multiple representations to coexist in PFC networks (Fig. 1), but none would be particularly strong. In this way, the PFC network is working with memories in a flexible manner, allowing multiple memory items to potentially control action. A strong D1 tone (state 2) would shift the processing mode so that a single goal state would be established strongly, and this goal state would be maintained even in the presence of alternative information or distracters. Thus, if a stimuli of behavioral significance is encountered and DA is released in PFC, networks within the PFC work with mnemonic information to consider either many options (state 1) or a single option (state 2) for action.

Certain situations and internal states evoke varying levels of DA in PFC and induce state 1 or state 2 network dynamics. Animals that are deprived of basic physiological needs, such as food or water, respond with a large increase in PFC DA levels when they subsequently encounter such stimuli (Feenstra et al., 1995; Feenstra and Botterblom, 1996; Taber and Fibiger, 1997; Ahn and Phillips, 1999). Likewise, stressors, such as food shock or restraint, also evoke significant increases in PFC DA (Horger and Roth, 1996; Finlay and Zigmiond, 1997; Feenstra, 2000). When the behavioral significance of the stimulus decreases, such as satiation in a hungry rat or

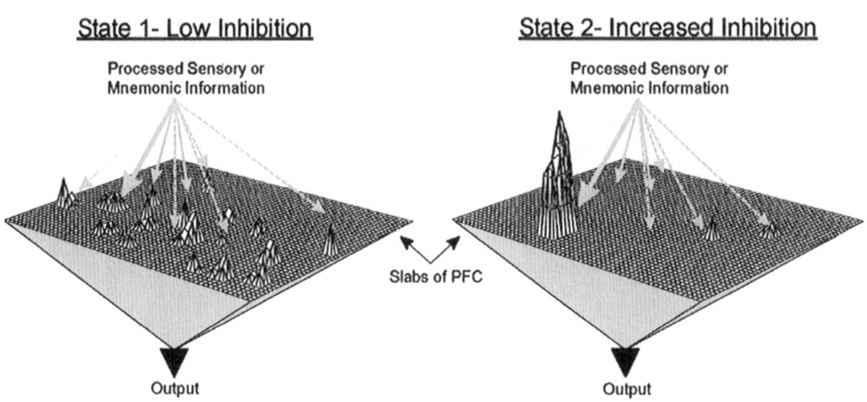

Figure 1: Theoretical two state dynamics for working memory buffers in the PFC. State 1 is characterized by low inhibition due to D2 down-regulation of GABAergic inhibition. With the simultaneous D2 down-regulation of excitability of pyramidal cells and NMDA currents in state 1, many assemblies are active but only weakly. State 2 is characterized by heightened inhibition due to D1-mediated up-regulation of GABAergic inhibition and excitability of interneurons. Coupled with the D1 mediated increase in pyramidal excitability and NMDA currents in state 2, a single representation is strongly activated and maintained in the face of competing inputs. Adopted from Seamans et al. (2001b) with permission.

multiple encounters with a familiar female, less DA release is typically observed (Fiorino et al., 1997; Ahn and Phillips, 1999). Perhaps the level of DA released in the PFC dictates which receptors are preferentially activated and which state is established. Indeed, it is evident from *in vivo* and *in vitro* data from the striatum and PFC that varying levels of DA exert opposing physiological actions via D1 versus D2 receptors on spiking and glutamate currents (Akaike et al., 1987; Hu and Wang, 1988; Willilams and Millar, 1990; Yang and Mogenson, 1990; Zheng et al., 1999). One possibility is that moderately significant stimuli (e.g. food to a satiated animal, or the presentation of a habituated stressor) would cause moderate activity of mesocortical DA system and set up a state 1 network dynamic. In this case, PFC networks work with memory related information in a flexible manner to determine the best course of action, based on experience. When highly significant stimuli are encountered (e.g. food to a hungry animal, or the presentation of a particularly stressful stimulus), a state 2 dynamic might occur and PFC networks work with memory-related information to establish a very fixed goal state that completely dominates PFC output even in the presence of distracters (i.e. the goal state representation is robust). Pushing the system even more, continual high DA loads as might occur with chronic exposure to drugs of abuse, may lock PFC into state 2. As a result, PFC networks work with memories in a very rigid manner, and an extremely limited number of goal state representations are established and maintained, but those that are completely control behavior. This would be the case in addictive behaviors where all cognitive resources are directed towards the attainment of the drug.

4. CONCLUSION

Working memory buffers in PFC do not simply hold memory information transiently but rather work with memories to guide action in a dynamic fashion according to internal and external stimuli. In conditions where highly important stimuli are encountered, PFC networks may establish a limited number of goal states perhaps via predominant activation of D1 receptors, at the expense of all competing information and goal states. In less stressful situations, PFC networks may deal flexibly with mnemonic information to guide forthcoming actions in manner that is less dire and more exploratory, perhaps via predominant activation of D2 receptors. The goal of future research will be to determine what types of stimuli and DA release events activate D1 versus D2 classes of DA receptors in PFC, and whether this varies on an individual or context dependent basis. Such information may provide a novel way to look at working memory processes in the PFC under normal and pathological conditions.

REFERENCES

Ahn S, Phillips AG (1999) Dopaminergic correlates of sensory-specific satiety in the medial prefrontal cortex and nucleus accumbens of the rat. J Neurosci 19: RC29.

Akaike A, Ohno Y, Sasa M, Takaori S (1987) Excitatory and inhibitory effects of dopamine on neuronal activity of the caudate nucleus neurons in vitro. Brain Res 418: 262-272.

Aujla H, Beninger RJ (2001) Hippocampal-prefrontocortical circuits: PKA inhibition in the prefrontal cortex impairs delayed nonmatching in the radial maze in rats. Behav Neurosci 115: 1204-1211.

Bachevalier J, Mishkin M (1986) Visual recognition impairment follows ventromedial but not dorsolateral prefrontal lesions in monkeys. Behav Brain Res 20: 249-261.

Baddeley A (1986) Working Memory, Oxford University Press.

Baddeley A, Della Sala S (1996) Working memory and executive control. Phil Trans Royal Soc Lond 351: 1397-1404.

Baddeley AD, Hitch G (1974) Working memory. In: The Psychology of Learning and Motivation. Advances in Research and Theory, (Bower GH, ed), pp 47-89, NY Academic Press.

Barnes CA (1995) Involvement of LTP in memory: Are we "searching under the street light"? Neuron 15: 751-754.

Batuev AS, Kurina NP, Shutov AP (1990) Unit activity of the medial wall of the frontal cortex during delayed performance in rats. Behav Brain Res 41: 95-102.

Brozowski TS, Brown RM, Rosvold HE, Goldman PS (1979) Cognitive deficits caused by regional depletion of dopamine in prefrontal cortex of Rhesus monkey. Science 205: 929-932.

Bruce CJ (1988) Single neuron activity in the monkey's prefrontal cortex. In: Neurobiology of Neocortex (Rakic P and Singer W, eds), pp 297-329, John Wiley and Sons.

Bunney BS, Aghajanian GK (1976) Dopamine and norepinephrine innervated cells in the rat prefrontal cortex: pharmacological differentiation using microiontophoretic techniques. Life Sci 19: 1783-1792.

Cépeda C, Radisavljevic Z, Peacock W, Levine MS, Buchwald NA (1992) Differential modulation by dopamine of responses evoked by excitatory amino acids in human cortex. Synapse 11: 330-341.

Chafee MV, Goldman-Rakic PS (1998) Matching patterns of activity in primate prefrontal area 8a and parietal area 7ip neurons during a spatial working memory task. J Neurophysiol 79: 2919-2940.

Chafee MV, Goldman-Rakic PS (2000) Inactivation of parietal and prefrontal cortex reveals interdependence of neural activity during memory-guided saccades. J Neurophysiol 83: 1550-1566.
Condé F, Marie-Lepoivre E, Audinat E, Crépel F (1995) Afferent connections of the medial frontal cortex of the rat. II. Cortical and subcortical afferents. J Comp Neurol 352: 567-593.
Constantinidis C, Franowicz MN, Goldman-Rakic PS (2001) The sensory nature of mnemonic representation in the primate prefrontal cortex. Nature Neurosci 4: 311-316.
Cook RG, Brown RF, Riley DA (1985) Flexible memory processing by rats: use of prospective and retrospective information in the radial arm maze. Anim Behav Proc 11: 453-469.
D'Esposito M, Postle BR (1999) The dependence of span and delayed-response performance on prefrontal cortex. Neuropsychologia 37: 1303-1315.
Diamond A, Goldman-Rakic PS (1989) Comparison of human infants and rhesus monkeys on Piaget's AB task: evidence for dependence on dorsolateral prefrontal cortex. Exp Brain Res 74: 24-40.
Diamond A, Zola-Morgan S, Squire LR (1989) Successful performance by monkeys with lesions of the hippocampal formation on AB and object retrieval, two tasks that mark developmental changes in human infants. Behav Neurosci 103: 526-537.
Durstewitz D, Seamans JK, Sejnowski TJ (2000) Dopamine-mediated stabilization of delay-period activity in a network model of prefrontal cortex. J Neurophysiol 83: 1733-1750.
Durstewitz D, Seamans JK (2002) The computational role of dopamine D1 receptors in working memory. Neural Netw 15: 561-572.
Feenstra MG (2000) Dopamine and noradrenaline release in the prefrontal cortex in relation to unconditioned and conditioned stress and reward. In: Progress in Brain Research, vol 126 (Uylings HBM, Van Eden CG, De Bruin JPC, Feestra MGP, and Pennartz CMA, eds), pp 133-163.
Feenstra MG, Botterblom MH (1996) Rapid sampling of extracellular dopamine in the rat prefrontal cortex during food consumption, handling and exposure to novelty. Brain Res 742: 17-24.
Feenstra MG, Botterblom MH, van Uum JF (1995) Novelty-induced increase in dopamine release in the rat prefrontal cortex in vivo: inhibition by diazepam. Neurosci Lett 189: 81-84.
Finlay JM, Zigmond MJ (1997) The effects of stress on central dopaminergic neurons: possible clinical implications. Neurochem Res 22: 1387-1394.

Fiorino DF, Coury A, Phillips AG (1997) Dynamic changes in nucleus accumbens dopamine efflux during the Coolidge effect in male rats. J Neurosci 17: 4849-4855.

Freedman DJ, Riesenhuber M, Poggio T, Miller EK (2002) Visual categorization and the primate prefrontal cortex: neurophysiology and behavior. J Neurophysiol 88: 929-94.

Funahashi S, Kubota K (1994) Working memory and prefrontal cortex. Neurosci Res 21: 1-11.

Funahashi S, Bruce CJ, Goldman-Rakic PS (1989) Mnemonic coding of visual space in the monkey's dorsolateral prefrontal cortex. J. Neurophysiol 61: 331-349.

Funahashi S, Chafee MV, Goldman-Rakic PS (1993) Prefrontal neuronal activity in rhesus monkeys performing a delayed anti-saccade task. Nature 365: 753-756.

Fuster JM (1973) Unit activity in prefrontal cortex during delayed-response performance: neuronal correlates of transient memory. J Neurophysiol 36: 61-78.

Fuster JM (1984) Behavioral electrophysiology of the prefrontal cortex. Trends Neurosci 7: 408-414.

Fuster JM (1990) Inferotempoal units in selective visual attention and short-term memory. J Neurophysiol 64: 681-697.

Fuster JM (1991) The prefrontal cortex and its relation to behavior. Prog Brain Res 87: 201-211.

Fuster JM (1993) Frontal Lobes. Curr Opn Neurobiol 3: 160-165.

Fuster JM (1995) Memory in the cerebral cortex: an empirical approach to neural networks in the human and nonhuman primate. MIT Press.

Fuster JM (2000) Executive frontal functions. Exp Brain Res 133: 66-70.

Fuster JM, Alexander GE (1971) Neuron activity related to short-term memory. Science 173: 652-654.

Fuster JM, Bauer RH, Jervey JP (1985) Functional interactions between inferotemporal and prefrontal cortex in a cognitive task. Brain Res 330: 299-307.

Gao WJ, Wang Y, Goldman-Rakic PS (2003) Dopamine modulation of perisomatic and peridendritic inhibition in prefrontal cortex. J Neurosci 23: 1622-1630.

Gaspar P, Bloch B, Le Moine C (1995) D1 and D2 receptor gene expression in the rat frontal cortex: cellular localization in different classes of efferent neurons. Eur J Neurosci 7: 1050-1063.

Godbout R, Mantz J, Pirot S, Glowinski J, Thierry A-M (1991) Inhibitory influence of the mesocortical dopaminergic neurons on their target cells: electrophysiological and pharmacological characterization. J Pharmacol Exp Therap 258: 728-738.

Goldman PS, Rosvold HE (1970) Localization of function within the dorsolateral prefrontal cortex of the rhesus monkey. Exp Neurol 27: 291-304.

Goldman-Rakic PS (1987) Circuitry of the prefrontal cortex and the regulation of behavior by representational knowledge. In: Handbook of Physiology (Plum F and Mountcastle V, eds), pp 373-417, American Physiological Society, Maryland.

Goldman-Rakic PS (1988) Topography of cognition: Parallel distributed networks in primate association cortex. Annu Rev Neurosci 11: 137-156.

Goldman-Rakic PS (1990) Cellular and circuit basis of working memory in prefrontal cortex of nonhuman primates. Prog Brain Res 85: 325-335.

Goldman-Rakic PS (1991) Prefrontal cortical dysfunction in schizophrenia: the relevance of working memory. In: Psychopathology and the Brain (Carroll BJ and Barrett JE, eds), pp 1-23, Raven Press.

Goldman-Rakic PS (1992) Dopamine-mediated mechanisms of the prefrontal cortex. The Neurosciences, 4: 149-159.

Goldman-Rakic PS (1995) Cellular basis of working memory. Neuron 14: 477-485.

Goldman-Rakic PS (1996) The prefrontal landscape: implications of functional architecture for understanding human mentation and the central exectutive. Phil Trans Royal Soc Lond 351: 1445-1453.

Gorelova N, Seamans JK, Yang CR (2002) Mechanisms of dopamine activation of fast-spiking interneurons that exert inhibition in rat prefrontal cortex. J Neurophysiol 88: 3150-3166.

Gorelova NA, Yang CR (2000) Dopamine D1/D5 receptor activation modulates a persistent sodium current in rat prefrontal cortical neurons in vitro. J Neurophysiol 84: 75-87.

Groenewegen HJ, Berendse HW, Wolters JG, Lohman AHM (1990) The anatomical relationship of the prefrontal cortex with the striatopallidal system, the thalamus and the amygdala: evidence for a parallel organization. Prog Brain Res 85: 95-118.

Gulledge AT, Jaffe DB (1998) Dopamine decreases the excitability of layer V pyramidal cells in the rat prefrontal cortex. J Neurosci 18:9139-9151.

Gulledge AT, Jaffe DB (2001) Multiple effects of dopamine on layer v pyramidal cell excitability in rat prefrontal cortex. J Neurophysiol 86: 586-95.

Hebb DO (1939) Intelligence in man after large removals of cerebral tissue: report of four left frontal lobe cases. J Gen Psychol 21: 73-87.

Hebb D (1977) The frontal lobe. CMA Journal 116: 1373-1374.

Henze DA, Gonzalez-Burgos GR, Urban NN, Lewis DA, Barrionuevo G (2000) Dopamine increases excitability of pyramidal neurons in primate prefrontal cortex. J Neurophysiol 84: 2799-2809.

Honig WK (1971) Animal Memory. Academic Press.
Horger BA, Roth RH (1996) The role of mesoprefrontal dopamine neurons in stress. Crit Rev Neurobiol 10: 395-418.
Hu XT, Wang RY (1988) Comparison of effects of D-1 and D-2 dopamine receptor agonists on neurons in the rat caudate putamen: an electrophysiological study. J Neurosci 8: 4340-4348.
Inoue M, Oomura Y, Auo S, Nishino H, Sikdar S (1985) Reward related neuronal activity in monkey dorsolateral prefrontal cortex during feeding behavior. Brain Res 326: 307-312.
Jay TM, Witter MP (1991) Distribution of hippocampal CA1 and subicular efferents in the prefrontal cortex of the rat studied by means of the anterograde transport of Phaseolus vulgaris leucoagglutinin. J Comp Neurol 313: 574-586.
Koch KW, Fuster JM (1989) Unit activity in monkey parietal cortex related to haptic perception and temporary memory. Exp Brain Res 76: 292-306.
Kojima S, Goldman-Rakic PS (1982) Delay-related activity of prefrontal neurons in rhesus monkeys performing delayed response. Brain Res 248: 43-49.
Kolb B (1984) Functions of the frontal cortex of the rat: a comparative review. Brain Res Rev 8: 65-98.
Konow A, Pribram KH (1970) Error recognition and utilization produced by injury to the frontal cortex in man. Neuropsychologia 8: 489-491.
Kubota K, Niki H (1971) Prefrontal cortical unit activity and delayed alternation performance in monkeys. J Neurophysiol 34: 337-347.
Lewis DA, Pierri JN, Volk DW, Melchitzky DS, Woo TU (1999) Altered GABA neurotransmission and prefrontal cortical dysfunction in schizophrenia. Biol Psychiatry 46: 616-626.
Ljungberg T, Apicella P, Schultz W (1992) Responses of monkey dopamine neurons during learning of behavioral reactions. J Neurophysiol 67: 145-163.
Manes F, Sahakian B, Clark L, Rogers R, Antoun N, Aitken M, Robbins T (2002) Decision-making processes following damage to the prefrontal cortex. Brain 125: 624-639.
Mantz J, Milla C, Glowinski J, Thierry AM (1988) Differential effects of ascending neurons containing dopamine and noradrenaline in the control of spontaneous activity and of evoked responses in the rat prefrontal cortex. Neuroscience 27: 517-526.
Miller EK, Cohen JD (2001) An integrative theory of prefrontal cortex function. Annu Rev Neurosci 24: 167-202.
Miller EK, Desimone R (1994) Parallel neuronal mechanisms for short-term memory. Science 263: 520-522.

Miller EK, Erickson CA, Desimone R (1996) Neural mechanisms of visual working memory in prefrontal cortex of the macaque. J Neurosci 16: 5154-5167.

Milner B (1963) Effects of different brain lesions on card sorting. The role of the frontal lobes. Arch Neurol 9: 90-100.

Milner B, Petrides M (1984) Behavioral effects of frontal-lobe lesions in man. Trends Neurosci 7: 403-407.

Mitchell BD, Cauller LJ (1997) Cortico-cortical and thalamocortical projections to layer I of the prefrontal/premotor neocortex in rats. Soc Neurosci Abstr 23: 1273.

Monchi O, Petrides M, Petre V, Worsley K, Dagher A (2001) Wisconsin Card Sorting revisited: distinct neural circuits participating in different stages of the task identified by event-related functional magnetic resonance imaging. J Neurosci 21: 7733-7741.

Mora F, Sweeney KF, Rolls ET, Sanguinetti AM (1976) Spontaneous firing rate of neurones in the prefrontal cortex of the rat: evidence for a dopamine inhibition. Brain Res 116:516-522.

Murphy BL, Arnsten AFT, Goldman-Rakic PS, Roth RH (1996) Increased dopamine turnover in the prefrontal cortex impairs spatial working memory performance in rats and monkeys. Proc Natl Acad Sci USA 93: 1325-1329.

Nauta WJH (1971) The problem of the frontal lobe: A reinterpretation. J Psychiat Res 8: 167-187.

Niki H, Watanabe M (1979) Prefrontal and cingulate unit activity during timing behavior in the monkey. Brain Res 171: 213-224.

Olton DS, Becker JT, Handleman GE (1979) Hippocampus, space and memory. Behav Brain Sci 2:313-365.

Orlov AA, Kurzina NP, Shutov AP (1988) Activity of medial wall neurons in frontal cortex of rat brain during delayed response reactions. Neurosci Behav Physiol 18:31-37.

Overton F (1897) Applied Physiology-Intermediate. pp 125-126. American Book Co. New York.

Owen AM, Downes JJ, Sahakian BJ, Polkey CE, Robbins TW (1990) Planning and spatial working memory following frontal lobe lesions in man. Neuropsychologia 28: 1021-1034.

Owen AM, Sahakian BJ, Semple J, Polkey CE, Robbins TW (1995) Visuospatial short term recognition memory and learning after temporal lobe excisions, frontal lobe excisions or amygdala hippocampectomy in man. Neuropsychologia 33: 1-24.

Owen AM, Doyon J, Petrides M, Evans AC (1996) Planning and spatial working memory: a positron emission tomography study in humans. Eur J Neurosci 8: 353-364.

Pandya DN, Yeterian EH (1990) Prefrontal cortex in relation to other cortical areas in the rhesus monkey: architecture and connections. In: Progress in Brain Research, vol 85 (Uylings H.B.M., van Eden C., de Bruin JPC, Corner MA, and Feenstra MGP, eds), pp 63-94, Elsevier, Amsterdam.

Passingham RE (1975) Delayed matching after selective prefrontal lesions in monkeys (Macac mulatta). Brain Res 92: 89-102.

Passingham RE (1993) The frontal lobes and voluntary action, Oxford University Press, Oxford.

Penit-Soria J, Audinat E, Crepel F (1987) Excitation of rat prefrontal cortical neurons by dopamine: an in vitro electrophysiological study. Brain Res 425: 263-274.

Petrides M (1989) Frontal lobes and memory. In: Handbook of Neuropsychology, vol 3 (Boller F and Graffman J, eds), pp 75-90, Elsevier, Amsterdam.

Petrides M (1994) Frontal lobes and behaviour. Curr Opn Neurobiol 4: 207-211.

Petrides M (1995) Functional Organization of the human frontal cortex for mnemonic processing: Evidence from neuroimaging studies. Ann NY Acad Sci 769: 85-96.

Petrides M (1996) Specialized systems for the processing of mnemonic information within the primate frontal cortex. Phil Trans Royal Soc Lond 351: 1455-1462.

Petrides M (2000a) Impairments in working memory after frontal cortical excisions. Adv Neurol 84: 111-118.

Petrides M (2000b) The role of the mid-dorsolateral prefrontal cortex in working memory. Exp Brain Res 133:44-54.

Petrides M, Milner B (1982) Deficits on subject-ordered tasks after frontal- and temporal-lobe lesions in man. Neuropsychologia 20: 249-262.

Pirot S, Godbout R, Mantz J, Tassin J-P, Glowinski J, Thierry A-M (1992) Inhibitory effects of ventral tegmental area stimulation on the activity of prefrontal cortical neurons: evidence for involvement of both dopaminergic and GABAergic components. Neuroscience 49: 857-865.

Pratt WE, Mizumori SJ (2001) Neurons in rat medial prefrontal cortex show anticipatory rate changes to predictable differential rewards in a spatial memory task. Behav Brain Res 123: 165-183.

Quintana J, Fuster JM (1992) Mnemonic and predictive functions of cortical neurons in a memory task. Neuroreport 3: 721-724.

Quintana J, Fuster JM, Yajeya J (1989) Effects of cooling parietal cortex on prefrontal units on delayed tasks. Brain Res 503: 100-110.

Rainer G, Miller EK (2002) Timecourse of object-related neural activity in the primate prefrontal cortex during a short-term memory task. Eur J Neurosci 15: 1244-1254.

Rainer G, Rao SC, Miller EK (1999) Prospective coding for objects in primate prefrontal cortex. J Neurosci 19: 5493-5505.

Redgrave P, Prescott TJ, Gurney K (1999) Is the short-latency dopamine response too short to signal reward error? Trends Neurosci 22: 146-151.

Rétaux S, Besson MJ, Penit-Soria J (1991) Opposing effects of dopamine D2 receptor stimulation on the spontaneous and the electrically-evoked release [^{3}H GABA] on rat prefrontal cortex slices. Neuroscience 42: 61-71.

Robbins TW (1996) Dissociating executive functions of the prefrontal cortex. Phil Trans Royal Soc Lond 351: 1463-1470.

Romo R, Shultz W (1990) Dopamine neurons of the monkey midbrain: contingencies of responses to active touch during self-initiated arm movements. J Neurophysiol 63: 592-606.

Rose JE, Woolsey CN (1948) Structure and relations of limbic cortex and anterior thalamic nuclei in rabbit and cat. J Comp Neurol 89: 279-347.

Sawaguchi T (1987) Catecholamine sensitivities neuron related to a visual reaction time task in the monkey prefrontal cortex. J Neurophysiol 48: 1100-1122.

Sawaguchi T, Goldman-Rakic PS (1994) The role of D1-dopamine receptor in working memory: local injections of dopamine antagonists into the prefrontal cortex of rhesus monkeys performing an oculomotor delayed-response task. J Neurophysiol 71: 515-528.

Sawaguchi T, Matsumura M (1985) Laminar distributions of neurons sensitive to acetylcholine, noradrenaline and dopamine in the dorsolateral prefrontal cortex of the monkey. Neurosci Res 2: 255-273.

Sawaguchi T, Matsumura M, Kubota K (1986) Dopamine modulates neuronal activities related to motor performance in the monkey prefrontal cortex. Brain Res 371, 404-408.

Sawaguchi T, Matsumura M, Kubota K (1988) Dopamine enhances the neuronal activity of spatial short-term memory performance in the primate prefrontal cortex. Neurosci Res 5: 465-473.

Sawaguchi T, Matsumura M, Kubota K (1990a) Catecholamine effects on neuronal activity related to a delayed response task in monkey prefrontal cortex. J Neurophysiol 63: 1385-1400.

Sawaguchi T, Matsumura M, Kubota K (1990b) Effects of dopamine antagonists on neuronal activity related to a delayed response task in monkey prefrontal cortex. J Neurophysiol 63: 1401-1412.

Schultz W (1992a) Predictive reward signal of dopamine neurons. J Neurophysiol 80: 1-27.

Schultz W (1992b) Activity of dopamine neurons in the behaving primate. The Neurosciences 4: 129-138.

Schultz W, Romo R (1990) Dopamine neurons of the monkey midbrain: contingencies of responses to stimuli eliciting immediate behavioral reactions. J Neurophysiol 63: 607-624.

Schultz W, Tremblay L, Holleman JR (1998) Reward prediction in primate basal ganglia and frontal cortex. Neuropharmacol 37, 421-429.

Seamans JK, Floresco SB, Phillips AG (1995) Functional differences between the prelimbic and anterior cingulate regions of the rat prefrontal cortex. Behav Neurosci 109: 1063-1073.

Seamans JK, Floresco SB, Phillips AG (1998) D1 receptor modulation of hippocampal-prefrontal cortical circuits integrating spatial memory with executive functions in the rat. J Neurosci 18:1613-1621.

Seamans JK, Durstewitz D, Christie BR, Stevens CF, Sejnowski TJ (2001a) Dopamine D1/D5 receptor modulation of excitatory synaptic inputs to layer V prefrontal cortex neurons. Proc Natl Acad Sci USA 98: 301-306.

Seamans JK, Gorelova N, Durstewitz D, Yang CR (2001b) Bidirectional dopamine modulation of GABAergic inhibition in prefrontal cortical pyramidal neurons. J Neurosci 21: 3628-3638.

Sesack SR, Bunney BS (1989) Pharmacological characterization of the receptor mediating electrophysiological responses to dopamine in rat medial prefrontal cortex: a microiontophoretic study. J Pharmacol Exp Therap 248: 1323-1333.

Sesack SR, Deutch AY, Roth RH, Bunney BS (1989) Topographical organization of the efferent projections of the medial prefrontal cortex in the rat: an anterograde tract-tracing study with phaseolus vulgaris leucoagglutinin. J Comp Neurol 290: 213-242.

Shallice T (1982) Specific impairments in planning. Phil Trans Royal Soc Lond 298: 199-209.

Shallice T, Burgess P (1996) The domain of supervisory processes and temporal organization of behaviour. Phil Trans Royal Soc Lond 351: 1405-1411.

Stuss DT, Benson DF (1986) The Frontal Lobes, Raven Press, New York.

Taber MT, Fibiger HC (1997) Activation of the mesocortical dopamine system by feeding: lack of a selective response to stress. Neuroscience 77: 295-298.

Uylings HBM, van Eden CG (1990) Qualitative and quantitative comparison of the prefrontal cortex in rat and in primates, including humans. In: Progress in Brain Research, vol 85 (Uylings H.B.M., van Eden C., de Bruin JPC, Corner MA, and Feenstra MGP, eds), pp 31-62, Elsevier, Amsterdam.

Verin M, Partiot A, Pillon B, Malapani C, Agid Y, Dubois B (1993) Delayed response tasks and prefrontal lesions in man--evidence for self generated patterns of behaviour with poor environmental modulation. Neuropsychologia 31: 1379-1396.

Vincent SL, Knan Y, Benes FM (1993) Cellular distribution of dopamine D1 and D2 receptors in rat medial prefrontal cortex. J Neurosci 13: 2551-2564.

Wallis JD, Anderson KC, Miller EK (2001) Single neurons in prefrontal cortex encode abstract rules. Nature 411: 953-956.

Watanabe M (1981) Prefrontal unit activity during delayed conditional discriminations in the monkey. Brain Res 225: 51-65.

Watanabe M (1986a) Prefrontal unit activity during delayed conditional go/no-go discrimination in the monkey. I. Relation to the stimulus. Brain Res 382: 1-14.

Watanabe M (1986b) Prefrontal unit activity during delayed conditional go/no-go discrimination in the monkey. II. Relation to go and no-go responses. Brain Res 382: 15-27.

Watanabe M (1990) Prefrontal unit activity during associative learning in the monkey. Exp Brain Res 80: 296-309.

Watanabe M (1996) Reward expectancy in primate prefrontal neurons. Nature 382: 629-632.

Watanabe M (1998) Cognitive and motivational operations in primate prefrontal neurons. Rev Neurosci 9: 225-241.

Watanabe T, Niki H (1985) Hippocampal unit activity and delayed response in the monkey. Brain Res 325: 241-254.

Watanabe M, Hikosaka K, Sakagami M, Shirakawa S (2002) Coding and monitoring of motivational context in the primate prefrontal cortex. J Neurosci 22: 2391-2400.

White IM, Wise SP (1999) Rule-dependent neuronal activity in the prefrontal cortex. Exp Brain Res 126: 315-335.

Williams GV, Millar J (1990) Differential Actions of Endogenous and Iontophoretic Dopamine in Rat Striatum. Eur J Neurosci 2: 658-661.

Williams GV, Goldman-Rakic PS (1995) Modulation of memory fields by dopamine D1 receptors in prefrontal cortex. Nature 376: 572-575.

Wise SP, Murray EA, Gerfen CR (1996) The frontal cortex-basal ganglia system in primates. Crit Rev Neurobiol 10: 317-356.

Yajeya J, Quintana J, Fuster J (1988) Prefrontal representation of stimulus attributes during delay tasks II. The role of behavioral significance. Brain Res 474: 222-230.

Yang CR, Mogenson GJ (1990) Dopaminergic modulation of cholinergic responses in rat medial prefrontal cortex: an electrophysiological study. Brain Res 524: 271-281.

Yang CR, Seamans JK (1996) Dopamine D1 receptor actions in layer v-vi rat prefrontal cortex neurons in vitro: Modulation of dendritic-somatic signal integration. J Neurosci 16: 1922-1935.

Zahrt J, Taylor JR, Mathew RG, Arnsten AFT (1997) Supranormal stimulation of D_1 dopamine receptors in the rodent prefrontal cortex impairs spatial working memory performance. J Neurosci 17:8528-8535.

Zheng P, Zhang XX, Bunney BS, Shi WX (1999) Opposite modulation of cortical N-methyl-D-aspartate receptor-mediated responses by low and high concentrations of dopamine. Neuroscience 91: 527-535.

Zhou FM, Hablitz JJ (1999) Dopamine modulation of membrane and synaptic properties of interneurons in rat cerebral cortex. J Neurophysiol 81: 967-976.

Chapter 3

DOPAMINE MODULATION OF PREFRONTAL CORTICAL NEURAL ENSEMBLES AND SYNAPTIC PLASTICITY:
Potential Involvement in Schizophrenia

Yukiori Goto, Kuei-Yuan Tseng, Barbara L. Lewis, and Patricio O'Donnell

Center for Neuropharmacology and Neuroscience, Albany Medical College, Albany, NY 12208

Keywords: Prefrontal cortex, dopamine, glutamate, membrane potential, ensemble coding, schizophrenia, animal model, synaptic plasticity, *in vivo* intracellular recording, i*n vitro* whole cell recording.

Abstract: The prefrontal cortex has been implicated in executive functions, and it can become dysfunctional in psychiatric disorders such as schizophrenia. Prefrontal pyramidal neurons exhibit dynamic membrane potential activity *in vivo*, which depends on local microcircuits and synaptic inputs from other brain structures and may define neural ensembles encoding information. Mesocortical dopamine modulates these membrane potential states, allowing for long-term synaptic plasticity in the prefrontal cortex. Dopamine-mediated ensemble coding reinforcement may therefore be important for associative learning and executive functions. Dysfunction of associative learning and neural plasticity induced by dopamine abnormalities in the prefrontal cortex may be central components in the pathophysiology of schizophrenia.

1. INTRODUCTION

The prefrontal cortex (PFC) has been recognized as a brain region mediating the highest cognitive functions. PFC damage in humans (Lewinsohn et al., 1972; Damasio et al., 1994; Muller et al., 2002) and animals (Glick and Greenstein, 1972; Shaw and Aggleton, 1993; Joel et al., 1997) typically disrupts executive functions. Electrophysiological recordings in primates and rodents reveal that PFC neurons exhibit electrical activity associated with working memory as well (Kubota, 1975; Goldberg et al., 1980; Funahashi et al., 1991; Mulder et al., 2000). Untangling the mechanisms of information processing in the PFC is important not only for understanding the neural basis of human cognitive functions, but also for the pathophysiology of schizophrenia, a disorder in which a PFC malfunction is a critical component (Weinberger et al., 1994; Andreasen et al., 1997).

2. ELECTROPHYSIOLOGICAL RECORDINGS FROM PREFRONTAL NEURONS

2.1 Membrane Potential Activity in PFC Neurons *In Vivo*

In vivo intracellular recordings from PFC pyramidal cells in anesthetized rodents reveal that their membrane potential fluctuates spontaneously between a very negative resting potential (DOWN state) and transient plateau depolarizations (UP state; Fig. 1) (Branchereau et al., 1996; Lewis and O'Donnell, 2000). Similar membrane potential activity has been reported in other cortical regions (Steriade et al., 1993), as well as in medium spiny neurons in the dorsal (Wilson, 1993) and ventral (O'Donnell and Grace, 1995; Goto and O'Donnell, 2001) striatum. Since such membrane potential fluctuations are not detected in the slice preparation unless some manipulation enhances synaptic activity (Sanchez-Vives and McCormick, 2000), excitatory synaptic inputs from other brain structures or microcircuits of the cortex are thought to mediate UP transitions in the PFC. Simultaneous *in vivo* intracellular and local field potential recordings exhibit synchronized UP transitions and field potential shifts, indicating that membrane potential fluctuations occur synchronously in populations of cortical neurons (Steriade, 2001a). Elimination of ventral hippocampal (VH) inputs has been shown to prevent UP transitions in the ventral striatum (O'Donnell and Grace, 1995) and in the PFC (O'Donnell et al., 2002). These results suggest that the amount of synchronous excitatory synaptic inputs from other cortical, limbic, or thalamic areas projecting to the PFC may be essential in driving plateau depolarizations.

2.2 Dopamine Effects on Membrane Potential Activity in the PFC

Mesocortical DA projections arising from the ventral tegmental area (VTA) (Phillipson, 1979) are important for PFC functions. Reciprocal connections between the PFC and VTA (Carr and Sesack, 2000b; Sesack

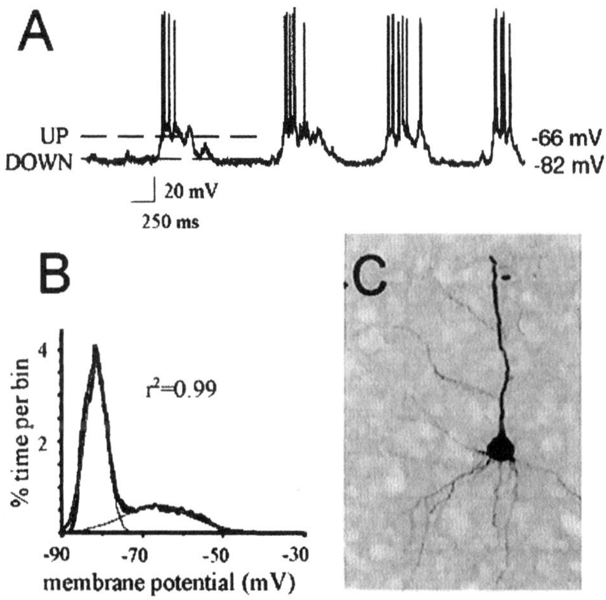

Figure 1: PFC pyramidal neurons exhibit membrane potential fluctuations from a very negative resting potential (DOWN) to transient plateau depolarizations (UP). A. Representative trace showing UP and DOWN transitions. B. Membrane potential distribution histogram of the trace shown in A. Two peaks corresponding to DOWN (-82 mV) and UP (-66 mV) states are observed, and the histogram is fitted to a bimodal Gaussian function (r^2=0.99). C. Neurobiotin injections revealed that recordings were from pyramidal neurons in deep layers of the PFC. Adopted from Lewis and O'Donnell (2000) with permission.

and Carr, 2002) may control DA release in the PFC. A DA modulation of PFC UP and DOWN membrane potential fluctuations was shown with electrical and chemical VTA activation (Lewis and O'Donnell, 2000). When the VTA is activated with trains of electrical pulses mimicking DA burst firing, a sustained membrane depolarization resembling the UP state and lasting for up to several seconds is typically evoked. The D_1 antagonist SCH23390 can reduce, but not block, the VTA-evoked membrane depolarization (Fig. 2), suggesting that D_1 receptor activation contributes to sustain the depolarization, although the transition to the depolarized state appears to be mediated by non-DA components. Because recent anatomical studies revealed that a substantial amount of VTA projection neurons to the PFC are not DA, but GABA cells (Carr and Sesack, 2000a), it is possible that GABA-mediated responses contribute to UP transitions. VTA GABA neurons exhibit slow frequency (~ 1 Hz) membrane potential fluctuations (Steffensen et al., 1998), and we have shown that PFC UP transitions are

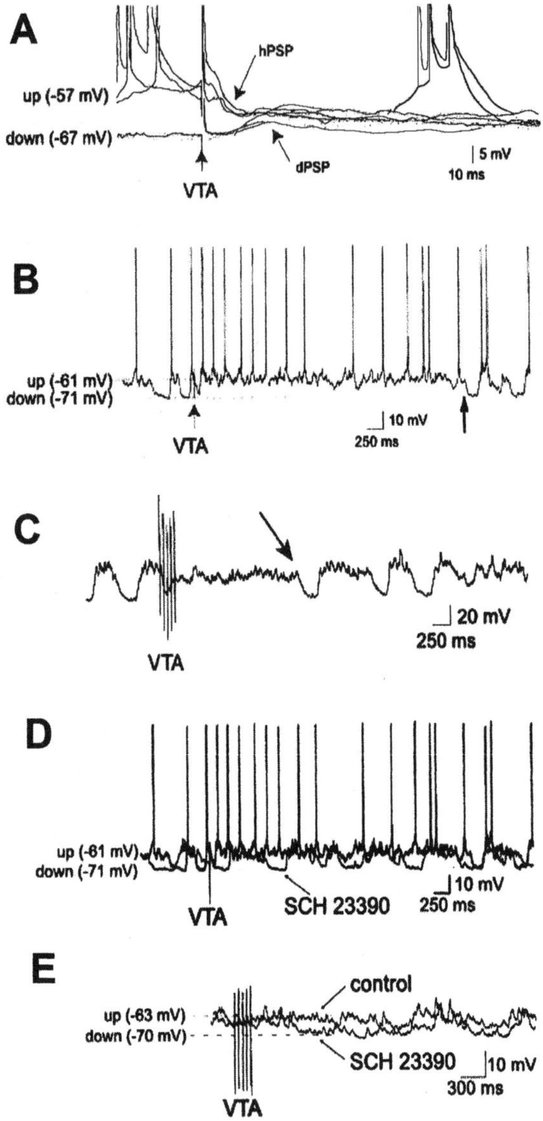

Figure 2: VTA stimulation induces sustained membrane potential depolarization requiring D_1 receptor activation. A. Overlay of traces showing variable response to VTA single-pulse stimulation. B. In some cases, single pulse stimulation evokes sustained membrane potential depolarization. C. With trains of pulses mimicking DA cell burst firing (5 pulses at 20 Hz), the plateau depolarization can be evoked more reliably. D, E. Plateau depolarizations evoked by VTA activation are shortened by the D_1 antagonist SCH23390. Adopted from Lewis and O'Donnell (2000) with permission.

correlated with local field potentials in the VTA (Peters et al., 2000). It is possible that VTA field potentials reflect GABA neuronal activity. It has been recently suggested that GABA could have an excitatory action when its spatial and temporal pattern in PFC neurons is paired with glutamatergic inputs (Gulledge and Stuart, 2003). However, transitions to the UP state do require glutamatergic inputs, since a VH lesion eliminates UP states (O'Donnell et al., 2002). Thus, D_1 receptors can sustain evoked depolarizations that depend primarily on glutamatergic inputs but may also involve activation of GABA receptors.

2.3 Dopamine-Glutamate Interactions in the PFC

The D_1 sustaining of plateau depolarizations may involve interactions with glutamate receptors. In the striatum, where expression of both D_1 and D_2 receptors is abundant, a number of studies have revealed a DA modulation of glutamate responses (Cepeda et al., 1993; Levine et al., 1996b). It has been shown that D_1 receptor activation enhances inward rectification, an effect blocked by potassium channel inactivation (Pacheco-Cano et al., 1996; Mermelstein et al., 1998). This indicates that D_1 receptors may facilitate inward rectifying potassium currents (K_{ir}). D_1 receptor activation also enhances calcium influx though L-type calcium channels (Hernández López et al., 1997) and NMDA (*N*-methyl-D-aspartate) currents (Levine et al., 1996a; Harvey and Lacey, 1997). The effect on K_{ir} would contribute to clamping the membrane potential to the DOWN state (Wilson, 1993; Wilson and Kawaguchi, 1996), and the other actions could contribute to a sustained depolarization. Thus, D_1 receptor can be both excitatory and inhibitory, depending on the membrane potential state (O'Donnell et al., 1999; Nicola et al., 2000). *In vitro* whole cell recordings from PFC pyramidal neurons have revealed similar DA-glutamate synergism (Wang and O'Donnell, 2001). Bath applications of either a D_1 agonist or NMDA alone at high concentrations increase cell excitability in PFC neurons. Low concentrations of a D_1 agonist and NMDA, which do not affect cell excitability when they are given separately, enhance cell excitability when they are co-applied to the bath. Such D_1-NMDA synergism can be prevented by pretreatment with a D_1 antagonist, PKA blockers, or by interruption of Ca^{2+} cascades (Wang and O'Donnell, 2001). These results suggest that D_1 activation enhances NMDA current through second messenger pathways involving calcium and PKA.

3. ENSEMBLE CODING AND SYNAPTIC PLASTICITY IN PFC

3.1 Neural Assemblies Defined by Membrane Potential States

Early electrophysiological studies have suggested that distributed networks (neural ensembles) of neurons may mediate information processing in the brain (Hebb, 1949; Kristan and Gerstein, 1970; Eccles, 1971). Recent simultaneous recordings from populations of neurons support this concept (Wilson and McNaughton, 1993; Deadwyler et al., 1996; Nicolelis et al., 1997). Since actual synchronization of action potential firing is either elusive or, at best, weak (Chang et al., 2000), it is possible that ensembles of active neurons are not defined by instantaneous synchronization of spike firing, but by whether a population of neurons is firing or not during a physiologically relevant period. If this is the case, subthreshold membrane potential activity may be a better strategy to define neural ensembles than action potential firing (O'Donnell, 1999, 2003). Thus, information in the PFC may be encoded with ensembles of neurons in their UP or DOWN membrane potential states (Fig. 3A). Since UP state transitions are dependent on excitatory synaptic inputs from other brain structures or cortical regions projecting to the PFC, ensembles of active neurons could be defined as integrating information from the thalamus, limbic structures (hippocampus and amygdala), and other cortical areas including the parietal cortex. The output of PFC neurons as action potential firing is further determined by the arrival of additional inputs during this period. In this sense, PFC neurons are both temporal integrators and detectors of coincident information. This combination renders the PFC suitable for temporal and cross-modal integration of information (Fuster, 1997; Fuster et al., 2000).

UP and DOWN membrane potential transitions have been studied in anesthetized animals. It is unclear whether cortical neurons in awake animals still exhibit such membrane potential fluctuations. The correlation between UP states and slow wave oscillation in the electroencephalogram (EEG) suggests that synchronous alterations between UP and DOWN states in cortical neurons are typical of slow-wave sleep (Steriade et al., 1993; Steriade and Amzica, 1998). Awake animals exhibit higher frequency components in their EEG. However, recent studies also provide indication that sustained depolarization and hyperpolarization can control information processing. For example, cortical activity measured with voltage-sensitive dyes reveals membrane hyperpolarization associated with oculomotor saccades (Seidemann et al., 2002). In addition, *in vivo* recordings from

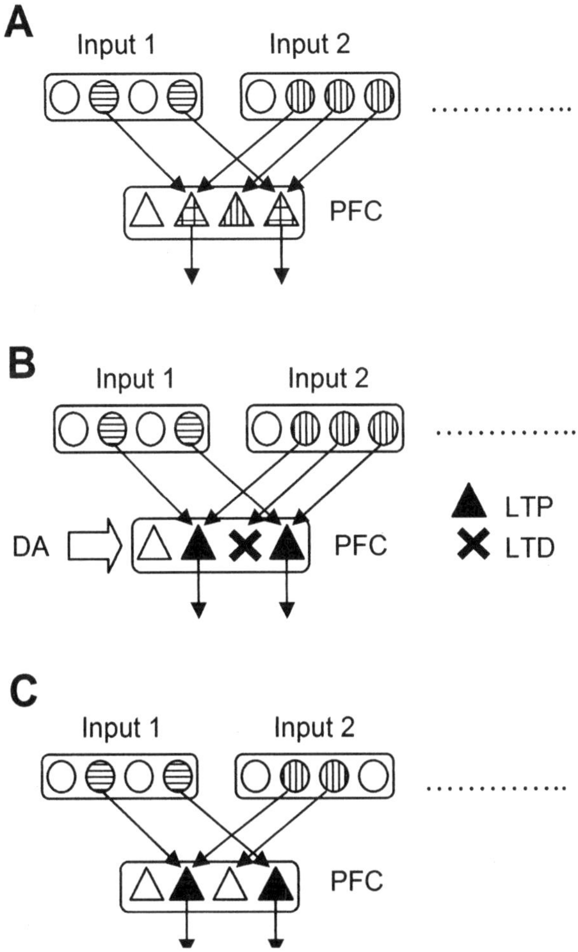

Figure 3: Neural ensemble coding and synaptic plasticity enabled by DA in the PFC. A. Synchronous excitatory synaptic inputs from other brain structures (inputs 1,2,3,....) initiate membrane depolarization in some neurons, although other neurons receiving asynchronous inputs may stay in the hyperpolarized state. The topology of neurons in the UP state may define an active neural ensemble encoding information in the PFC. B. DA activation, typically in response to salient and behaviorally relevant stimuli, can reinforce the ongoing neural ensembles with induction of LTP. On the other hand, LTD may be induced in neurons in the DOWN state. C. Neural ensembles reinforced by synaptic plasticity can be easily evoked in similar situations.

striatal neurons in awake monkeys (Kitano et al., 2002) and unanesthetized rats (Wilson and Groves, 1981) indicate the existence of bistable membrane potentials. UP-DOWN membrane potential alternations in anesthetizedanimals resemble slow-wave sleep conditions. Even in those conditions, information processing during UP states may be important for learning and plasticity mechanisms (Steriade, 2001a,b; Lee and Wilson, 2002). It is possible that in awake animals, neuronal populations loose synchrony of membrane potential fluctuations, resulting in disappearance of slow components in the electroencephalogram. In the presence of behaviorally relevant stimuli that activate the mesocortical pathway, a large number of neural ensembles could be set into a persistent UP state (O'Donnell, 2003).

3.2 DA Modulation of Neural Ensembles and Synaptic Plasticity

The D_1 facilitation of UP states may contribute to working memory. A membrane depolarization prolonged by D_1 receptor activation can explain the sustained action potential firing typically observed in PFC neurons during working memory tasks in primates. Indeed, D_1, but not D_2, receptor blockade disrupts sustained spike firing in PFC neurons and working memory performance (Goldman-Rakic, 1995, 1999).

DA may also affect plasticity in the PFC by sustaining UP states. Long-term potentiation (LTP) (Gurden et al., 1999, 2000) and long-term depression (LTD) (Otani et al., 1998; Takita et al., 1999) have been reported in the PFC. DA is known to modulate synaptic plasticity via D_1 receptor activation, since both inactivation of the mesocortical projection and D_1 receptor blockade disrupt LTP induction in the hippocampal–PFC pathway (Gurden et al., 1999, 2000). A facilitation of synaptic plasticity by DA may be due to D_1 receptors sustaining UP states and thereby facilitating NMDA responses by bringing these receptors out of their inactive voltage range. By reinforcing LTP, D_1 receptors may ensure the reproducibility of a given ensemble of PFC neurons in the UP state. It is possible that a DA reinforcement of LTD is also voltage-dependent. LTD is more commonly induced in the PFC using the slice preparation (Law-Tho et al., 1995; Otani et al., 1998), in which PFC neuron membrane potential is within the range of the *in vivo* DOWN state. Although speculative, in the presence of DA and its resulting state-stabilization, LTP may be enhanced only on cells in the UP state, whereas LTD would be the plasticity mechanism enhanced in neurons in the DOWN state. This may be related to pre- and postsynaptic spike timing determining LTP or LTD induction (Markram et al., 1997; Bi and Poo, 1999, 2001). The possibility of DA supporting either LTP or LTD in a given system is supported by recent evidence that a first DA application may enhance LTD, whereas a second DA application results in LTP induction (Blond et al., 2002). Such a dual effect of DA would certainly contribute to strengthening the pattern of network activity associated with salient stimuli, resulting in the learning reinforcement function that has been proposed for DA (Schultz, 1998, 2002). A combination of synaptic response enhancement during UP states and input attenuation during DOWN states can result in a filtering mechanism by which only strong stimuli (perhaps those effectively reinforced by plasticity) can overcome the "inhibition"; in other words, an increase in the signal-to-noise ratio. The outcome would be that the network of neurons in the UP state during a salient event is both strengthened and filtered of irrelevant information by the multiple facets of DA actions. Memories could be retrieved by the relative ease of reproducing a similar ensemble in conditions resembling the initial context (Fig. 3B, C).

4. PFC ENSEMBLES AND SYNAPTIC PLASTICITY MAY BE ALTERED IN SCHIZOPHRENIA

4.1 Alteration of Prefrontal Response to Dopamine in a Developmental Animal Model of Schizophrenia

A neonatal VH lesion in rodents and primates has been proposed as a developmental animal model of schizophrenia. These animals exhibit abnormal behaviors such as exaggerated locomotion in response to DA agonists (Lipska et al., 1993), NMDA antagonists (Al-Amin et al., 2001), or stress (Lipska et al., 1995), but only after puberty. This time course is similar to what is observed in the onset of symptoms in schizophrenia (Weinberger, 1995). In addition, cognitive deficits in working memory (Lipska et al., 2002), latent inhibition (Grecksch et al., 1999), or sensory gating (Lipska et al., 1996), and reduction of social interactions (Sams Dodd et al., 1997) are commonly observed in animals with neonatal VH lesion as well as in schizophrenia patients. Thus, this animal model stresses the link between early-life limbic compromise (Lipska and Weinberger, 2000) and delayed symptom onset in schizophrenia. Because the VH has a massive projection to the PFC (Jay et al., 1989; Jay and Witter, 1991), it is expected that PFC function is also changed in these animals. Reduced *N*-acetyl aspartate (NAA) (Bertolino et al., 1997) and reduced expression of GAD67 (Lipska and Weinberger, 2000) and BDNF (Lipska et al., 2001b; Ashe et al., 2002) mRNAs are reported in the PFC of animals with a neonatal VH lesion. We have investigated whether a neonatal VH lesion affects PFC neuron physiology with *in vivo* intracellular recordings (Fig. 4) (O'Donnell et al., 2002). Surprisingly, although an adult VH lesion eliminated UP transitions,

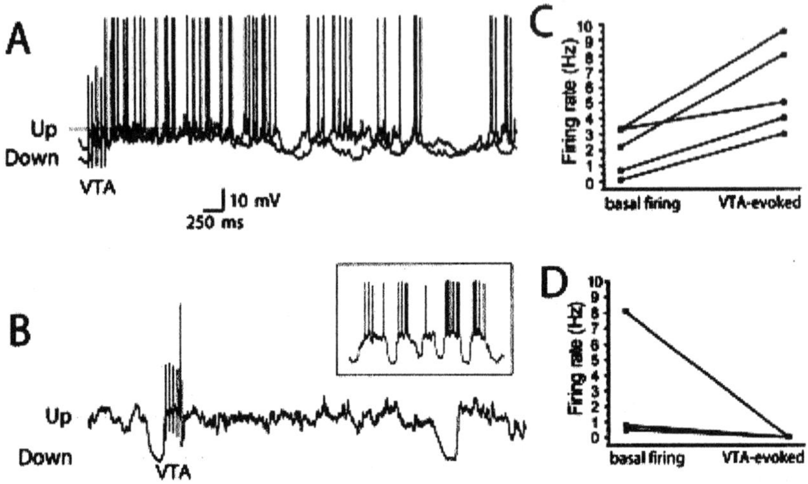

Figure 4: A developmental alteration in PFC neurons in response to VTA activation. A, C. Trains of VTA stimulation evoke sustained membrane depolarization with increased spike firing in adult animals with a neonatal VH lesion. B, D. Trains of VTA stimulation evoke sustained plateau depolarizations in animals with a neonatal sham treatment, but decreased spike firing as observed during the response, which is similar to that of naïve animals. Adopted from O'Donnell et al (2002) with permission.

PFC neurons in animals with a neonatal VH lesion still exhibit spontaneous membrane potential oscillations. This indicates that VH inputs are an important set of glutamatergic afferents that contribute to UP states in PFC pyramidal neurons. The persistence of UP states in animals with a neonatal VH lesion suggests that other brain structures can replace the VH inputs in this ability to drive PFC pyramidal neurons if they are eliminated in early development. The primary alteration in PFC neurons from animals with a neonatal VH lesion is their response to mesocortical activation. When the VTA is stimulated with a train of electrical pulses mimicking DA cell burst firing, a membrane depolarization with suppressed spike firing is observed in naïve or sham treated animals. However, increased spike firing is observed during the VTA-evoked membrane depolarization in animals with a neonatal VH lesion. This enhanced response to VTA stimulation can only be observed after puberty. Although the mechanisms for such developmental changes are not understood, this result indicates that altered DA response in the PFC may be responsible for at least some behavioral abnormalities in this animal model.

4.2 Possible Network Mechanism of Prefrontal Dysfunction in Schizophrenia

Hypofrontality is a major component in schizophrenia pathophysiology (Weinberger et al., 1994). It has been linked to the severity of "negative symptoms" (i.e. social withdrawal and lack of affect) (Wolkin et al., 1992), as well as to a variety of cognitive deficits observed in this disorder (Carter et al., 1998). But what really is hypofrontality? Traditionally, it is viewed as lack of PFC activation during tasks that would normally engage this brain region, measured as changes in regional cerebral blood flow (Fig. 5A) (Weinberger et al., 1994; Andreasen et al., 1997). This causes deficits in working memory resembling those seen in PFC lesions (Muller et al., 2002). Our finding of enhanced firing in PFC neurons during VTA-evoked depolarizations in animals with a neonatal VH lesion would suggest that in those animals, the PFC becomes hyper-, but not hypo-, active upon DA activation.

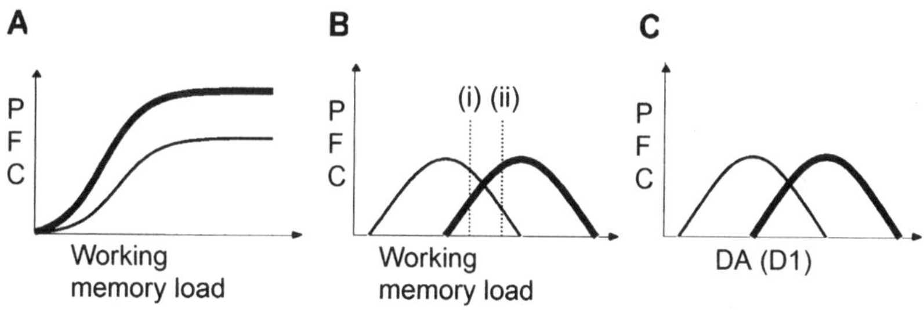

Figure 5: Relations between working memory capacity or D_1 receptor activation and PFC activation. A. Schizophrenia hypofrontality is traditionally considered as an impaired PFC activation during working memory tasks. Bold and thin lines indicate PFC activation with increasing working memory demands in normal and schizophrenia cases, respectively. B. Manoach suggests that highest PFC activation in schizophrenia patients is identical to normal subjects, but lower degree of working memory capacity results in highest activation of the PFC with low demands compared to normal, and high cognitive loads result in inactivation of the PFC (Manoach et al., 2003). Thus, in this model, low cognitive demands induce higher PFC activity in schizophrenia than normal subject (indicated by (i); inverted correlation), although high working memory demands (which is the level of tasks that most of experiments report hypofrontality in schizophrenia) results in higher PFC activation in normal subject than schizophrenia (indicated by (ii)). C. Working memory is also known to be disrupted by insufficient and excessive activation of D_1 receptors in the PFC, resulting in similar inverted U-shape relation between D_1 activation and performance (Lidow et al., 1998). Thus, schizophrenia may involve enhanced PFC neural activity at similar DA levels as normal subjects.

Recent clinical studies have challenged the concept of a hypoactive PFC in hypofrontality. It appears now that, when working memory performance is adjusted to equal level, schizophrenia patients exhibit even higher PFC activation than normal subjects (Manoach et al., 1999, 2000; Callicott et al., 2000; Ramsey et al., 2002; Manoach, 2003). This finding suggests an insufficient PFC activity with poor task outcome rather than PFC hypofunction. It has also been suggested that working memory capacity is reduced in the schizophrenia (Callicott et al., 2000; Manoach et al., 2000). Since PFC activity is related to working memory load (Cohen et al., 1997; Manoach, 2003), it is conceivable that poor outcome of working memory performance in schizophrenia is related to an overload of a limited PFC capacity (Fig. 5B). An inverted U-shape in the correlation between PFC activation and working memory load is similar to the correlation between working memory performance and activation of D_1 receptors in the PFC (Fig. 5C; Lidow et al., 1998). Thus, it is possible that the increased firing of PFC neurons following mesocortical activation in animals with a neonatal VH lesion is related to a DA-dependent PFC overload that yields poor performance. A hyper-responsive DA system has been suggested to underlie positive symptoms in schizophrenia (i.e. hallucinations and delusions). It is possible that the neural bases of both negative and positive symptoms are linked, as VTA DA cell activity and PFC cell firing are interdependent.

A number of mechanisms could result in abnormal PFC cell firing in response to mesocortical activation. This altered PFC response to DA can be understood by considering tonic/phasic DA release in the mesocorticolimbic network. A recent human imaging study in schizophrenia patients reveals increased D_1 receptor expression in the PFC (Abi-Dargham et al., 2002), suggesting upregulation of D_1 receptors. Upregulation of a D_1 receptor interacting protein, calcyon, has also been reported in schizophrenia PFC (Koh et al., 2003). These are likely due to reduced tonic DA release, which depends on regular DA cell firing and determines the basal levels of DA (O'Donnell and Grace, 1998). A reduced tonic DA release may be related to the reduction in DA innervation of the PFC in schizophrenia, as evidenced by fewer tyrosine hydroxylase positive terminals than normal subjects (Akil et al., 1999). In this hypotonic DA condition, PFC is rendered hypoactive, yielding negative symptoms. Since autoreceptors present in DA terminals control phasic DA release, it is likely that reduced tonic DA levels will yield enhanced phasic DA release with DA cell burst firing (Grace, 1991). This may cause an excessive D_1 receptor activation, overloading a reduced-capacity PFC system. As a consequence, DA-dependent plasticity mechanisms and learning are impaired in the PFC, causing deficits in switching strategies and response selection. An enhanced PFC response to phasic DA release would in turn increase mesolimbic activity and DA release in the ventral striatum, resulting in positive symptoms.

Another possibility that could explain an abnormal PFC response to VTA activation in animals with a neonatal VH lesion is a deficit in activation of local circuit interneurons in the PFC. Cortical interneurons are important modulators of pyramidal cell activity. Post mortem studies have revealed a loss of specific population of GABA interneurons in several cortical regions in schizophrenia (Benes, 1995; Volk et al., 2000). Since interneurons receive DA innervation (Sesack et al., 1998), it is likely that in the presence of a reduced local network of interneurons, pyramidal PFC neurons would be excessively activated. This effect could be compounded with the exaggerated D_1 response, contributing to an ineffective PFC function.

4.3 Associative Learning and Prefrontal Synaptic Plasticity Dysfunction in Schizophrenia

The altered glutamatergic and DA transmission proposed for schizophrenia may affect synaptic plasticity mechanisms in the PFC. A reduction of dendritic spines in PFC pyramidal neurons has been reported in schizophrenia brains (Glantz and Lewis, 2000) as well as in animals with a neonatal VH lesion (Lipska et al., 2001a), suggesting reduced excitatory synaptic inputs in this area. Decreases of *N*-acetyl aspartate (NAA) in schizophrenia (Bertolino et al., 1999) and in this animal model (Bertolino et al., 1997) also indicate that excitatory inputs to the PFC are reduced. Administration of NMDA antagonists such as MK-801 or phencyclidine (PCP) induces schizophrenia-like symptoms (Luby et al., 1959; Heresco-Levy and Javitt, 1998; Jentsch and Roth, 1999). These conditions would result in an impairment of LTP and LTD induction in the PFC. A number of factors known to modulate plasticity are affected both in schizophrenia and in animals with a neonatal VH lesion. For example, brain-derived neurotrophic factor (BDNF) is identified as an important regulator of synaptic plasticity (Balkowiec and Katz, 2002; Kovalchuk et al., 2002; Messaoudi et al., 2002). BDNF is affected in schizophrenia (Wassink et al., 1999; Krebs et al., 2000; Virgos et al., 2001) as well as in animals with a neonatal VH lesion, in which there is reduced expression of BDNF mRNA in the PFC (Lipska et al., 2001b; Ashe et al., 2002). Antipsychotics increase BDNF mRNA expression (Chlan-Fourney et al., 2002) and have been suggested to induce synaptic plasticity (Konradi and Heckers, 2001). In addition, BDNF modulates DA systems (Guillin et al., 2001). Thus, cortical synaptic plasticity may be altered in schizophrenia.

A synaptic plasticity deficit would in turn cause dysfunction of neural ensemble formation and alteration of neural transmission in the ventral

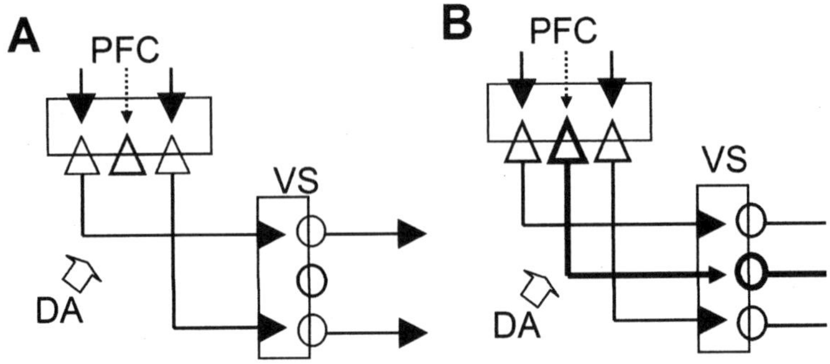

Figure 6: Abnormal PFC-ventral striatum neural ensembles formed with abnormal synaptic plasticity may be responsible for schizophrenia positive symptoms. A. In normal conditions, weak irrelevant inputs (dashed arrow) are filtered away by decrease in signal-to-noise ratio with phasic DA release. Units that receive strong inputs are reinforced by LTP, and units that receive weak information are depressed by LTD. This allows the selection of proper PFC neural ensembles, which may drive activity in selected ventral striatum (VS) neurons. B. Developmental compromise in the hippocampus alters DA response in PFC pyramidal neurons at adulthood. In this condition, irrelevant information is processed in the PFC neurons by disruption of signal-to-noise ratio with phasic DA release. Such information irrelevant inputs can form synaptic plasticity, resulting abnormal neural ensemble formation in the PFC and, as a consequence, in the VS. Such abnormal neural ensemble formation in the PFC–VS pathway may induce psychosis.

striatum (Fig. 6). Indeed, animals with a neonatal VH lesion also show altered responses to the VTA stimulation in the ventral striatum (Goto and O'Donnell, 2002). These abnormal responses are not observed when the PFC is lesioned (Goto and O'Donnell, 2003), suggesting that excessive glutamate release from the PFC in response to DA activation affects basal ganglia responses. The word "schizophrenia" as defined by Eugen Bleuler originates from the Greek words *Schizein*, "to split" and *phren*, "mind" (Bleuler, 1952). Thus, he identified the key symptom of schizophrenia as dissociative thinking. Its converse, associative learning, requires limbic-PFC interactions. Context-related information processed by the hippocampus must be incorporated into the cortico-basal ganglia networks to select the appropriate set of behavioral responses. Thus, deficits in limbic-PFC flow of information will disrupt goal-directed behaviors. Recent studies by Earl Miller have shown that the PFC indeed processes associative learning (Miller et al., 1996; Asaad et al., 1998; Miller, 2000; Miller and Cohen, 2001). There is also evidence that this is disrupted in schizophrenia (Gold et al., 2000; Martins Serra et al., 2001). Abnormal NMDA and DA activity resulting in impaired plasticity may be responsible for such cognitive deficits in schizophrenia. Further studies in the role of synaptic plasticity in the PFC and its role in the formation of neural ensembles, which may mediate associative learning, can yield more insight for the central components responsible for schizophrenia pathophysiology.

REFERENCES

Abi-Dargham A, Mawlawi O, Lombardo I, Gil R, Martinez D, Huang Y, Hwang DR. Keilp J, Kochan L, Van Heertum R, Gorman JM, Laruelle M (2002) Prefrontal dopamine D1 receptors and working memory in schizophrenia. J Neurosci 22:3708-3719.

Akil M, Pierri JN, Whitehead RE, Edgar CL, Mohila C, Sampson AR, Lewis DA (1999) Lamina-specific alterations in the dopamine innervation of the prefrontal cortex in schizophrenic subjects. Am J Psychiatry 156:1580-1589.

Al-Amin HA, Shannon Weickert C, Weinberger DR. Lipska BK (2001) Delayed onset of enhanced MK-801-induced motor hyperactivity after neonatal lesions of the rat ventral hippocampus. Biol Psychiatry 49:528-539.

Andreasen NC, O'Leary DS, Flaum M, Nopoulos P, Watkins GL, Boles Ponto LL, Hichwa RD (1997) Hypofrontality in schizophrenia: distributed dysfunctional circuits in neuroleptic-naive patients. Lancet 349:1730-1734.

Asaad WF, Rainer G, Miller EK, (1998) Neural activity in the primate prefrontal cortex during associative learning. Neuron 21:1399-1407.

Ashe PC, Chlan-Fourney J, Juorio AV, Li XM (2002) Brain-derived neurotrophic factor (BDNF) mRNA in rats with neonatal ibotenic acid lesions of the ventral hippocampus. Brain Res 956:126-135.

Balkowiec A, Katz DM (2002) Cellular mechanisms regulating activity-dependent release of native brain-derived neurotrophic factor from hippocampal neurons. J Neurosci 22:10399-10407.

Benes FM (1995) Altered glutamatergic and GABAergic mechanisms in the cingulate cortex of the schizophrenic brain. Arch Gen Psychiatry 12:1019-1024.

Bertolino A, Saunders RC, Mattay VS, Bachevalier J, Frank JA, Weinberger DR (1997) Altered development of prefrontal neurons in rhesus monkeys with neonatal mesial temporo-limbic lesions: a proton magnetic resonance spectroscopic imaging study. Cereb Cortex 7:740-748.

Bertolino A, Knable MB, Saunders RC, Callicott JH, Kolachana B, Mattay VS, Bachevalier J, Frank JA, Egan M, Weinberger DR (1999) The relationship between dorsolateral prefrontal N-acetylaspartate measures and striatal dopamine activity in schizophrenia. Biol Psychiatry 45:660-667.

Bi G, Poo M (2001) Synaptic modification by correlated activity: Hebb's postulate revisited. Annu. Rev. Neurosci. 24:139-166.

Bi G-Q, Poo M-M (1999) Distributed Synaptic Modification in Neural Networks Induced by Patterned Stimulation. Nature 401:792-796.

Bleuler E (1952) Dementia praecox; or, The group of schizophrenias. International universities press, New York.

Blond O, Crepel F, Otani S (2002) Long-term potentiation in rat prefrontal slices facilitated by phased application of dopamine. Eur. J. Pharmacol. 438:115-116.

Branchereau P, Van Bockstaele EJ, Chan J, Pickel VM (1996) Pyramidal neurons in rat prefrontal cortex show a complex synaptic response to single electrical stimulation of the locus coeruleus region: evidence for antidromic activation and GABAergic inhibition using in vivo intracellular recording and electron microscopy. Synapse 22:313-331.

Callicott JH, Bertolino A, Mattay VS, Langheim FJ, Duyn J, Coppola R, Goldberg TE, Weinberger DR (2000) Physiological dysfunction of the dorsolateral prefrontal cortex in schizophrenia revisited. Cereb Cortex 10:1078-1092.

Carr DB, Sesack SR (2000a) GABA-containing neurons in the rat ventral tegmental area project to the prefrontal cortex. Synapse 38:114-123.

Carr DB, Sesack SR (2000b) Projections from the rat prefrontal cortex to the ventral tegmental area: target specificity in the synaptic associations with mesoaccumbens and mesocortical neurons. J. Neurosci. 20:3864-3873.

Carter CS, Perlstein W, Ganguli R, Brar J, Mintun M, Cohen, JD (1998) Functional hypofrontality and working memory dysfunction in schizophrenia. Am J Psychiatry 155:1285-1287.

Cepeda C, Buchwald NA, Levine MS (1993) Neuromodulatory actions of dopamine in the neostriatum are dependent upon the excitatory amino acid receptor subtypes activated. Proc Natl Acad Sci USA 90:9576-9580.

Chang JY, Janak PH, Woodward DJ (2000) Neuronal and behavioral correlations in the medial prefrontal cortex and nucleus accumbens during cocaine self-administration by rats. Neuroscience 99:433-443.

Chlan-Fourney J, Ashe P, Nylen K, Juorio AV, Li XM (2002) Differential regulation of hippocampal BDNF mRNA by typical and atypical antipsychotic administration. Brain Res. 954:11-20.

Cohen JD, Perlstein WM, Braver TS, Nystrom LE, Noll DC, Jonides J, Smith EE (1997) Temporal dynamics of brain activation during a working memory task. Nature 386:604-608.

Damasio H, Grabowski T, Frank R, Galaburda AM, Damasio A (1994) The return of Phineas Gage: clues about the brain from the skull of a famous patient. Science 264:1102-1105.

Deadwyler SA, Bunn T, Hampson RE (1996) Hippocampal ensemble activity during spatial delayed-nonmatch-to-sample performance in rats. J Neurosci 16:354-372.

Eccles JC (1971) Functional significance of arrangement of neurones in cell assemblies. Arch Psychiatr Nervenkr 215:92-106.

Funahashi S, Bruce CJ, Goldman-Rakic PS (1991) Neuronal activity related to saccadic eye movements in the monkey's dorsolateral prefrontal cortex. J Neurophysiol 65:1464-1483.

Fuster JM (1997) The Prefrontal Cortex: Anatomy, Physiology, and Neuropsychology of the Frontal Lobe (3rd ed), Lippincott-Raven, Philadelphia.

Fuster JM, Bodner M, Kroger JK (2000) Cross-modal and cross-temporal association in neurons of frontal cortex. Nature 405:347-351.

Glantz LA, Lewis DA (2000) Decreased dendritic spine density on prefrontal cortical pyramidal neurons in schizophrenia. Arch Gen Psychiatry 57:65-73.

Glick SD, Greenstein S (1972) Amnesia following cortical brain damage in mice. Behav Biol 7:573-583.

Gold JM, Bish JA, Iannone VN, Hobart MP, Queern CA, Buchanan RW (2000) Effects of contextual processing on visual conditional associative learning in schizophrenia. Biol Psychiatry 48:406-414.

Goldberg RB, Fuster JM, Alvarez-Pelaez R (1980) Frontal cell activity during delayed response performance in squirrel monkey (Saimiri sciureus). Physiol Behav 25:425-432.

Goldman-Rakic PS (1995) Cellular basis of working memory. Neuron 14:477-485.

Goldman-Rakic PS (1999) The Physiological Approach: Functional Architecture of Working Memory and Disordered Cognition in Schizophrenia. Biol Psychiatry 46:650-661.

Goto Y, O'Donnell P (2001) Network synchrony in the nucleus accumbens in vivo. J Neurosci 21:4498-4504.

Goto Y, O'Donnell P (2002) Delayed mesolimbic system alteration in a developmental animal model of schizophrenia. J Neurosci 22:9070-9077.

Goto Y, O'Donnell P (2003) Altered prefrontal cortex - nucleus accumbens information processing in a developmental animal model of schizophrenia. Ann N Y Acad Sci (in press)

Grace AA (1991) Phasic versus tonic dopamine release and the modulation of dopamine system responsivity: a hypothesis for the etiology of schizophrenia. Neuroscience 41:1-24.

Grecksch G, Bernstein HG, Becker A, Höllt V, Bogerts B (1999) Disruption of latent inhibition in rats with postnatal hippocampal lesions. Neuropsychopharmacol 20:525-532.

Guillin O, Diaz J, Carroll P, Griffon N, Schwartz JC, Sokoloff P (2001) BDNF controls dopamine D3 receptor expression and triggers behavioural sensitization. Nature 411:86-89.

Gulledge AT, Stuart GJ (2003) Excitatory actions of GABA in the cortex. Neuron 37:299-309.

Gurden H, Tassin J-P, Jay TM (1999) Integrity of the mesocortical dopaminergic system is necessary for complete expression of in vivo hippocampal-prefrontal cortex long-term potentiation. Neuroscience 94:1019-1027.

Gurden H, Takita M, Jay TM (2000) Essential role of D1 but not D2 receptors in the NMDA receptor- dependent long-term potentiation at hippocampal-prefrontal cortex synapses in vivo. J Neurosci 20:RC106.

Harvey J, Lacey MG (1997) A postsynaptic interaction between dopamine D1 and NMDA receptors promotes presynaptic inhibition in the rat nucleus accumbens via adenosine release. J Neurosci 17:5271-5280.

Hebb DO (1949) The Organization of Behavior: A Neuropsychological Theory. John Wiley and Sons, New York.

Heresco-Levy U, Javitt DC (1998) The role of N-methyl-D-aspartate (NMDA) receptor-mediated neurotransmission in the pathophysiology and therapeutics of psychiatric syndromes. Eur Neuropsychopharmacol 8:141-152.

Hernández López S, Bargas J, Surmeier DJ, Reyes AD, Galarraga E (1997) D1 receptor activation enhances evoked discharge in neostriatal medium spiny neurons by modulating an L-type Ca^{2+} conductance. J Neurosci 17:3334-3342.

Jay TM, Witter MP (1991) Distribution of hippocampal CA1 and subicular efferents in the prefrontal cortex of the rat studied by means of anterograde transport of Phaseolus vulgaris-leucoagglutinin. J Comp Neurol 313:574-586.

Jay TM, Glowinski J, Thierry AM (1989) Selectivity of the hippocampal projection to the prelimbic area of the prefrontal cortex in the rat. Brain Res 505:337-340.

Jentsch JD, Roth RH (1999) The neuropsychopharmacology of phencyclidine: from NMDA receptor hypofunction to the dopamine hypothesis of schizophrenia. Neuropsychopharmacol 20:201-225.

Joel D, Tarrasch R, Feldon J, Weiner I (1997) Effects of electrolytic lesions of the medial prefrontal cortex or its subfields on 4-arm baited, 8-arm radial maze, two-way active avoidance and conditioned fear tasks in the rat. Brain Res 765:37-50.

Kitano K, Cateau H, Kaneda K, Nambu A, Takada M, Fukai T (2002) Two-state membrane potential transitions of striatal spiny neurons as evidenced by numerical simulations and electrophysiological recordings in awake monkeys. J Neurosci 22:RC230.

Koh PO, Bergson C, Undie AS, Goldman-Rakic PS, Lidow MS (2003) Up-regulation of the D_1 dopamine receptor-interacting protein, Calcyon, in patients with schizophrenia. Arch Gen Psychiatry 60: 311-319.

Konradi C, Heckers S (2001) Antipsychotic drugs and neuroplasticity: insights into the treatment and neurobiology of schizophrenia. Biol Psychiatry 50:729-742.

Kovalchuk Y, Hanse E, Kafitz KW, Konnerth A (2002) Postsynaptic induction of BDNF-mediated long-term potentiation. Science 295:1729-1734.

Krebs MO, Guillin O, Bourdell MC, Schwartz JC, Olie JP, Poirier MF, Sokoloff P (2000) Brain derived neurotrophic factor (BDNF) gene variants association with age at onset and therapeutic response in schizophrenia. Mol Psychiatry 5:558-562.

Kristan WB Jr, Gerstein GL (1970) Plasticity of synchronous activity in a small neural net. Science 169:1336-1339.

Kubota K (1975) Prefrontal unit activity during delayed-response and delayed-alternation performances. Jpn J Physiol 25:481-493..

Law-Tho D, Desce JM, Crepel F (1995) Dopamine favours the emergence of long-term depression versus long-term potentiation in slices of rat prefrontal cortex. Neurosci Lett 188:125-128.

Lee AK, Wilson MA (2002) Memory of sequential experience in the hippocampus during slow wave sleep. Neuron 36:1183-1194.

Levine MS, Altemus KL, Cepeda C, Cromwell HC, Crawford C, Ariano MA, Drago J, Sibley DR, Westphal H (1996a) Modulatory actions of dopamine on NMDA receptor-mediated responses are reduced in D1A-deficient mutant mice. J Neurosci 16:5870-5882.

Levine MS, Li Z, Cepeda C, Cromwell HC, Altemus KL (1996b) Neuromodulatory actions of dopamine on synaptically-evoked neostriatal responses in slices. Synapse 24:65-78.

Lewinsohn PM, Zieler RE, Libet J, Eyeberg S, Nielson G (1972) Short-term memory: a comparison between frontal and nonfrontal right- and left-hemisphere brain-damaged patients. J Comp Physiol Psychol 81:248-255.

Lewis BL, O'Donnell P (2000) Ventral tegmental area afferents to the prefrontal cortex maintain membrane potential 'up' states in pyramidal neurons via D_1 dopamine receptors. Cereb Cortex 10:1168-1175.

Lidow MS, Williams GV, Goldman-Rakic PS (1998) The cerebral cortex: a case for a common site of action of antipsychotics. Trends Pharmacol Sci 19:136-140.

Lipska BK, Weinberger DR (2000) To model a psychiatric disorder in animals: schizophrenia as a reality test. Neuropsychopharmacol 23:223-239.

Lipska BK, Jaskiw GE, Weinberger DR (1993) Postpubertal emergence of hyperresponsiveness to stress and to amphetamine after neonatal excitotoxic hippocampal damage: a potential animal model of schizophrenia. Neuropsychopharmacol 9:67-75.

Lipska BK, Chrapusta SJ, Egan MF, Weinberger DR (1995) Neonatal excitotoxic ventral hippocampal damage alters dopamine response to mild repeated stress and to chronic haloperidol. Synapse 20:125-130.

Lipska BK, Swerdlow NR, Geyer MA, Jaskiw GE, Braff DL, Weinberger DR (1996) Neonatal excitotoxic hippocampal damage in rats causes post-pubertal changes in prepulse inhibition of startle and its disruption by apomorphine. Psychopharmacol (Berl) 122:35-43.

Lipska BK, Kolb B, Halim ND, Weinberger DR (2001a) Synaptic abnormalities in prefrontal cortex and nuclues accumbens of adult rats with neonatal hippocampal damage. Schizophr Res 49:47.

Lipska BK, Khaing ZZ, Weickert CS, Weinberger DR (2001b) BDNF mRNA expression in rat hippocampus and prefrontal cortex: effects of neonatal ventral hippocampal damage and antipsychotic drugs. Eur J Neurosci 14:135-144.

Lipska BK, Aultman JM, Verma A, Weinberger DR, Moghaddam,B (2002) Neonatal damage of the ventral hippocampus impairs working memory in the rat. Neuropsychopharmacol 27:47-54.

Luby ED, Cohen BD, Rosenbaum G, Domino EF (1959) Study of a new schizophrenomimetic drug- Sernyl. Arch Gen Psychiatry 81:363-369.

Manoach DS (2003) Prefrontal cortex dysfunction during working memory performance in schizophrenia: reconciling discrepant findings. Schizophr Res 60:285-298.

Manoach DS, Press DZ, Thangaraj V, Searl MM, Goff DC, Halpern E, Saper CB, Warach S (1999) Schizophrenic subjects activate dorsolateral prefrontal cortex during a working memory task, as measured by fMRI. Biol Psychiatry 45:1128-1137.

Manoach DS, Gollub RL, Benson ES, Searl MM, Goff DC, Halpern E, Saper CB, Rauch SL (2000) Schizophrenic subjects show aberrant fMRI activation of dorsolateral prefrontal cortex and basal ganglia during working memory performance. Biol Psychiatry 48:99-109.

Markram H, Lubke J, Frotscher M, Sakmann B (1997) Regulation of synaptic efficacy by coincidence of postsynaptic APs and EPSPs. Science 275:213-215.

Martins Serra A, Jones SH, Toone B, Gray JA (2001) Impaired associative learning in chronic schizophrenics and their first-degree relatives: a study of latent inhibition and the Kamin blocking effect. Schizophr Res 48:273-289.

Mermelstein PG, Song WJ, Tkatch T, Yan Z, Surmeier DJ (1998) Inwardly rectifying potassium (IRK) currents are correlated with IRK subunit expression in rat nucleus accumbens medium spiny neurons. J Neurosci 18:6650-6661.

Messaoudi E, Ying SW, Kanhema T, Croll SD, Bramham CR (2002) Brain-derived neurotrophic factor triggers transcription-dependent, late phase long-term potentiation in vivo. J Neurosci 22:7453-7461.

Miller EK (2000) The prefrontal cortex and cognitive control. Nat Rev Neurosci 1:59-65.

Miller EK, Cohen JD (2001) An integrative theory of prefrontal cortex function. Annu Rev Neurosci 24:167-202.

Miller EK, Erickson CA, Desimone R (1996) Neural mechanisms of visual working memory in prefrontal cortex of the macaque. J Neurosci 16:5154-5167.

Mulder AB, Nordquist R, Orgut O, Pennartz CM (2000) Plasticity of neuronal firing in deep layers of the medial prefrontal cortex in rats engaged in operant conditioning. In: Progress in Brain Research, vol 126 (Uylings HBM, Van Eden CG, De Bruin JPC, Feestra MGP, and Pennartz CMA, eds), pp 287-301, Elsevier, Amsterdam.

Muller NG, Machado L, Knight RT (2002) Contributions of subregions of the prefrontal cortex to working memory: evidence from brain lesions in humans. J Cogn Neurosci 14:673-686.

Nicola SM, Surmeier J, Malenka RC (2000) Dopaminergic modulation of neuronal excitability in the striatum and nucleus accumbens. Annu Rev Neurosci 23:185-215.

Nicolelis MA, Ghazanfar AA, Faggin BM, Votaw S, Oliveira LM (1997) Reconstructing the engram: simultaneous, multisite, many single neuron recordings. Neuron 18:529-537.

O'Donnell P (1999) Ensemble coding in the Nucleus Accumbens. Psychobiol 27:187-197.

O'Donnell P (2003) Dopamine gating of forebrain neural ensembles. Eur J Neurosci 17:1-7.

O'Donnell P, Grace AA (1995) Synaptic interactions among excitatory afferents to nucleus accumbens neurons: hippocampal gating of prefrontal cortical input. J Neurosci 15:3622-3639.

O'Donnell P, Grace AA (1998) Dysfunctions in multiple interrelated systems as the neurobiological bases of schizophrenic symptom clusters. Schizophr Bull 24:267-283.

O'Donnell P, Greene J, Pabello N, Lewis BL, Grace AA (1999) Modulation of cell firing in the nucleus accumbens. Ann N Y Acad Sci 877:157-175.

O'Donnell P, Lewis BL, Weinberger DR, Lipska BK (2002) Neonatal hippocampal damage alters electrophysiological properties of prefrontal cortical neurons in adult rats. Cereb Cortex 12:975-982.
Otani S, Blond O, Desce JM, Crepel F (1998) Dopamine facilitates long-term depression of glutamatergic transmission in rat prefrontal cortex. Neuroscience 85:669-676.
Pacheco-Cano MT, Bargas J, Hernandez-Lopez S, Tapia D, Galarraga E (1996) Inhibitory action of dopamine involves a subthreshold Cs(+)-sensitive conductance in neostriatal neurons. Exp Brain Res 110:205-211.
Peters YM, Lewis BL, O'Donnell P (2000) Synchronous activity in the ventral tegmental area and prefrontal cortex. Ann N Y Acad Sci 909:267-269.
Phillipson OT (1979) Afferent projections to the ventral tegmental area of Tsai and interfascicular nucleus: a horseradish peroxidase study in the rat. J Comp Neurol 187:117-143.
Ramsey NF, Koning HA, Welles P, Cahn W, van der Linden JA, Kahn RS (2002) Excessive recruitment of neural systems subserving logical reasoning in schizophrenia. Brain 125:1793-1807.
Sams Dodd F, Lipska BK, Weinberger DR (1997) Neonatal lesions of the rat ventral hippocampus result in hyperlocomotion and deficits in social behaviour in adulthood. Psychopharmacol (Berl) 132:303-310.
Sanchez-Vives MV, McCormick DA (2000) Cellular and network mechanisms of rhythmic recurrent activity in neocortex. Nat Neurosci 3:1027-1034.
Schultz W (1998) Predictive reward signal of dopamine neurons. J Neurophysiol 80:1-27.
Schultz W (2002) Getting formal with dopamine and reward. Neuron 36:241-263.
Seidemann E, Arieli A, Grinvald A, Slovin H (2002) Dynamics of depolarization and hyperpolarization in the frontal cortex and saccade goal. Science 295:862-865.
Sesack SR, Carr DB (2002) Selective prefrontal cortex inputs to dopamine cells: implications for schizophrenia. Physiol Behav 77:513-517.
Sesack SR, Hawrylak VA, Melchitzky DS, Lewis DA (1998) Dopamine innervation of a subclass of local circuit neurons in monkey prefrontal cortex: ultrastructural analysis of tyrosine hydroxylase and parvalbumin immunoreactive structures. Cereb Cortex 8:614-622.
Shaw C, Aggleton JP (1993) The effects of fornix and medial prefrontal lesions on delayed non-matching-to-sample by rats. Behav Brain Res 54: 91-102.

Steffensen SC, Svingos AL, Pickel VM, Henriksen SJ (1998) Electrophysiological characterization of GABAergic neurons in the ventral tegmental area. J Neurosci 18:8003-8015.

Steriade M (2001a) The intact and sliced brain. The MIT press, Cambridge.

Steriade M (2001b) Active neocortical processes during quiescent sleep. Arch Ital Biol 139:37-51.

Steriade M, Amzica F (1998) Coalescence of sleep rhythms and their chronology in corticothalamic networks. Sleep Res Online 1:1-10.

Steriade M, Nuñez A, Amzica F (1993) Intracellular analysis of relations between the slow (< 1 Hz) neocortical oscillation and other sleep rhythms of the electroencephalogram. J Neurosci 13:3266-3283.

Takita M, Izaki Y, Jay TM, Kaneko H, Suzuki SS (1999) Induction of stable long-term depression in vivo in the hippocampal-prefrontal cortex pathway. Eur J Neurosci 11:4145-4148.

Virgos C, Martorell L, Valero J, Figuera L, Civeira F, Joven J, Labad A, Vilella E (2001) Association study of schizophrenia with polymorphisms at six candidate genes. Schizophr Res 49:65-71.

Volk DW, Austin MC, Pierri JN, Sampson AR, Lewis DA (2000) Decreased glutamic acid decarboxylase67 messenger RNA expression in a subset of prefrontal cortical gamma-aminobutyric acid neurons in subjects with schizophrenia. Arch Gen Psychiatry 57:237-245.

Wang J, O'Donnell P (2001) D_1 dopamine receptors potentiate NMDA-mediated excitability increase in layer V prefrontal cortical pyramidal neurons. Cereb Cortex 11:452-462.

Wassink TH, Nelson JJ, Crowe RR, Andreasen NC (1999) Heritability of BDNF alleles and their effect on brain morphology in schizophrenia. Am J Med Genet 88:724-728.

Weinberger DR (1995) From neuropathology to neurodevelopment. Lancet 346:552-557.

Weinberger DR, Aloia MS, Goldberg TE, Berman KF (1994) The frontal lobes and schizophrenia. J Neuropsychiatry Clin Neurosci 6:419-427.

Wilson CJ (1993) The generation of natural firing patterns in neostriatal neurons. Prog Brain Res 99:277-297.

Wilson CJ, Groves PM (1981) Spontaneous firing patterns of identified spiny neurons in the rat neostriatum. Brain Res 220:67-80.

Wilson CJ, Kawaguchi Y (1996) The origins of two-state spontaneous membrane potential fluctuations of neostriatal spiny neurons. J Neurosci 16:2397-2410.

Wilson MA, McNaughton BL (1993) Dynamics of the hippocampal ensemble code for space. Science 261:1055-1058.

Wolkin A, Sanfilipo M, Wolf AP, Angrist B, Brodie JD, Rotrosen J (1992) Negative symptoms and hypofrontality in chronic schizophrenia. Arch Gen Psychiatry 49:959-965.

Acknowledgements

This work was supported by USPHS grants MH57683, MH60131, DA14020, and a NARSAD Independent Investigator Award to P. O'D.

Chapter 4

INDUCTION PROPERTIES OF SYNAPTIC PLASTICITY IN RAT PREFRONTAL NEURONS

Satoru Otani and Bogdan Kolomiets
Neurobiologie des Processus Adaptatifs CNRS-UMR 7102, Université Paris VI, Paris, France

Keywords: Long-term potentiation, long-term depression, dopamine, postsynaptic depolarization, prelimbic area, memory model, cognitive function.

Abstract: In rat prelimbic (prefrontal) slices, layer I-II to layer V pyramidal neuron glutamatergic synapses show long-term depression (LTD) and potentiation (LTP) of synaptic strength. First, LTD is induced by high-frequency synaptic stimuli (100 pulses at 50 Hz, 4 times) in the presence of dopamine. Our analyses show that the synaptic responses and postsynaptic depolarization during high-frequency stimuli are larger in the presence of dopamine than in its absence. These dopamine effects are *N*-methyl-D-aspartate (NMDA) receptor-dependent. Second, LTP is induced by the same stimuli, in the presence of dopamine, if the synapses are previously exposed to dopamine. Interestingly, the synaptic responses and depolarization during the LTP-inducing high-frequency stimuli are smaller than those in control and LTD conditions. Third, LTP can also be induced by short burst-type stimuli without dopamine (5–10 pulses at 50 Hz, 5–6 times). With the short burst stimuli, prelimbic neurons show little inter-burst synaptic fatigue and fire to each burst episode, a property not seen in any of the groups in which the long trains of stimuli were applied. These data suggest that there are separate induction mechanisms for synaptic plasticity in rat prefrontal neurons. We will then discuss on functional implications of these LTP and LTD.

1. INTRODUCTION

Synaptic plasticity has been observed in every brain region so far examined. The brain may use the plastic nature of the synapse as a tool to store information. Among plastic changes, long-term potentiation (LTP) and long-term depression (LTD) have been most widely studied. LTP and LTD are candidate cellular substrates for various forms of memory and behavioral modifications (e.g. Bannerman et al., 1995; Kirkwood et al., 1996; Rogan et al., 1997; Reynolds et al., 2001; Ungless et al., 2001). LTP and LTD are observed also in rodent prefrontal cortex (PFC; the prelimbic area of medial frontal cortex, see Chapter 1) (Hirsch and Crepel, 1990; Law-Tho et al., 1995; Vickery et al., 1997; Otani et al., 1998; Gurden et al., 1999; Takita et al., 1999; Blond et al., 2002; Herry and Garcia, 2002). A notable characteristic of prefrontal LTP/LTD is the strong modulation by neuromodulator dopamine. This fact appears relevant to the numerous behavioral and clinical studies that indicate dopamine as a critical factor for prefrontal-related normal and abnormal behaviors (e.g. Robbins and Everitt, 2002). Particularly intriguing to us is the fact that the PFC is not only important for short-term memory, but is also involved in certain forms of long-term memory (Otani, 2002; see also Chapter 12). In this chapter, we focus on LTP and LTD in rat prefrontal neurons maintained *in vitro*. We will explain our already published findings but will extend our discussion in several ways. 1) We will show our thorough re-analyses of the synaptic responses occurring during LTP/LTD-inducing stimuli. 2) We will show our new findings on LTP whose induction does not need dopaminergic modulation. 3) We will propose simple cellular models for certain physiological and pathological processes involving the PFC.

2. BASIS AND THE PRESENT FOCUS

2.1 General Methods

Brain slices containing the prelimbic area of the medial frontal cortex (2.2–3.7mm from bregma) are prepared from young (23-30 days old) male Sprague-Dawley rats (Otani et al., 1998, 1999, 2002). We place a bipolar tungsten stimulating electrode on layer I-II to stimulate presynaptic axons by single mono-phasic square pulses (100 μs width, 0.033 Hz). Evoked excitatory postsynaptic potentials (EPSPs) are recorded from cell body of the layer V pyramidal neurons by the use of a glass micro-pipette filled with 3 M potassium acetate. In all experiments, fast GABA-A (γ-amino-butyric acid-A) receptor-mediated synaptic inhibition is reduced by bicuculline methiodide (1 μM) added in bathing medium. A schematic drawing of the experimental preparation is shown in Figure 1A.

In all experiments, synaptic responses of about 10 mV amplitude are evoked to the 0.033 Hz single test stimuli in order to acquire baseline level (Fig. 1). After at least a 15 min recording, the synapses are conditioned by high-frequency stimuli applied to the presynaptic axons. In the conventional protocol, the stimuli consisted of four episodes (10 sec interval) of a 2 sec train (100 pulses at 50 Hz). In short-burst protocol, stimuli consisted of five or six episodes (8 – 10 sec interval) of 80 or 180 ms duration of stimuli (5 or 10 pulses at 50 Hz). An increase or decrease of the synaptic responses was expressed as a percent change of the initial rising slope (≤1 ms period from response onset) from baseline level. Synaptic responses evoked during the conditioning stimuli were recorded in magnetic tape for off-line analyses.

2.2 LTD: Previous Findings

First, the application of the conditioning stimuli (100 pulses at 50 Hz, x 4 at 0.1 Hz), in control medium, does not induce lasting synaptic plasticity (Otani et al., 1998, 1999; Fig. 1B). The same stimuli, however, induce LTD when delivered at the end of a 10-15 min bath-application of dopamine (100 μM in 20 μM ascorbic acid) (Otani et al., 1998, 1999; Fig. 1C). A series of our in-depth studies revealed the underlying mechanisms of this dopamine-facilitated LTD as follows (see also Otani et al., 2003 for review).

1. Dopamine acts on both D1-like and D2-like receptors. Stimulation of either class of receptors appears sufficient to facilitate LTD induction.
2. The dopamine-facilitated LTD is NMDA (*N*-methyl-D-aspartate) receptor-independent.
3. Induction of the dopamine-facilitated LTD requires postsynaptic depolarization during conditioning. Indeed, dopamine enhances postsynaptic responses during high-frequency conditioning stimuli.
4. Induction of the dopamine-facilitated LTD requires synaptic activation of both group I and group II metabotropic glutamate receptors (mGluRs) during conditioning.
5. A mechanism of the cooperation between dopamine receptors and the mGluRs is convergent postsynaptic activation of mitogen-activated protein kinases (MAP kinases).
6. Mechanisms of group II mGluR involvement include postsynaptic activation of phospholipase C and consequential protein kinase C activation and internal Ca^{2+} release (Otani et al., 2002).

2.3 LTP: Previous Findings

In our slice condition, the conditioning stimuli coupled to a dopamine bath-application always induced LTD and never induced LTP. By contrast, in the anesthetized rats, release of dopamine in the PFC by ventral tegmental stimulation facilitates LTP (Gurden et al., 1999, 2000). We reasoned that the lack of baseline dopamine receptor stimulation in our slice condition may be a source for the discrepancy. For example, in freely moving rats, ventral

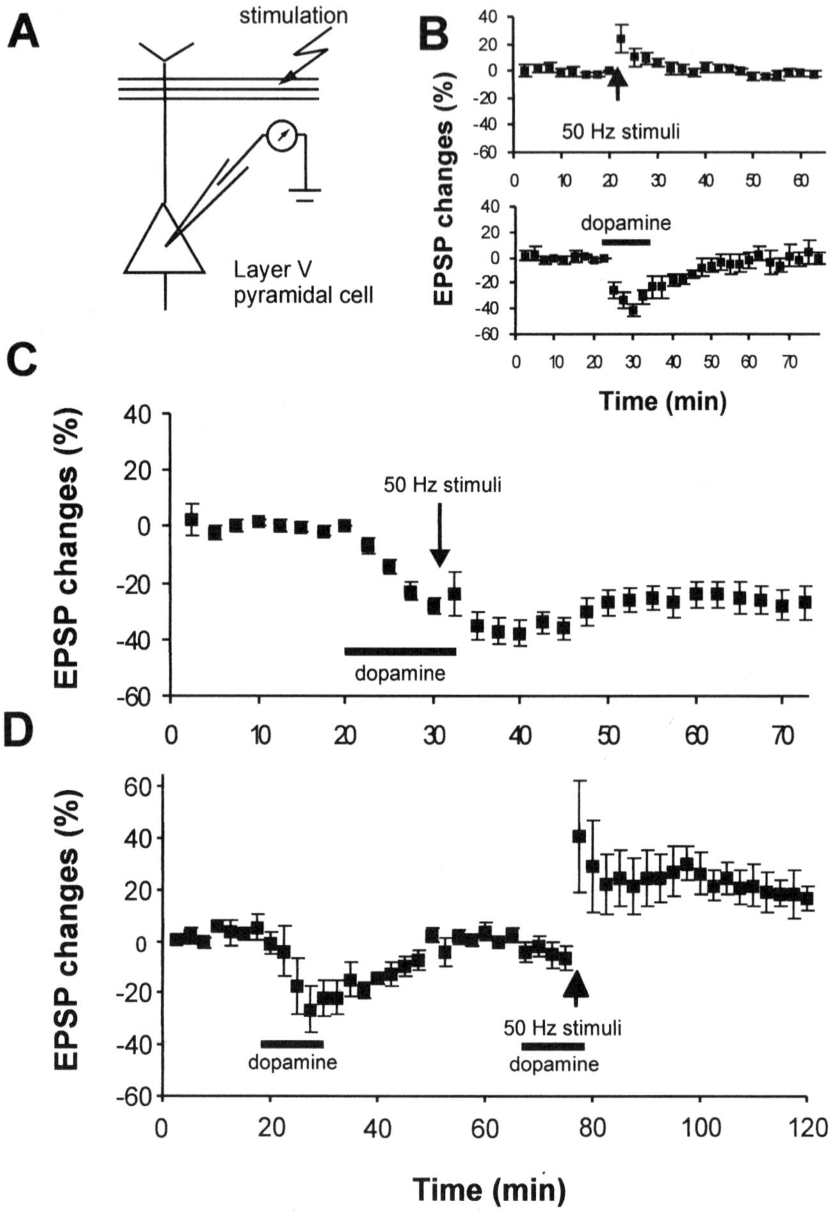

Figure 1: LTD and LTP induction in rat prelimbic neurons. A. Schematic drawing of our experimental preparation. B. Neither 50 Hz conditioning alone (upper) or 100 μM dopamine bath-application alone (lower) induces lasting changes of the slope of the EPSP. C. However, a combination of these two results in synaptic depression that lasts longer than 40 min (i.e. LTD). D. The same combination as in C induces LTP if the neurons have been "primed" by prior sole bath-application of dopamine.

tegmental neurons discharge spontaneously and raise their firing rate phasically at various behavioral occasions (Kosobud et al., 1994; Kiyatkin and Rebec, 1998). In fact, dopamine concentration in rat PFC is tonically maintained (Takahata and Moghaddam, 2000) and increases phasically during conditioned and unconditioned appetitive tasks (Bassareo and Di Chiara, 1997) as well as during conditioned and unconditioned aversive tasks (Feenstra et al., 2001). Under anesthesia also, there is always a basal level of dopamine in the PFC (Gurden et al., 2000). In contrast, we usually allow brain slices to recover from the dissection insult for 3 hours or longer. Because the axons of dopaminergic afferents are severed in our condition, the dopamine receptors are largely left unstimulated during this period. Routinely, we detect no effects of dopamine receptor antagonists on baseline synaptic responses, which should otherwise occur if dopamine receptors are stimulated, since the presence of dopamine reduces synaptic responses (Otani et al., 1998).

We tested therefore in our slice preparation the effect of prior application of dopamine on later plasticity induction (Blond et al., 2002). First, we bath-applied dopamine identically as in our previous studies. When the responses recovered from the acute transient depression by the dopamine, dopamine was applied for the second time, and this second dopamine was coupled to 50 Hz stimuli. This procedure induced LTP (Fig. 1D). The inhibitory effect of the second dopamine on the synaptic responses is always smaller than the first dopamine (–4.3 ± 4.6% vs –33 ± 5.5%, p<0.0025), suggesting that the first dopamine triggers some lasing "priming" intracellular effects (see Otani et al., 2003).

2.4 Present Focus

In this chapter, based on the following reasons, we will focus on the point 3 in **2.2**, i.e. the dopamine effect on postsynaptic responses during conditioning stimuli. We included some newly conducted experiments.

1. This dopamine effect has not been thoroughly analyzed in our past studies (Otani et al., 1998, 1999).
2. This effect suggests that signal-to-noise ratio might be exaggerated by dopamine. Thus, dopamine severely reduces (by about 40%; Otani et al., 1998, 1999) synaptic responses to single 0.033 Hz stimuli (Fig. 1B). A reduction of the low-frequency synaptic responses and an enhancement of high-frequency synaptic responses might mean that in the presence of dopamine, "background noise" is filtered while "significant" events are amplified.

3. Other studies report dopaminergic enhancement of postsynaptic depolarization through the modulation of tetrodotoxin-sensitive, slowly-inactivating persistent Na^+ current (I_{NaP}) (Yang and Seamans, 1996; Gorelova and Yang, 2000; but see Geijo-Barrientos and Pastore, 1995; Gulledge and Jaffe, 1998). However, in these studies, the effect was tested by direct postsynaptic current injection. Detailed analyses of the dopamine effect with synaptically-applied input will add important information to the literature.

3. EFFECTS OF DOPAMINE ON POSTSYNAPTIC RESPONSES

To re-analyze dopaminergic enhancement of postsynaptic responses during conditioning stimuli, we measured the following parameters: 1) the number of spikes per train episode, 2) decay of postsynaptic depolarization during a train, and 3) the size of the EPSPs during a train.

3.1 LTD Inducing-Stimuli

3.1.1 Spike numbers

On the contrary to our previous description (Otani et al., 1998), we found no statistically significant enhancement of spike numbers per train episode by dopamine (Fig. 2C). This result is supported further by the fact that mean peak depolarization during a train episode is not different between the dopamine and control groups (not shown; spike threshold was taken as the peak when a spike(s) was present). Thus, while dopamine increased spike number during a depolarizing current step (Yang and Seamans, 1996; Gorelova and Yang, 2000), this effect was not seen with our 50 Hz synaptic drive. However, this may be due to the moderate amount of depolarization evoked by the synaptic stimuli.

3.1.2 Decay of postsynaptic depolarization during conditioning

By contrast, dopamine prolonged postsynaptic depolarization during 50 Hz stimulus train. We computed the amount of depolarization every 100 ms following a train onset (calculated in 10 ms windows). This generates twenty voltage values from one neuron for one train episode. We then drew mean decay curve of the voltage values in each experimental group for each of the four train episodes. Figure 2B shows that dopamine prolonged postsynaptic depolarization in the first train episode compared to control (two-way ANOVA, $F(19,342)=1,743$, $p<0.03$). The effect is relatively small but consistent. Significant interactions were not seen in later train episodes, but two-tailed t-test revealed that the depolarization 400 ms after train onset in

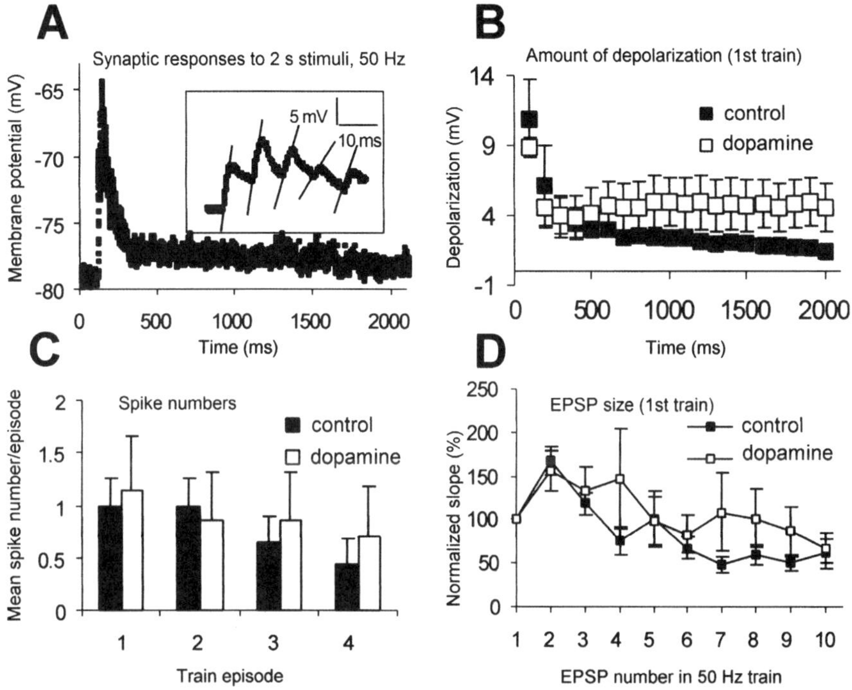

Figure 2: LTD induction in the presence of dopamine is accompanied by enhanced postsynaptic depolarization and EPSP sizes during conditioning stimuli. A. An example of synaptic responses during 2 sec 50 Hz train. This was taken from control non-dopamine condition. The inset shows in different scales the first five EPSPs evoked to the 50 Hz pulses, and how the slope of the responses were measured. B. Mean postsynaptic depolarization in control and LTD-inducing dopamine conditions. The data are taken from the first of four train episodes. Dopamine prolonged postsynaptic depolarization. C. Mean spike numbers in the control and dopamine groups at each of the four train episodes. There was no statistical difference. D. Mean normalized slope values for the first 10 EPSPs in the first train episode. Dopamine enhanced the EPSPs during the train.

the third episode in dopamine group is significantly larger than control group (p<0.03). Comparison at 100 ms in this episode shows a tendency (p<0.08).

Interestingly, the dopaminergic prolongation of depolarization was blocked by NMDA receptor antagonist DL-2-amino-5-phosphonopentanoic acid (DL-AP5; 100 μM) (Fig. 3A). In Figure 3B, we plotted AP5-sensitive portions of the depolarization under the dopamine and the non-dopamine control conditions (calculated as: Δ(dopamine)-Δ(dopamine+AP5), and Δ(control)-Δ(control+AP5)). During the first train episode, the NMDA receptor-mediated portion grew towards the end of the 2 sec train in the presence of dopamine, while in its absence, it rapidly decayed ($F(19,342)=2,572$, $p<0.0005$). In later episodes, the portion was larger mainly in the early periods in the presence of dopamine (third train, $F(19,323)=1,968$, $p<0.01$).

3.1.3 Size of the EPSPs during conditioning

We also measured the slope of the first 10 EPSPs during 50 Hz trains (note: later EPSPs were often very small to measure efficiently). We then normalized them relative to the first EPSP in a given train. Dopamine enhanced the EPSPs during the first train episode (two-way ANOVA, $F(12,216)=1,946$, $p<0.035$; Fig. 2D). During the third and fourth episodes, the EPSPs were larger in the presence of dopamine, although the differences were statistically marginal ($p<0.075$).

Again, the increases in the EPSP slope under dopamine condition are mediated by NMDA receptors. In the presence of DL-AP5, the EPSP sizes were similar between the dopamine and the non-dopamine control conditions (not shown).

3.1.4 Role of D1 receptors

Seamans et al. (2001) showed an AP5-sensitive late enhancement of depolarization during 20 Hz synaptic stimuli by dopamine D1 agonist. Wang and O'Donnell (2001) showed the enhancement of depolarization upon current injection by synergistic action of NMDA and D1 receptors. Therefore, we re-analyzed the postsynaptic responses in the D1 agonist SKF 38393-treated group (Otani et al., 1998). SKF 38393 was present in the bath for 10 – 15 min before 50 Hz stimuli were applied. SKF 38393 enhanced postsynaptic depolarization in the second ($F(19,228)=1,832$, $p<0.025$), the third ($F(19,228)=3,351$, $p<0.0001$), and the fourth train episodes ($F(19,228)=3,950$, $p<0.0001$). Figure 3C shows the enhancement seen in the third train. It was noted that SKF 38393 enhanced mainly the early depolarization (up to about 1 s), whereas dopamine affected also late depolarization. The interaction between the SKF 38393 and dopamine groups in the third train episode is significant ($F(19,209)=1.638$, $p<0.05$), raising the possibility that, although D1 agonist enhances postsynaptic depolarization during 50 Hz stimuli, the mechanism may slightly differ from that of dopamine. Interestingly, SKF 38393 did not enhance the slope of the EPSPs, unlike dopamine (not shown).

At this stage, we have to note that the induction of dopamine-facilitated LTD is *independent* of NMDA receptors (Otani et al., 1998). Therefore, the prolongation of postsynaptic depolarization and the increases of the EPSPs by dopamine are not causally related to the induction of this homosynaptic LTD. Nevertheless, postsynaptic hyperpolarization blocks dopamine-

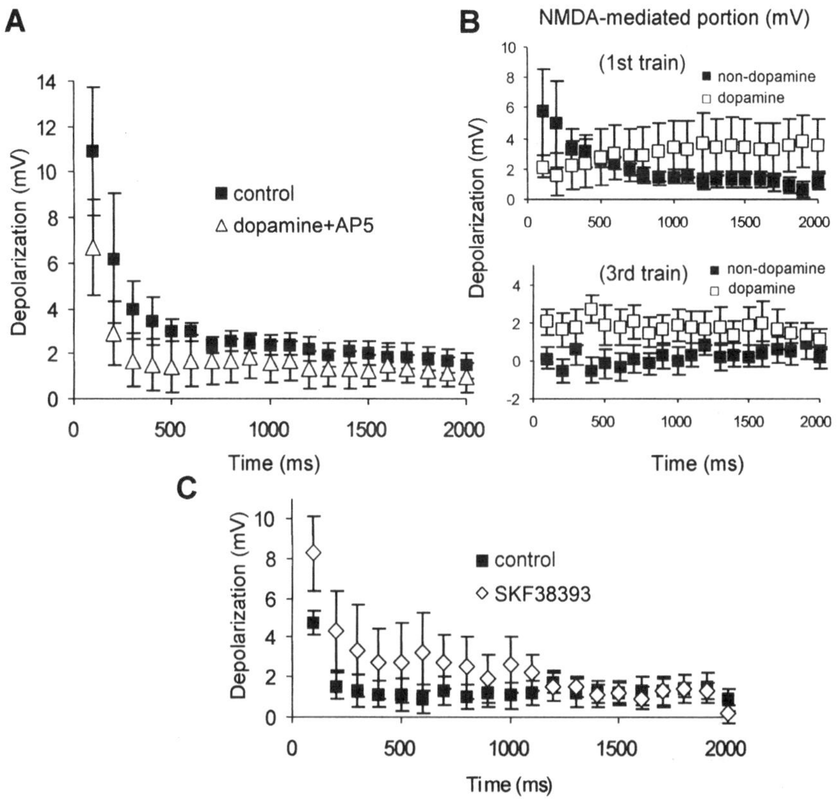

Figure 3: Depolarization enhancement by dopamine during 50 Hz stimuli is NMDA receptor-mediated and may be mimicked by D1 agonist. A. The presence of the NMDA receptor antagonist DL-AP5 blocked the enhancement of postsynaptic depolarization by dopamine (see Fig. 2B). B. We plotted NMDA receptor-mediated portion of depolarization in non-dopamine and dopamine conditions for the first and third train episodes. It is clear that dopamine enhances NMDA-mediated depolarization. Dopamine, by contrast, did not affect NMDA-independent portions of depolarization (not shown). C. Dopamine D1 receptor agonist SKF 38393 enhanced postsynaptic depolarization. This is the third train episode.

facilitated LTD (Otani et al., 1998). Thus, for this LTD, depolarization achieved in the control and the AP5 conditions must be sufficient.

3.2 LTP Inducing-Stimuli

We previously noted that the postsynaptic depolarization during LTP-inducing 50 Hz stimuli is *smaller* than that during LTD (Blond et al., 2002). This is an interesting observation, since LTP induction is believed to require larger postsynaptic depolarization and, consequently, larger intracellular calcium increases than LTD (Bienenstock et al., 1982; Dudek and Bear, 1992; Artola and Singer, 1993; Hansel et al., 1996; Otani and Connor, 1998; Cormier et al., 2001). We thus thoroughly analyzed the postsynaptic responses during LTP-inducing conditioning stimuli.

3.2.1 Spike numbers

There was no statistical difference in the number of spikes per train episode between LTP condition and control or LTD condition. Similarly, no difference was found in the peak depolarization between these groups.

3.2.2 Decay of postsynaptic depolarization during conditioning

By contrast, the postsynaptic depolarization during conditioning was smaller in LTP condition than control condition in the first ($F(19,323)=2,828$, $p<0.0001$) and the second train episodes ($F(19,304)=2.039$, $p<0.01$) (Fig. 4). These effects are due to the differences in the initial 400 – 500 ms periods. Comparison between LTP and LTD conditions revealed a significant interaction in the fourth train episode ($F(19,266)=3.103$, $p<0.001$), although in all episodes, LTP was accompanied steadily by smaller (albeit marginally significant) postsynaptic depolarization than LTD. Taken together, we suggest that dopamine-facilitated LTP does not require enhanced postsynaptic depolarization for the induction.

3.2.3 The size of the EPSPs during conditioning

Normalized EPSP slopes in LTP condition were not different from those in control condition, but were smaller than LTD condition in the first and third train episodes (not shown). In the first episode, there was a significant groups x time interaction ($F(12,144)=1,991$, $p<0.03$), while there was a significant group effect in the third ($F(1,13)=8,222$, $p<0.02$).

3.3 Discussion: Dopamine Effects during LTD and LTP Induction

We summarize our investigation on dopamine-facilitated LTD and LTP as follows.

1. When dopamine receptors are in "non-primed" state, glutamatergic high-frequency synaptic inputs coincident with dopamine receptor stimulation induce homosynaptic LTD.

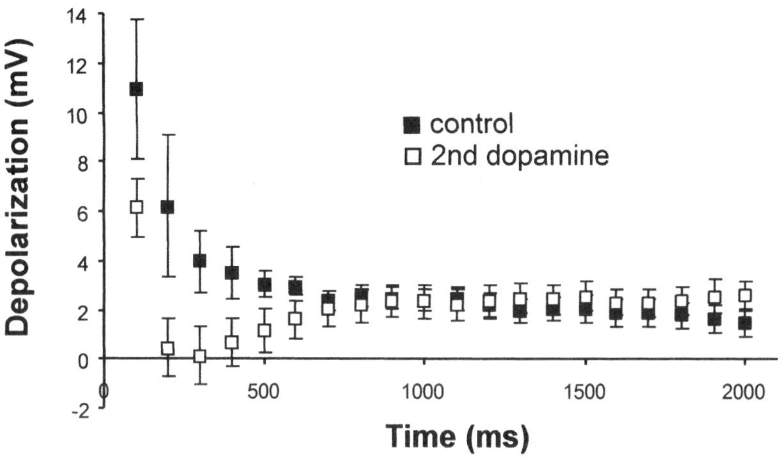

Figure 4: LTP-inducing stimulation in the presence of dopamine is accompanied by smaller postsynaptic depolarization than control. This was taken from the fist train episode.

2. During the LTD-inducing high-frequency stimuli, dopamine, perhaps through D1 receptors, enhances postsynaptic depolarization. This effect is mediated by NMDA receptors.
3. During the same stimuli, dopamine enhances the slope of the evoked EPSPs. This effect is also NMDA-mediated, but not mediated by D1 receptors.
4. When dopamine receptors are in "primed" state, the high-frequency inputs coincident with dopamine receptor stimulation induce homosynaptic LTP, not LTD.
5. This LTP induction involves smaller postsynaptic depolarization than LTD.

There are two main points of discussion.

1) Dopamine-facilitated homosynaptic LTD does not require NMDA receptors (Otani et al., 1998). Then, what role, if any, do the NMDA-mediated depolarization and increased EPSPs play for synaptic plasticity?

It may be that NMDA receptor-dependent homosynaptic plasticity does occur, but is masked by massive LTD (Hirsch and Crepel, 1991). Supporting this view, LTD induction in the presence of AP5 seems slightly more consistent than LTD induction in its absence (LTD>10%, 7/7 in the presence of AP5 vs 9/14 in the absence, $p<0.1$, χ^2 test). The second possibility is that the increased depolarization serves to facilitate Hebbian associative processes (Levy and Steward, 1979, 1983). Clark and Collingridge (1996) showed that even a few mV synaptic input effectively affects adjacent synaptic region to enhance NMDA receptor-mediated transmission at that region. Thus, dopaminergic enhancement of postsynaptic depolarization may facilitate NMDA-dependent associative processes. In addition, the enhanced depolarization lowers requirement for additional associated inputs to fire the neuron to transmit the information.

2) What are the possible mechanisms for the dopamine "metaplasticity" that converts LTD to LTP?

Gorelova and Yang (2000) showed that D1 agonist SKF81297 lastingly decreases the first spike latency upon current injection. Seamans et al. (2001) showed that SKF81297 lastingly potentiates NMDA receptor-mediated synaptic transmission. These effects may facilitate the induction of synaptic plasticity. However, in our case, neither the increased spiking nor the NMDA-mediated enhanced postsynaptic responses were seen during LTP induction (see 3.2.). At this stage, we could not provide answers as regards dopamine metaplasticity. Studies are currently underway in our laboratory to address this question.

4. LTP BY BRIEF BURST STIMULI

4.1 Prelimbic Neuronal Activity during Behavior

The 50 Hz 2 sec conditioning stimuli were used in early studies (Hirsch and Crepel, 1990; Law-Tho et al., 1995). This type of long trains of pulses have been used to induce LTP in many hippocampal studies, since hippocampal neurons can show relatively long high-frequency discharge during, for example, spatial exploration (O'Keefe and Nadel, 1978). Stimulus patterns mimicking theta rhythm are a more adapted version (Larson et al., 1986). Because ventral hippocampus sends monosynaptic projection to the prelimbic area in the rat (Jay and Witter, 1991) and because the hippocampal–prelimbic connection is important for spatial working memory in the rat (Floresco et al., 1997), theta burst-like stimuli (e.g. 4 pulses at 100 Hz, repeated 10 times at 5 Hz) were adopted in prelimbic studies and shown to induce LTP (Vickery et al., 1997; Morris et al., 1999). It is likely that other monosynaptic projection fibers also converge to prelimbic area and convey sensory information (Perez-Jaranay and Vives, 1991; Herry et al., 1999).

Several authors did find some increased firings of relatively long duration in the prelimbic area (Batuev et al., 1990; Kosobud et al., 1994; Mulder et al., 2000). However, these authors also report another type of activity that is much shorter in duration. Mulder et al. (2000) found a sharp transient increase (at about 30 Hz for about 250 ms) that occurs in response to a conditioned sensory stimulus. A similar sharp increase to a conditioned stimulus was seen by Batuev et al. (1990). Kosobud et al. (1994) found a sharp firing increase at a lever-press, which was the conditioned response to initiate a reward delivery in their task. An interesting point is that, in two of the three studies (Batuev et al., 1990; Mulder et al., 2000), the sharp increases were seen even in early-learning periods. As learning progressed, more persistent neuronal activity in response to the conditioned stimuli were detected. It may be then useful to test whether these brief activities can induce lasting synaptic plasticity in prelimbic neurons.

4.2 Short Bursts Induce LTP in Prelimbic Neurons without Dopamine

We fixed stimulus frequency at 50 Hz for the purpose of comparison with earlier studies. Duration of one train (burst) was 80 – 180 ms (i.e. 5 or 10 pulses). We repeated it 5 or 6 times in every 8 or 10 s. All other methods were identical to the previous experiments.

The burst stimuli, in control medium, induced LTP in 4 of 8 neurons tested (mean LTP in the four cases 17 ± 4.4% at 30–40 min). One example is shown in Figure 5A. We emphasize that the conventional long trains (50 Hz 2 sec x 4 in every 10 s) do not induce LTP or LTD without dopamine (0/12 cases, $p<0.05$, χ^2 test). (Note: in this case, the number of episode is 4, versus 5 or 6 in the short burst protocol. But our additional preliminary data suggest that application of the long train 6 times still does not induce LTP. It is therefore unlikely that the absence of LTP in the long train condition is due to the smaller number of episodes in this condition).

In a separate group, we applied the short burst in the presence of dopamine (the first application). In 3 of 4 cases, LTD was induced (not shown). The remaining neuron showed no lasting changes. Mean change (-18 ± 8,9%, n=4) is significantly smaller than mean change calculated from the 8 neurons tested without dopamine (7.8 ± 4.5%, $p<0.02$).

4.3 Possible Cellular Mechanism for LTP by Short Bursts

We noted that during the repeated delivery of a short burst, the prelimbic neurons discharged to each episode. Such a sustained firing was not seen with the long stimulus trains. In Figure 5B, we plotted mean spike number per burst for the first four burst episodes to compare with the long trains. As shown, with the short burst, the neurons show little inter-episode fatigue and fire on average at each of the burst episodes. In contrast, firing to a 2 sec train gradually decays across episodes. There was a significant groups x episodes interaction (F(3,63)=3,185, $p<0.03$). Interestingly, the same interaction was found when the short burst group was compared to the dopamine-facilitated LTP group (F(3,45)=3,332, $p<0.03$, not shown), suggesting that these two forms of LTP may not share common postsynaptic mechanisms.

The analysis of the EPSPs within a burst/train shows that the ratio of the second EPSP/the first EPSP is larger in the short burst protocol than the long train protocol (Fig. 5C). This enhanced response contributed to the sustained, across-episodes spiking in the short burst condition. This effect on the EPSP was also evident when compared to the dopamine-facilitated LTP group ($p<0.01$ in the third train).

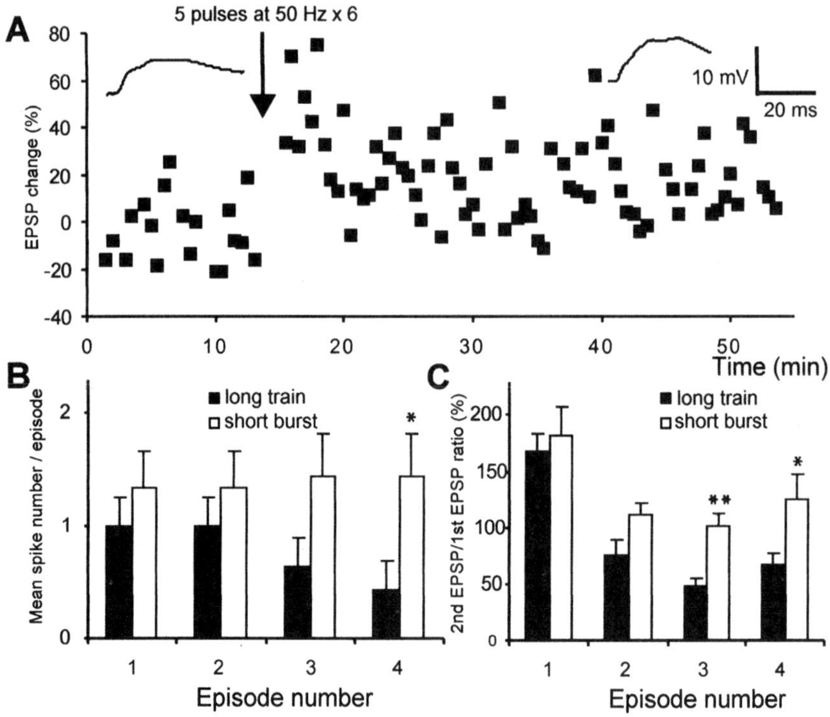

Figure 5: Short burst stimuli, unlike long trains of stimuli (see Fig. 1B), can induce LTP without dopamine. A. An example of LTP induction by short burst stimuli. Similar observation was made in 4 out of 8 neurons tested. Insets are averaged EPSPs taken from baseline period and 40 min after the stimuli. B. With short burst stimuli, there was no inter-episode reduction of spiking, unlike with conventional long trains. * $p<0.03$ (t-test). C. During short burst stimuli, the ratio of the second EPSP/the first EPSP is larger in the third and fourth episodes than long trains, that contributed to the sustained inter-episode spiking. * $p<0.025$, ** $p<0.002$.

4.4 Discussion: A Way to Induce LTP without Dopamine

We summarize our findings on LTP induced by short burst stimuli.

1. Short burst stimuli (5 – 10 pulses at 50 Hz x 5 – 6) can induce LTP in the absence of dopamine. This effect was not seen with conventional long trains of stimuli (100 pulses at 50 Hz x 4).
2. Prelimbic neurons fire to each of the burst episodes. This effect was not seen with the long trains.
3. The ratio of the second EPSP/the first EPSP is larger within a short burst than within a long train. This effect contributes to the sustained firing across episodes in the former protocol.
4. When dopamine is present in the bath (the first application), the short burst stimuli induce LTD.

Even though rat prelimbic neurons probably receive high-frequency inputs of relatively long duration via projection fibers, these stimuli alone may not induce lasting synaptic plasticity. These stimuli may be effective in inducing LTP or LTD when dopamine (and probably other neuromodulators) is present at/around the synaptic sites. On the contrary, short burst inputs alone can lastingly potentiate prelimbic synapses. When dopamine is present, the same short burst stimuli induce LTD (3/4 cases). Our preliminary data suggest that when synapses are "primed" by prior dopamine, short bursts also induce LTP in the presence of (second) dopamine.

We do not know how spiking is inhibited across long train episodes. We speculate that synaptic stimulation of group II mGluRs might be involved in the inhibition. When group II mGluR antagonist MSOPPE ((RS)-α-methylserine-O-phosphate monophenyl ester) is present in the bath, the conventional long trains successfully discharge prelimbic neurons at each of the four episodes as with the short bursts ($p<0.05$ compared to control, data not shown). This effect was not seen with group I mGluR antagonist AIDA ((RS)-1-aminoindan-1,5-dicarboxylic acid). It may be that short burst stimuli minimally activate the perisynaptically located mGluRs and minimize synaptic fatigue that would be otherwise induced by group II mGluRs.

5. DOES SYNAPTIC PLASTICITY HAVE ROLES IN PREFRONTAL FUNCTIONS?

5.2 Prefrontal Memory and LTP?

It is known that under behavioral conditions, prefrontal dopamine concentration is tonically maintained by ventral tegmental activity (Takahara and Moghaddam, 2000) and that it elevates repeatedly at a presentation of reward (Bassareo and Di Chiara, 1997). According to our *in vitro* results, under such conditions, coincident activity enhancement of dopaminergic and glutamatergic fibers converging on the same postsynaptic neuron induces LTP in the glutamatergic synapses on that neuron. The results obtained in the anesthetized condition, in which there is always baseline dopamine concentration in the PFC, support this view (Gurden et al., 1999, 2000).

We propose that coincidental dopaminergic and glutamatergic activity occurs in the phase of conscious reward anticipation and that a resulted LTP may serve as a memory trace. For example, the dialysis study in rats showed elevations of prefrontal dopamine concentration in response to food consumption and to the presentation of conditioned stimulus signaling the food (Bassareo and Di Chiara, 1997). In another study, increased firing of ventral tegmental neuron was seen at a lever-press to initiate a reward delivery, at the reward delivery, and just after the reward consumption (Kosobud et al., 1994). In this operant task, a reward consumption was immediately followed by another reward-seeking action. Therefore, the end of reward consumption was actually when next reward anticipation started. Kosobud et al. (1994) further showed that portion of prelimbic neurons coincidentally increase their firing at these behavioral phases. Firing of prelimbic neurons upon reward anticipation (a food tray entry) was seen also by Mulder et al. (2000). Other authors reported also increased prefrontal firings in other anticipatory phases (Batuev et al., 1990; Pratt and Mizumori, 2001).

Importantly, these rat studies are consistent with key monkey studies: i.e. the studies that showed phasic firing of dopamine neurons to a primary

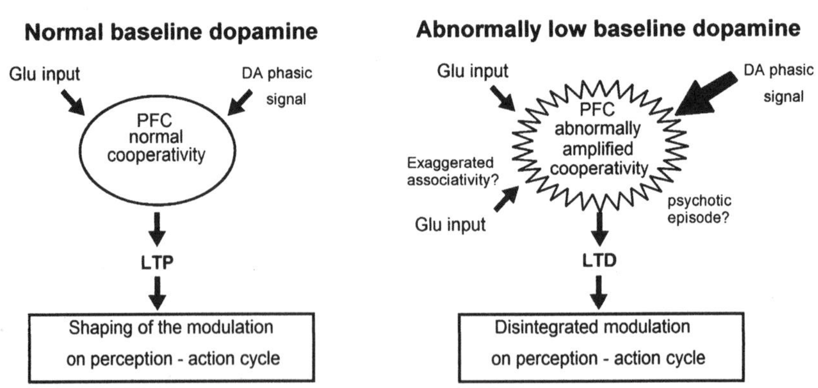

Figure 6: Possible functional significance of dopamine-facilitated LTP and LTD. See text for details.

reward or to a conditioned stimulus signaling the reward (Schultz, 2002) and the studies that showed firing of dorsolateral prefrontal neurons in "delay period" (Goldman-Rakic, 1995). Note that during the delay period, animals attend to a conditioned sensory stimulus anticipating a reward (Watanabe, 1996).

Thus, as shown in Figure 6 (left), we propose that during these motivational phases, coincident activity of dopamine fibers and glutamatergic projection fibers (from the structures such as the amygdala and hippocampus) may induce LTP at the glutamatergic synapses. LTP may serve as a trace that helps later conscious behavioral guidance in the same or similar situations (Otani, 2002) by means of the "biasing" (Miller and Cohen, 2001) on habitual perception–action cycles that involve the striatum.

Possible role for the dopamine-independent LTP by short burst stimuli is still unclear. It might serve as a priming signal for dopamine-dependent LTP, as indicated by the fact that short bursts occur in early-learning periods (Batuev et al., 1990; Mulder et al., 2000). Alternatively, it might occur in intrinsic connections in order to reinforce local networks that have been potentiated by the dopamine–glutamate coincident activity.

5.3 Prefrontal Pathology and LTD?

LTD facilitated by dopamine in "non-primed" naïve neurons might best model certain pathological processes (Fig. 6, right). For example, Moghaddam, Adams, and their colleagues (Moghaddam et al., 1997; Adams and Moghaddam, 1998; Moghaddam and Adams, 1998) showed in rats that the injection of phencyclidine (PCP), a psychotomimetic NMDA antagonist that causes schizophrenia-like symptoms in humans, induces some 600% increases of prefrontal dopamine efflux as well as two-fold increases of glutamate (also Suzuki et al., 2002). A pharmacological block of the glutamate increase blocks PCP-induced working memory impairment (Moghaddam and Adams, 1998), suggesting that the coincident efflux of dopamine and glutamate is necessary for the memory impairment. Note that prefrontal LTD appears to require strong dopaminergic input coincident with glutamatergic input and occurs in the presence of a NMDA antagonist (Otani et al., 1998). Human schizophrenia may involve two pathological phases of dopaminergic innervation: i.e. abnormally low and abnormally high (Yang et al., 1999). If prefrontal dopamine system is somehow down-regulated in schizophrenics (Akil et al., 1999) to set up postsynaptic receptor hypersensitivity (Abi-Dargham et al., 2002), an event-related phasic release of dopamine (Fig. 6 right, thick arrow) coincident with glutamatergic input may result in abnormally amplified biochemical and electrophysiological cooperativity. Such an event may be accompanied by exaggerated synaptic association due to the increased postsynaptic depolarization (Fig. 2B). This state might correspond to a psychotic episode, and a consequence of this might be a lasting reduction of glutamatergic transmission. LTD might then correspond to some cognitive abnormalities of schizophrenics such as reduced ability of working memory.

REFERENCES

Abi-Dargham A, Mawlawi O, Lombardo I, Gil R, Martinez D, Huang Y, Hwang D-R, Keilp J, Kochan L, Van Heertum R, Gorman JM, Laruelle M (2002) Prefrontal dopamine D1 receptors and working memory in schizophrenia. J Neurosci 22:3708-3719.

Adams B, Moghaddam B (1998) Corticolimbic dopamine neurotransmission is temporally dissociated from the cognitive and locomotor effects of phencyclidine. J Neurosci 18:5545-5554.

Akil M, Pierri JN, Whitehead RE, Edgar CL, Mohila C, Sampson AR, Lewis DA (1999) Lamina-specific alterations in the dopamine innervation of the prefrontal cortex in schizophrenic subjects. Am J Psychiatry 156:1580-1589.

Artola A, Singer W (1993) Long-term depression of excitatory synaptic transmission and its relationship to long-term potentiation. Trends Neurosci 16:480-487.

Bannerman DM, Good MA, Butcher SP, Ramsay M, Morris RGM (1995) Distinct components of spatial learning revealed by prior training and NMDA receptor blockade. Nature 378:182-186.

Bassareo V, Di Chiara G (1997) Differential influence of associative and nonassociative learning mechanisms on the responsiveness of prefrontal and accumbal dopamine transmission to food stimuli in rats fed *ad libitum.* J Neurosci 17:851-861.

Batuev AS, Kursina NP, Shutov AP (1990) Unit activity of the medial wall of the frontal cortex during delayed performance in rats. Behav Brain Res 41:95-102.

Bienenstock EL, Cooper LN, Munro PW (1982) Theory for the development of neuron selectivity: orientation specificity and binocular interaction in visual cortex. J Neurosci 2:32-48.

Blond O, Crepel F, Otani S (2002) Long-term potentiation in rat prefrontal slices facilitated by phased application of dopamine. Eur J Pharmacl 438: 115-116.

Clark KA, Collingridge GL (1996) Evidence that heterosynaptic depolarization underlies associativity of long-term potentiation in rat hippocampus. J Physiol 490:455-462.

Cormier RL, Grennwood AC, Connor JA (2001) Bidirectional synaptic plasticity correlated with the magnitude of dendritic calcium transients above a threshold. J Neurophysiol 85:399-406.

Dudek SM, Bear MF (1992) Homosynaptic long-term depression in area CA1 of hippocampus and effects of N-methyl-D-aspartate receptor blockade. Proc Natl Acad Sci USA 89:4363-4367.

Feenstra MGP, Vogel M., Botterblom MHA, Joosten RNJMA, de Bruin JPC (2001) Dopamine and noradrenaline efflux in the rat prefrontal cortex after classical aversive conditioning to an auditory cue. Eur J Neurosci 13:1051-1054.

Floresco SB, Seamans JK, Phillips AG (1997) Selective roles for hippocampal, prefrontal cortical, and ventral striatal circuits in radial-arm maze tasks with or without a delay. J Neurosci 17:1880-1890.

Geijo-Barrientos E, Pastore C (1995) The effects of dopamine on the subthreshold electrophysiological responses of rat prefrontal cortex neurons in vitro. Eur J Neurosci 7:358-366.

Goldman-Rakic PS (1995) Cellular basis of working memory. Neuron 14:477-485.

Gorelova NA, Yang CR (2000) Dopamine D1/D5 receptor activation modulates a persistent sodium current in rat prefrontal cortical neurons in vitro. J Neurophysiol 84:75-87.

Gresch PJ, Sved AF, Zigmond MJ, Finlay JM (1994) Stress-induced sensitization of dopamine and norepinephrine efflux in medial prefrontal cortex of the rat. J Neurochem 63:575-583.

Gulledge AT, Jaffe DB (1998) Dopamine decreases the excitability of layer V pyramidal cells in the rat prefrontal cortex. J Neurosci 18:9139-9151.

Gurden H, Tassin JP, Jay TM (1999) Integrity of the mesocortical DA system is necessary for complete expression of in vivo hippocampal – prefrontal cortex long-term potentiation. Neuroscience 94:1019-1027.

Gurden H, Takita M, Jay TM (2000) Essential role of D1 but not D2 receptors in the NMDA receptor-dependent long-term potentiation at hippocampal – prefrontal cortex synapses *in vivo*. J Neurosci 20:RC106 (1-5).

Hansel C, Artola A, Singer W (1996) Relation between dendritic Ca2+ levels and the polarity of synaptic long-term modifications in rat visual cortex neurons. Eur J Neurosci 9:2309-2322.

Herry, C. & Garcia, R. (2002) Prefrontal cortex long-term potentiation, but not long-term depression, is associated with the maintenance of extinction of learned fear in mice. J Neurosci 22:577-583.

Herry C, Vouimba R-M, Garcia R (1999) Plasticity in the mediodorsal thalamo-prefrontal cortical transmission in behaving mice. J Neurophysiol 82:2827-2832.

Hirsch JC, Crepel F (1991) Blockade of NMDA receptors unmasks a long-term depression in synaptic efficacy in rat prefrontal neurons in vitro. Exp. Brain Res 85:621-624.

Jay TM, Witter MP (1991) Distribution of hippocampal CA1 and subicular efferents in the prefrontal cortex of the rat studied by means of anterograde transport of Phaseolus vulgaris-leucoagglutinin. J Comp Neurol 313:574-586.

Jung MW, Qin Y, McNaughton BL, Barnes CA (1998) Firing characteristics of deep layer neurons in prefrontal cortex in rats performing spatial working memory tasks. Cereb Cortex 8:437-450.

Kirkwood A, Rioult MG, Bear MF (1996) Experience-dependent modification of synaptic plasticity in visual cortex. Nature 381:526-528.

Kiyatkin EA, Rebec GV (1998) Heterogeneity of ventral tegmental area neurons: single-unit recording and iontophoresis in awake, unrestrained rats. Neuroscience 85:1285-1309.

Kosobud AEK, Harris GC, Chapin JK (1994) Behavioral associations of neuronal activity in the ventral tegmental area of the rat. J Neurosci 14:7117-7129.

Larson J, Wong D, Lynch G (1986) Patterned stimulation at the theta frequency is optimal for the induction of hippocampal long-term potentiation. Brain Res 368:347-350.

Law-Tho D, Desce JM, Crepel F (1995) Dopamine favours the emergence of long-term depression versus long-term potentiation in slices of rat prefrontal cortex. Neurosci Lett 188:125-128.

Levy WB, Steward O (1979) Synapses as associative memory elements in the hippocampal formation. Brain Res 175:233-245.

Levy WB, Steward O (1983) Temporal contiguity requirements for long-term associative potentiation/depression in the hippocampus. Neuroscience 8:791-797.

Miller E.K. & Cohen, J.D. (2001) An integrative theory of prefrontal cortex function. Annu Rev Neurosci 24:167-202.

Mizoguchi K, Yuzurihara M, Ishige A, Sasaki H, Chui D-H, Tabira T (2000) Chronic stress induces impairment of spatial working memory because of prefrontal dopaminergic dysfunction. J Neurosci 15:1568-1574.

Moghaddam B, Adams BW (1998) Reversal of phencyclidine effects by a group II metabotropic glutamate receptor agonist in rats. Science 281:1349-1352.

Moghaddam B, Adams BW, Verma A, Daly D (1997) Activation of glutamatergic neurotransmission by ketamine: a novel step in the pathway from NMDA receptor blockade to dopaminergic and cognitive disruptions associated with the prefrontal cortex. J Neurosci 17:2921-2927.

Morris SH, Knevett S, Lerner EG, Bindman LJ (1999) Group I mGluR agonist DHPG facilitates the induction of LTP in rat prelimbic cortex in vitro. J Neurophysiol 82:1927-1933.

Mulder AB, Nordquist R, Örgüt O, Pennartz CMA (2000) Plasticity of neuronal firing in deep layers of the medial prefrontal cortex in rats engaged in operant conditioning. In: Progress in Brain Research, vol 126 (Uylings HBM, Van Eden CG, De Bruin JPC, Feestra MGP, and Pennartz CMA, eds), pp 287-301, Elsevier, Amsterdam.

O'Keefe J, Nadel L (1978) The Hippocampus as a Cognitive Map, Clarendon Press, Oxford.

Otani S. (2002) Memory trace in prefrontal cortex: theory for the cognitive switch. Biol Rev 77:563 - 577.

Otani S, Connor JA (1998) Requirement of rapid Ca^{2+} entry and synaptic activation of metabotropic glutamate receptors for the induction of long-term depression in adult rat hippocampus. J Physiol (Lond) 511:761-770.

Otani S, Blond O, Desce J-M, Crépel F (1998) Dopamine facilitates long-term depression of glutamatergic transmission in rat prefrontal cortex. Neuroscience 85:669-676.

Otani S, Auclair N, Desce J-M, Roisin M-P, Crépel F (1999) Dopamine receptors and groups I and II mGluRs cooperate for long-term depression induction in rat prefrontal cortex through converging postsynaptic activation of MAP kinases. J Neurosci 19:9788-9802.

Otani S, Daniel H, Takita M, Crepel F (2002). Long-term depression induced by postsynaptic group II mGluRs linked to phospholipase C and intracellular calcium rises in rat prefrontal cortex. J Neurosci 22:3434-3444.

Otani S, Daniel H, Roisin M-P, Crepel F (2003) Dopaminergic modulation of long-term synaptic plasticity in rat prefrontal neurons. Cereb Cortex (in press).

Perez-Jaranay JM, Vives F (1991) Electrophysiological study of the response of medial prefrontal cortex neurons to stimulation of the basolateral nucleus of the amygdala in the rat. Brain Res 8:97-101.

Pratt WE, Mizumori SJY (2001) Neurons in rat medial prefrontal cortex show anticipatory rate changes to predictable differential rewards in a spatial memory task. Behav Brain Res 123:165-183.

Reynolds JJ, Hyland BL, Wickens JR (2001) A cellular mechanisms of reward-related learning. Nature 413:67-70.

Robbins TW and Everitt BJ (2002) Dopamine – its role in behaviour and cognition in experimental animals and humans. In: Handbook of Experimental Pharmacology, vol 154/II, Chap 19 (Di Chiara G, ed), pp 173-211.

Rogan MT, Stäubli UV, LeDoux JE (1997) Fear conditioning induces associative long-term potentiation in the amygdala. Nature 390:604-607.

Rossi S, Cappa SF, Babiloni C, Pasqualetti P, Miniussi C, Carducci F, Babiloni F, Rossini PM (2001) Prefrontal cortex in long-term memory: an "interference" approach using magnetic stimulation. Nature Neurosci 4:948-952.

Schultz W (2002) Getting formal with dopamine and reward. Neuron 36:241-263.

Seamans JK, Durstewitz D, Christie B.R, Stevens CF, Sejonowski TJ (2001) Dopamine D1/D5 receptor modulation of excitatory synaptic inputs to layer V prefrontal cortex neurons. Proc Natl Acad Sci USA 98:301-306.

Suzuki Y, Jodo E, Takeuchi S, Niwa S, Kayama Y (2002) Acute administration of phencyclidine induces tonic activation of medial prefrontal cortex neurons in freely moving rats. Neuroscience 114:769-779.

Takahata R., Moghaddam B. (2000) target-specific glutamate regulation of dopamine neurons in the ventral tegmental area. J Neurochem 75:1775-1778.

Takita M, Izaki Y, Jay TM, Kaneko H, Suzuki SS (1999) Induction of stable long-term depression in vivo in the hippocampal-prefrontal cortex pathway; Eur J Neurosci 11:4145-4148.

Ungless MA, Whistler JL, Malenka RC, Bonci A (2001) Single cocaine exposure in vivo induces long-term potentiation in dopamine neurons. Nature 411:583-587.

Vickery RM, Morris SH, Bindman LJ (1997) Metabotropic glutamate receptors are involved in long-term potentiation in isolated slices of rat medial frontal cortex. J Neurophysiol 78:3039-3046.

Wang J, O'Donnell P (2001) D1 dopamine receptors potentiate NMDA-mediated excitability increase in layer V prefrontal cortical pyramidal neurons. Cereb Cortex 11:452-462.

Watanabe M (1996) Reward expectancy in primate prefrontal neurons. Nature 382:629-632.

Yang CR, Seamans JK (1996) Dopamine D1 receptor actions in layer v-vi rat prefrontal cortex neurons in vitro: Modulation of dendritic-somatic signal integration. J Neurosci 16: 1922-1935.

Yang CR, Seamans JK, Gorelova N (1999) Developing a neuronal model for the pathophysiology of schizophrenia based on the nature of electrophysiological actions of dopamine in the prefrontal cortex. Neuropsychopharmacol 21:161-194.

Chapter 5

UP AND DOWN REGULATION OF SYNAPTIC STRENGTH AT HIPPOCAMPAL TO PREFRONTAL CORTEX SYNAPSES

Thérèse M. Jay[1], Hirac Gurden[2], Cyril Rocher[2], Maïté Hotte[2], and Michael Spedding[2]

[1]Physiopathologie des Maladies Psychiatriques, INSERM EMI 0117, Paris, France, and [2]Neurobiologie de l'Apprentissage et de la Mémoire, CNRS UMR 8620, Université Paris XI, Orsay, France

Keywords: Long-term potentiation, long-term depression, prelimbic area, dopamine D_1, dopamine D_2, PKA, working memory, stress.

Abstract: Specific patterns of stimulation applied in the ventral hippocampus produce long-term potentiation (LTP) or long-term depression of stimulated synapses in the prefrontal cortex (PFC, or prelimbic area) *in vivo*, and these different forms of plasticity are reversible. LTP induction is dependent on NMDA receptors and the activation of the cAMP-dependent kinase, PKA. The mesocortical dopamine input is an important determinant in synaptic plasticity at the hippocampal to PFC synapses. An increase in prefrontal dopamine as well as a local stimulation of D_1 receptors is able to induce a long-lasting enhancement of hippocampal–prefrontal LTP, whereas a significant cortical dopamine depletion or a specific blockade of D_1, but not D_2, receptors results in a dramatic impairment of cortical LTP. Together, these data demonstrate that DA and D_1 receptors are necessary for the expression of synaptic plasticity in PFC. In addition, the stimulation paradigm used to induce this NMDA-dependent LTP causes an increase in dopamine release in PFC, suggesting a direct role of dopamine in the induction mechanisms. We propose a cooperative action of D_1 and NMDA receptors in the induction mechanisms of prefrontal LTP involving mostly PKA-dependent mechanisms. These results are significant for current understanding of prefrontal memory mechanisms and their abnormalities in schizophrenia. In recent studies, we have shown that LTP at hippocampal to PFC synapses is dramatically impaired by stress, suggesting a role of this limbic/cortical circuit in depression.

1. INTRODUCTION

The prefrontal cortex (PFC) is an heterogeneous neocortical region known to be involved in highly processed sensory information through afferents from the parietal and temporal regions of the cerebral cortex and in higher cognitive functions because of its association with the limbic structures (for review, see Groenewegen and Uylings, 2000). Additionally, the PFC is also implicated in visceral functions by its direct reciprocal connections with the hypothalamus and brainstem structures.

The highly specific hippocampal–prefrontal network provides the PFC with the possibility to gain access to memory processes. Based on behavioral and physiological data, the hippocampus and PFC are in a cooperative relationship for working memory. However, even if these two structures are part of a common network, they also subserve different functions in cognitive processes, and the hippocampus contribution may be preeminent when information needs to be associated with long-term memory. The direct hippocampal to PFC pathway and its changes in synaptic plasticity is a useful framework for investigating the functional operations of hippocampal–PFC communication in cognitive functions.

The aim of the present chapter is to provide an overview on the hippocampal–prefrontal circuit, the regulation of synaptic efficacy in the prelimbic cortex, and its modulation by the dopaminergic system with the incidence of environmental factors. The review concludes with a presumptive functional role of the hippocampal–prefrontal network in the pathophysiology of schizophrenia and depression.

2. THE HIPPOCAMPAL–PREFRONTAL CIRCUIT

2.1 Anatomical Organization

The PFC in the rat is directly connected with two output structures of the hippocampal formation, area CA1 and the subiculum, and this pathway is topographically organized along the longitudinal and transverse axis of the hippocampus (Jay and Witter, 1991). CA1 neurons, except in the most dorsal part of the hippocampus, and the neurons of the entire dorsoventral extent of the proximal region of the subiculum (close to CA1), innervate the PFC. There is a selectivity of the hippocampal projection to the presumed non-motor related sub-areas of the PFC (Jay et al., 1989). The CA1 region and the subiculum project to both the medial PFC, i.e. the prelimbic, infralimbic and medial orbital areas (Jay and Witter, 1991) and the lateral PFC (Verwer et al., 1997). The hippocampal fibers course through the fimbria and fornix to reach the lateral septum, nucleus accumbens, and the different subdivisions of the PFC. In these areas, hippocampal fibers and terminals are present in all cell layers, but are more densely distributed in layers V and VI of the ventral region of the prelimbic cortex. It is important to note that this pathway is ipsilateral and unidirectional, two characteristics that are of particular relevance to prefrontal functions. Unlike other neocortical areas such as the perirhinal or entorhinal cortices which are recipocally connected to the hippocampus (Witter et al., 1989), area CA1 and subiculum do not, in return, receive direct projections from the PFC (Sesack et al., 1989).

Although there are considerable variations across species, it was recently reported that the orbital and medial networks of the PFC are relatively comparable across species because of similarities in the position and connections of these two subareas (Öngür and Price, 2000). The hippocampal–prefrontal network is in agreement with this hypothesis at least when comparing rats and monkeys. In primates, hippocampal fibers primarily originate from the rostral hippocampus (region at the border between CA1 and the subiculum) and terminate in the medial and orbital areas of the PFC (Barbas and Blatt, 1995; Carmichael and Price, 1995). The rostral hippocampus in primates being considered equivalent to the ventral hippocampus of the rat, we can conclude that a similar organization of the hippocampal–PFC projections is observed in both rats and monkeys. Additionally, the strongest influence of the PFC on the hippocampus in primates is also not direct but goes through the parahippocampal cortices (Goldman-Rakic et al., 1984; Carmichael and Price, 1995).

The transfer of hippocampal information to the PFC is then conveyed to cortical and subcortical targets. Interestingly, this excitatory input identified as projecting to the nucleus accumbens and ventral tegmental area (Jay et al., 1995b) indicates a possible involvement in different functional aspects. Future investigation is required to identify the other efferent populations of prefrontal neurons that are synaptically driven by the hippocampus.

2.2 A Monosynaptic Glutamatergic Pathway

Electrophysiological studies have first demonstrated that neurons in the hippocampal formation exert an excitatory influence onto pyramidal neurons in the PFC and that this pathway uses glutamate as a transmitter (Ferino et al, 1987; Laroche et al., 1990; Jay et al., 1992; Mulder et al., 1997). Following single pulse stimulation of the hippocampus, the excitatory responses have a relatively long latency (18 ms) but compatible with the latency of antidromic spikes recorded in CA1 cells after stimulation of the prelimbic cortex (Ferino et al., 1987). Several ultrastructural and electrophysiological studies have confirmed that the hippocampal–prefrontal circuit is a monosynaptic pathway that utilizes an excitatory amino acid transmitter. Hippocampal terminals in the PFC form almost exclusively asymmetric synapses (Carr and Sesack, 1996). Synaptic contacts are formed with dendritic spines from pyramidal neurons but also with dendritic shafts of local circuit GABA (γ-aminobutyric acid) neurons (Gabbott et al., 2002). Furthermore, intracellular recordings of prefrontal neurons have shown excitatory postsynaptic potentials (EPSPs) evoked by hippocampal stimulation that occurred with a latency of around 18 ms followed by synaptic events which are probably the result of an activation of local circuit neurons (Thierry et al., 2000).

PFC neurons that respond to hippocampal stimulation are strongly activated by the agonists of the AMPA (α-amino-3-hydroxy-5-methyl-4-isoxazolepropionic acid) and NMDA (*N*-methyl-D-aspartic acid) glutamate receptor subtypes (Jay et al., 1992). The excitatory responses of PFC neurons evoked by hippocampal stimulation are blocked by the selective AMPA antagonist, CNQX (6-cyano-7-nitroquinoxaline-2,3-dione), indicating that normal neurotransmission at hippocampal to PFC synapses is AMPA receptor-dependent (Jay et al., 1992; Gigg et al., 1994). Conversely, the NMDA receptor antagonist AP5 (D-(-)-2-amino-5-phosphonopentanoic acid) does not affect the excitatory responses to low-frequency stimulation of the hippocampus.

3. Different Forms of Synaptic Plasticity

The last two decades have seen the initial investigations of experimentally induced synaptic plasticity, long-term potentiation (LTP) and long-term depression (LTD) extended from the hippocampus to neocortex. However, most of the studies on cortical plasticity have been carried out on *in vitro* slice preparation of different neocortical regions in immature animals. It is only recently that the neocortex in adult animals was observed to be still capable of plastic changes.

LTP can be induced at the synapses between the hippocampal input and the neurons in the prelimbic cortex *in vivo* (Laroche et al., 1990). High-frequency stimulation (250 Hz pulses, 200 ms duration, 10 trains at 0.1 Hz) applied in the ventral CA1-subicular region is able to elicit a rapid increase in the amplitude of the field potential which lasts several hours in the anesthetized animals. The increase in synaptic strength persists for several days in the awake freely-moving rats (Jay et al., 1996a). Interestingly, while tetanic stimulation of the fornix, i.e. the pathway of hippocampal fibers to the PFC, results in LTP simultaneously in at least two of the target structures, the prelimbic cortex and the nucleus accumbens (Mulder et al., 1997), LTP in the prelimbic cortex is long-lasting whereas that in the nucleus accumbens is limited to a short time. LTP can also be induced *in vivo* both in the prelimbic cortex by the stimulation of mediodorsal thalamus (Herry et al., 1999) and in the lateral PFC by the stimulation of basolateral amygdala (Escobar et al., 1998).

Activation of the NMDA receptors is required to elicit LTP in the PFC *in vivo* by stimulation of ventral hippocampus (Fig. 1; Jay et al. 1995a). When the NMDA receptor antagonist AP5 is infused locally in the prelimbic cortex, LTP is completely blocked while normal synaptic transmission is not affected. However, once induced, the maintenance of LTP is NMDA receptor independent. Thus, the NMDA receptor is critically involved in the induction of LTP at the hippocampal to PFC synapses. Although different synapses are investigated on the *in vitro* slices from rat PFC, repeated bursts of stimulation in layer II of the prelimbic cortex induce LTP in deep layer neurons, which involves glutamatergic activation of metabotropic receptors (Vickery et al., 1997). These results suggest a contribution of different glutamate receptors in synaptic plasticity in the PFC, although AMPA receptors have not yet been investigated on *in vitro* or *in vivo* preparations.

Studies into the mechanisms of the expression of *in vivo* hippocampal–prefrontal LTP have just begun to reveal that the cAMP-dependent protein kinase A (PKA) is a regulator of the increase in synaptic strength during LTP (Fig. 1). A rapid NMDA-dependent activation of cytosolic PKA in the PFC was found within minutes after the induction of LTP (Jay et al., 1998). The decay time of PKA activation in the minute range implies that the activated PKA is an induction switch rather than a device for late LTP as largely suggested for the hippocampus, even though similar data have been reported during the induction of LTP in area CA1 (Roberson and Sweatt, 1996). In addition, blocking the downstream PKA cascade by the injection of a PKA inhibitor (Rp-cAMP) in the PFC resulted in a dramatic impairment of the early and late stages of cortical LTP (Gurden et al., 2000). Although it is always assumed that mechanisms underlying synaptic plasticity in the neocortex are similar to those in the hippocampus, here is a major difference on the temporal dynamics of one key molecule, PKA. PKA is able to trigger LTP through up-regulation of NMDA receptor activation by phosphorylation (Blank et al., 1997), through direct phosphorylation of specific AMPA and NMDA receptors subunits (Leonard and Hell, 1997; Lee et al., 2000), and through direct stimulation of the expression of genes by phosphorylation of the cAMP response element-binding protein (CREB) (Impey et al., 1996; Moore et al., 1996).

LTD, the flip side of the synaptic plasticity, has been largely explored in the neocortex but was first established in brain slices (Artola et al., 1990). Whereas neocortical preparations have shown to be resistant to the induction of LTP, the standard low-frequency stimulation protocol (1Hz, 15 min) is

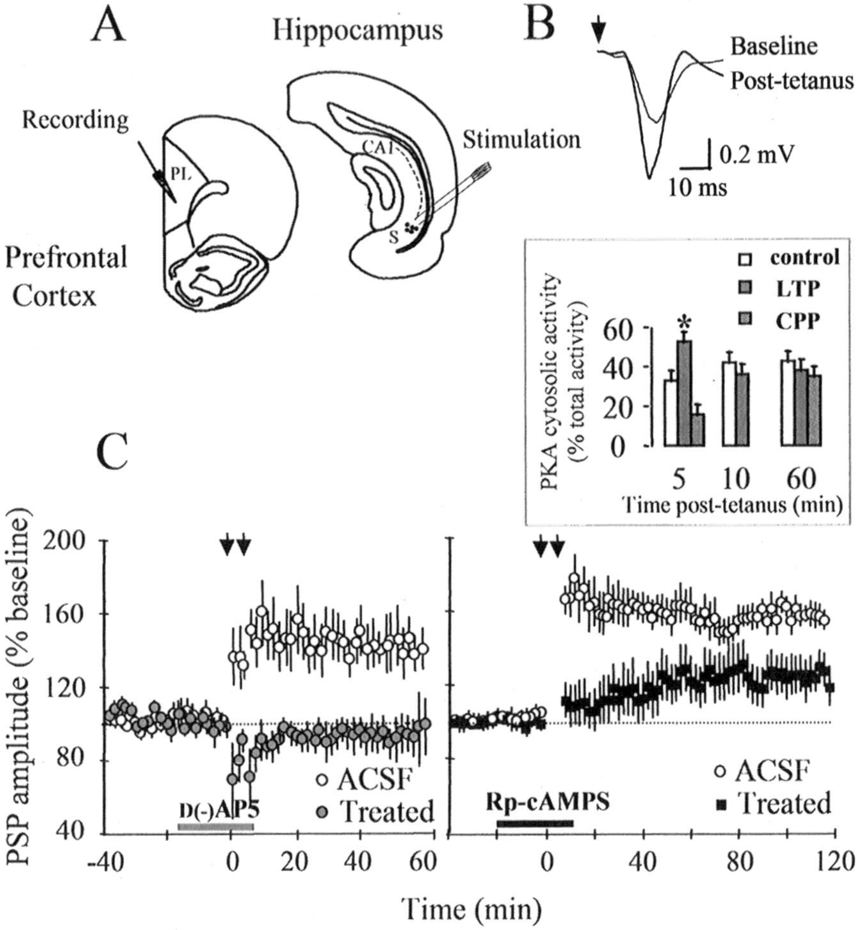

Figure 1: NMDA receptor-dependent LTP induced *in vivo* at hippocampal to PFC synapses. A. Schematic diagram representing the location of stimulating and recording electrodes in ventral hippocampus (S, subiculum) and prelimbic cortex (PL). B. Examples of pre- and post-tetanus hippocampal–prefrontal synaptic responses characterized by a large negative component with a peak at around 18 ms latency. Measurements of the amplitude of the synaptic responses are taken by drawing a line from a tangent drawn between the negative onset and offset with the use of an A/Dvance software. C. Hippocampal–prefrontal LTP is dependent on NMDA receptors and the activation of PKA. Each point represents mean ± SEM of averaged responses to four stimuli given at 30 sec intervals. Values are expressed as percentage changes relative to baseline. Tetanic stimulation (250 Hz, 200 ms, 2 series) is indicated by arrowheads. Left graph shows a blockade of LTP by the NMDA antagonist D-AP5 (adopted from Jay et al., 1995a). Right graph shows that inhibition of PKA by local infusion of Rp-cAMPS dramatically decreases LTP amplitude during at least 2 h after tetanus (adopted from Gurden et al., 2000). *Inlet*, The rapid increase in PKA activity during LTP is NMDA receptor-dependent.

able to induce LTD in a number of cortical regions. However, this stimulation protocol was found to be ineffective at hippocampal to prefrontal synapses in anaesthetized rats (Burette et al., 1997). Instead, depotentiation was induced by low-frequency trains consisting of 2-pulse bursts. An appropriate induction protocol to elicit LTD at those synapses is made of 900 stimulus trains delivered at 1 Hz, where each train consists of five 250 Hz pulses (Takita et al., 1999). The duration of LTD *in vivo* depends on the number of the low-frequency trains delivered. Thus, multiple low-frequency stimulation sessions are required to reliably induce LTD *in vivo*, and this was also observed on LTD in the sensorimotor cortex of freely-moving rats (Froc et al., 2000). The mechanisms underlying depotentiation and LTD at the hippocampal to PFC synapses have not been explored yet.

Taken together, the PFC can support bi-directional and reversible adjustment of synaptic strength *in vivo*, depending on the stimulation paradigm, and these rules are consistent with theoretical models of learning and memory, where both LTP and LTD, i.e. stronger and weaker synapses, contribute to the storage of information.

4. DOPAMINE REGULATES SYNAPTIC PLASTICITY IN THE PREFRONTAL CORTEX *IN VIVO*

The mesocortical dopamine system which arises from cell bodies in the ventral tegmental area (VTA) terminates primarily in the PFC and targets the dendritic spines and shafts of layers V-VI pyramidal neurons (Van Eden et al., 1987, Sesack et al., 1995). In view of the role of dopamine in working memory and other aspects of cognitive function, a number of electrophysiological studies have explored the action of dopamine on prefrontal neurons and evoked synaptic responses, but failed to give a consistent picture of dopamine actions. Considering that the dopamine system does not function on its own but rather interacts with afferent projections and modulates their ability to integrate a proper input, we studied the effects of dopamine on hippocampal–prefrontal synaptic plasticity and investigated the role of dopamine receptor subtypes using multiple approaches.

4.1 Dopamine Controls LTP Induction and Expression

Infusion of dopamine (1 mM) in the PFC through reverse microdialysis (see Section 4.2) prior to the application of tetanus in the hippocampus increases the magnitude (two folds) of LTP in the PFC, and this greater LTP is maintained over time. Comparable results can also be obtained with the application of nomifensine (5μM), a dopamine reuptake blocker known to increase and prolong dopamine release, and these effects are pronounced at very low concentration (Jay et al., 1996a, b).

4.1.1 Endogenous dopamine enhances LTP

Then, the interest was to study the effects of physiologically released dopamine from VTA neurons on hippocampal–PFC synaptic plasticity. In order to increase local dopamine in the PFC, transient stimulation of the VTA (50 Hz, 2 s), a protocol known to increase the release capacity for mesocortical dopamine neurons (Garris et al., 1993), was applied prior to tetanic stimulation to the hippocampus. This short stimulation of dopamine cells is sufficient to produce a long-lasting increase in the magnitude of LTP in the hippocampal–PFC synapses (Fig. 2; Gurden et al., 1999). In contrast, VTA stimulation applied after tetanus-induced LTP did not produce any further increases in the potentiation. Thus, it is a sequential application of the two paradigms which is able to induce the synaptic enhancement. Together, these results suggest a contribution of dopamine in the induction and expression mechanisms of hippocampal–PFC LTP.

On the other hand, contrasting effects of VTA stimulation on the activity of PFC neurons receiving hippocampal input (baseline responses) have been observed, and these results were later confirmed with field potential responses. Following single pulse stimulation of the VTA, PFC neurons decrease their firing rate (Jay et al., 1995b), and single pulse or burst stimulation of the VTA both produces a transient depression in PFC synaptic responses to hippocampal stimulation (Gurden et al., 1999). There is an abundant literature on the effects of dopamine from the VTA on PFC individual neurons. With intracellular *in vivo* recordings, it was observed that PFC neurons spontaneously oscillate between two membrane potential states: a down-state (very negative) and an up-state (less negative) membrane potentials (Lewis and O'Donnell, 2000; also see Chapter 3 of this volume). These authors have recently shown that burst stimulation of the VTA evoked more prolonged up-state in prefrontal neurons (Lewis and O'Donnell, 2000) that could provide a postsynaptic depolarization (increasing NMDA function) sufficient to facilitate LTP at hippocampal to PFC synapses. At the same time, they also found a reduction in prefrontal neuron firing after VTA stimulation, which could explain the transient depression observed in our preparation. This bi-directional modulatory influence of dopamine from the VTA on the hippocampal–prefrontal responses and LTP could also be explained by the different nature of the evoked excitatory synaptic responses, i.e. a normal synaptic transmission that is mostly dependent on AMPA receptors and an LTP strongly dependent on NMDA receptors.

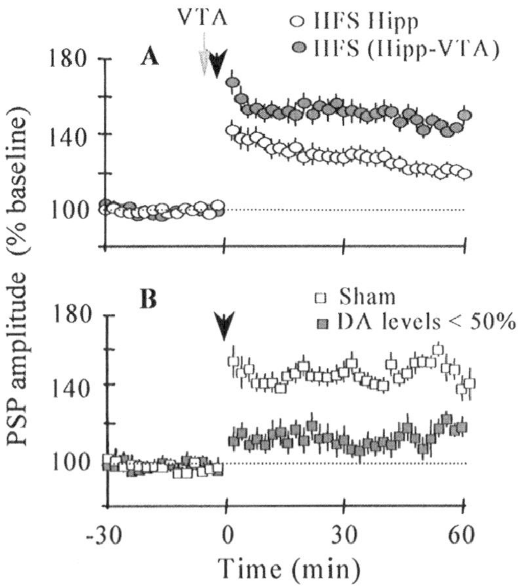

Figure 2: Involvement of mesoprefrontal dopamine neurons in the expression of hippocampal–prefrontal LTP. A. Long-lasting enhancement of LTP produced by a transient high frequency stimulation (HFS, 50 Hz, 2 sec) of the VTA applied prior to hippocampal HFS. B. LTP is severely affected in VTA-lesioned rats that show decreases of cortical dopamine levels greater than 50 % of the sham-operated animals.

However, we cannot exclude a GABAergic component that would be present in the evoked response after low frequency stimulation and disappear on the induction of LTP; a similar disinhibition is known to occur with LTP in the hippocampus (Davies et al., 1991).

4.1.2 A critical level of dopamine is essential for LTP

On the contrary to the long-term enhancement of LTP by dopamine, a reduction of dopamine in the PFC induces a dramatic decrease in the magnitude of LTP at hippocampal–PFC synapses (Fig. 2). VTA-lesioned rats with a major cortical dopamine depletion (50% compared to sham-operated animals) showed almost no LTP, and, interestingly, by pooling all the data (n=14 with different levels of dopamine), a significant correlation was found between the magnitude of cortical dopamine depletion and the disruption of the hippocampal–prefrontal LTP (Gurden et al., 1999). Thus, these results strengthened the evidence of a functional role for dopamine in the regulation of prefrontal LTP and demonstrate that LTP at the hippocampal–prefrontal synapses requires the integrity of the mesocortical dopamine system. Dopamine neurotransmission influences the expression of changes in postsynaptic plasticity and may even induce these changes.

Comparable *in vivo* studies investigating the interactions of dopamine inputs from the substantia nigra and the excitatory inputs from the cerebral cortex have also pointed out an up-regulation by dopamine of corticostriatal

LTD (Reynolds and Wickens, 2000). This form of synaptic plasticity that follows high-frequency stimulation is prevented or reversed by concurrent stimulation of the substantia nigra. Other investigators, using PFC slices, have shown that dopamine favors the emergence of LTD over LTP and produces LTD when tetanic stimulation alone does not induce any synaptic plasticity (Law-Tho et al., 1995; Otani et al., 1998). However, in a recent study, these authors reported that a second application of dopamine when coupled to high-frequency stimuli induces LTP instead of LTD on the layer I-II to layer V glutamatergic synapses (Blond et al., 2002; also see Chapter 4 of this volume). Thus, dopamine in the PFC can induce either LTD or LTP, at least *in vitro*. The effects of dopamine on *in vivo* hippocampal–prefrontal LTD still remain to be investigated.

4.2 D_1 Receptor Activation is a Necessary Requirement for LTP

Further investigation into the role of specific dopamine receptor subtypes in hippocampal–prefrontal LTP was accomplished using the same procedure as for the infusion of dopamine (i.e. reverse microdialysis). The advantage of using such a procedure is that it allows controlling of the perfusion of drugs and, simultaneously, recording of the synaptic responses. Control animals received artificial cerebral spinal fluid (ACSF) continuously while treated animals received drugs (at different doses) for 30 min starting 20 min before tetanus. The main finding was that dopamine D_1 receptors exert a clear facilitating effect on LTP but also that D_1 receptor activation is necessary for the LTP at hippocampal–prefrontal synapses (Fig. 3; Gurden et al., 2000). LTP is significantly higher when the full D_1 agonist SKF81297 is locally infused in the PFC prior to tetanus. The increase in LTP amplitude is significantly larger at certain doses tested when compared to ACSF-controls, demonstrating that an optimal range of D_1 receptor activation is necessary to induce sustained enhancement of prefrontal LTP. Conversely, application of the D_1 receptor antagonist SCH23390 at different doses in the PFC dose-dependently impaired LTP at the hippocampal to PFC synapses (Fig. 3). In addition, we found that the D_2 receptor antagonist sulpiride did not affect the cortical LTP. Thus, D_1 but not D_2 receptors play an essential role on the expression of LTP in these synapses.

D_1-like receptors are more widespread in the hippocampal formation and predominantly represented by the D5 receptors, and the first evidence for a role of D_1 receptors in LTP was reported in the CA1 region (Frey et al., 1991). Specific inhibitors of D_1 but not D_2 receptors have been shown to prevent late stages of LTP (longer than 1 to 2 h) without effect on early LTP, and these results were later confirmed by other *in vitro* studies and, more recently, by a study in awake animals (Swanson-Park et al., 1999). D_1

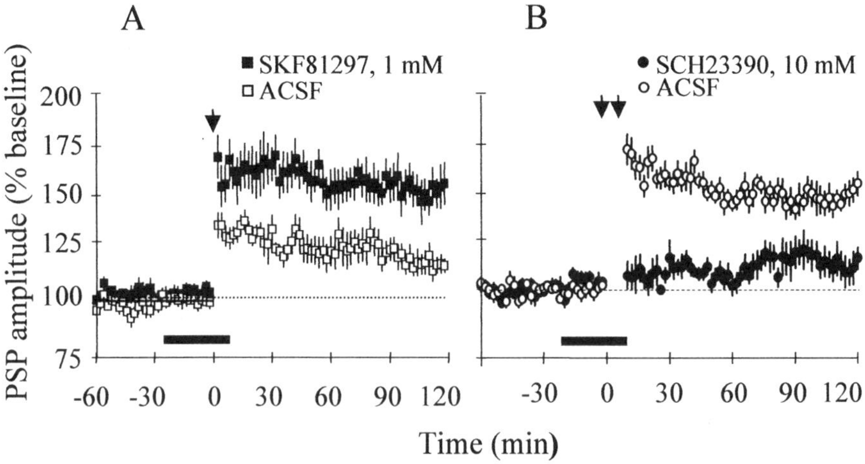

Figure 3: Effects of D_1 agonist and antagonist on hippocampal–prefrontal LTP. A. Infusion of D1 agonist SKF 81297 induces a persistent facilitation of LTP. B. Infusion of D1 antagonist SCH 23390 produces a sustained decrease in LTP amplitude.

receptors are also strongly implicated in corticostriatal LTP under Mg^{2+}-free condition (Kerr and Wickens, 2001) with a time course comparable to prefrontal rather than hippocampal LTP.

4.3 LTP Causes Release of Dopamine

Given the presence of dopaminergic fibers close to the dendrites of pyramidal cells and hippocampal terminals (Carr and Sesack, 1996), it was conceivable that the tetanus could exert its potentiation by trans-synaptically exciting dopamine neurons that release dopamine to the synaptic cleft. To test this hypothesis, we measured dopamine release in the PFC (direct microdialysis) before and after induction of LTP and found that potentiation was followed by a significant but transient increase in dopamine release in the PFC (Gurden et al., 2000). Therefore, hippocampal tetanization that induces LTP could act by triggering dopamine release. This result is consistent with the detrimental effects on LTP caused by the presence of a D_1 antagonist.

One possible mechanism could be that repetitive stimulation of the hippocampus activates indirectly VTA cells that project to prefrontal neurons, increasing dopamine release: this hypothesis could be tested through inactivation of the VTA. An alternative explanation is that LTP induction favors dopamine release through activation of NMDA receptors presumably located on dopamine terminals.

4.4 Proposed Interactions at the Postsynaptic Level

The direct role of dopamine and D_1 receptors on the induction of the NMDA receptor-dependent LTP at hippocampal to PFC synapses suggests a cooperative action of dopamine (mostly D_1) and glutamate (NMDA and AMPA) receptors occurring at the postsynaptic level (Fig. 4). *In vivo* induction and expression of PFC LTP is strongly dependent on PKA activity, and infusion of an adenylate cyclase activator, forskolin, mimicks the D_1 agonist effect (Gurden et al., 2000). Therefore, stimulation of the dopamine D_1 receptor during LTP induction probably activates adenylate cyclase that increases intracellular cAMP level which in turn activates PKA and, through PKA-dependent mechanisms, facilitates LTP. Indeed, dopamine activation is known to increase phosphorylation of specific AMPA receptor subunits (GluR1, Snyder et al., 2000), and the state of phosphorylation at this PKA-site has been correlated with changes in synaptic strength (Lee et al., 2000). In addition, the rate of phosphorylation/dephosphorylation of the NR1 subunit which is required for a functional NMDA receptor is also controlled by dopamine (Snyder et al., 1998). At the cellular level, the cAMP-regulated phosphoprotein (DARPP-32) is a signal transduction molecule that regulates the efficacy of dopamine signaling, and D_1 receptors require DARPP-32 to mediate their action. DARPP-32 activation occurs through the cascade involving cAMP/PKA, and once phosphorylated by PKA, DARPP-32 is a potent inhibitor of the protein phosphatase 1 (PP1) (Hemmings et al., 1984). Therefore, the control of PP1 through DARPP-32, a key regulator of dopamine transmission, is likely to have a significant effect on the regulation of the strength of synaptic plasticity in the PFC.

Dopamine and glutamate could also act cooperatively at the transcriptional level, and D_1 and NMDA receptors activate neuron-specific program of immediate early genes. PKA stimulates the transcription of a number of genes by catalyzing the phosphorylation of CREB, and PP1 retains the ability of CREB to stimulate transcription. Here is another step where dopamine receptors could control the kinetics and duration of phosphorylation of CREB through the PKA/PP1 signaling complex. Given the role of the transcription factor CREB in long lasting forms of synaptic plasticity, these interactions could explain the strong impact of dopamine through D_1 receptors on the duration of LTP.

This hypothetic role of dopamine through PKA-dependent mechanisms is in agreement with several studies demonstrating how PKA activity can serve

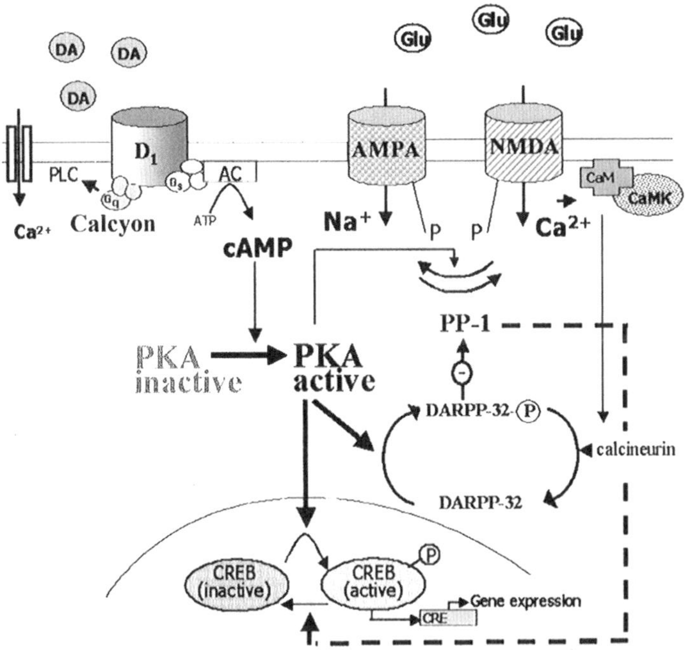

Figure 4: Scheme for the proposed cooperative action of dopamine (mostly D_1) and glutamate (NMDA and AMPA) receptors in the induction of hippocampal–prefrontal LTP, leading to a persistent enhancement of LTP. A postsynaptic convergence on regulatory pathways at the PKA level involving different substrates of PKA (ionotropic glutamate receptors, phosphoprotein DARPP-32, and CREB) could account for the control of synaptic strength. The D1 receptor positively coupled to adenylate cyclase (AC) increases AC activity leading to the formation of cAMP which subsequently activates the cAMP-dependent PKA by binding to the regulatory subunits. The release of the active catalytic subunit permits phosphorylation of specific target substrates localized in different cellular compartments (cytosol, membrane, nucleus). (1) Activated PKA phosphorylates both AMPA and NMDA receptors and DARPP-32. Once phosphorylated, DARPP-32 is converted to a potent inhibitor of PP-1 that will promote phosphorylation. (2) Concurrently, activation of NMDA receptors increases the influx of Ca^{2+} into the cell and the Ca^{2+}/calmodulin (CaM) complex activates different CaM kinases (II and IV) and a CaM dependent protein phosphatase, calcineurin (PP2B), leading to increased dephosphorylation of DARPP-32 and activation of PP-1. (3) PKA also phosphorylates CREB, switching this transcription factor from its inactive to active state. Conversely, PP-1 specifically dephosphorylates CREB. Another possibility may be the D_1 receptor-stimulated Ca^{2+} release involving a crosstalk between Gs- and Gq-coupled receptors and the protein calcyon.

as a gate for synaptic plasticity by modulating calmodulin-dependent protein kinase II (CaMKII) through PP1, suggesting a major role of the dynamic balance between these kinase and phosphatase activities to set the synaptic strength (Blitzer et al., 1998, Lisman and Zhabotinsky, 2001). Other studies have also indicated that PKA activity, through phosphorylation of CREB and activation of CRE-driven gene products, controls late stages of synaptic plasticity that depend on protein synthesis, contributing to the stabilization of synapses or the recruitment of new synapses (Bolshakov et al., 1997).

Alternatively, dopamine could act by elevating intracellular calcium concentration (Surmeier et al., 1995), and the importance of the second messenger Ca^{2+} has also been questioned in the dopamine modulation of synaptic plasticity. Recently, a protein calcyon was shown to confer the ability to stimulate intracellular Ca^{2+} release on D_1 receptors by potentiating a crosstalk between Gs- and Gq-coupled receptors (Lezcano et al., 2000). What we can conclude from these studies is that D_1 receptors likely modulate synaptic plasticity through both Ca^{2+} and cAMP and are able to integrate multiple signals to produce maximal cAMP signals which play a critical role in LTPs.

4.5 Behavioral Significance of Dopamine Control of LTP

From the data summarized in this chapter, it appears that dopamine is involved in the selective gating of information flow from the hippocampus to the PFC. In both rats and primates, dopamine or the mesocortical dopaminergic system is known to be important for learning in delay-dependent tasks requiring efficient working memory (Brozoski et al., 1979; Simon et al., 1980; Sawaguchi and Goldman-Rakic, 1994). A certain stage of dopamine with an optimal level of D_1 receptor activation appears to be essential for the cellular mechanisms of working memory, and interestingly, similar requirements are also true for a proper hippocampal–prefrontal LTP expression (Williams and Goldman-Rakic, 1995; Gurden et al., 2000). The hippocampal input to the PFC could be one important target of dopamine modulation in this process.

Indeed, the functional implication of the hippocampal–PFC pathway has been explored in short-term memory processes and the performance in a spatial delay-interposed task (radial maze) shown to be specifically dependent on the hippocampal–prefrontal network and D_1 receptor modulation of hippocampal inputs to the PFC (Floresco et al., 1997; Seamans et al., 1998). On top of these experiments, using an electrophysiological approach to describe the type of synaptic modifications occurring in this pathway during a comparable delayed spatial task, we observed a depression of the response starting during the delay and extending to the period of learning until the performance of the animals was reached (Burette et al., 2000). These data are in agreement with the behavioral studies mentioned above reporting an implication of the network linking the hippocampus to the PFC in delayed spatial tasks. These short-term memory experiments used a longer delay (30 min) which implicates the recall of spatial information and its integration into a prospective response strategy rather than the short-term retention of information. Therefore, dopamine plays a critical role in selecting the appropriate strategies. The occurrence of a depression on the hippocampal–prefrontal circuit during these delayed spatial tasks could also signify a major influence of dopamine. Our results and those of others have demonstrated the potential of mesocortical dopamine neurons to bi-directionally modulate excitability of prefrontal neurons depending on the ongoing activity of the converging inputs. In this context, it would be worthwhile to identify the contribution of mesoprefrontal dopamine neurons on the hippocampal–prefrontal synaptic response.

Although considered as global neuromodulatory systems, dopamine neurons are capable to deliver precisely timed information to specific target structures and influence a number of cognitive functions. As recently suggested by Shultz (see review, Shultz, 2002), the dopamine signal progresses by a very rapid and brief firing, through a wave of activity to the PFC, and creates and/or "fixes" the plasticity of ongoing glutamatergic activity. The modification occurs only when the dopamine signal is active at about the same time as the cortical glutamatergic input.

5. IMPAIRMENT OF LTP IN RESPONSE TO STRESS

Stress can be defined as a threatening and inescapable situation or event that can promote physiological and behavioral disturbances ranging from psychiatric (particularly mood) disorders to immunological dysfunctions. Specifically, the dramatic increase in corticosteroid hormones (mainly corticosterone in rodents, cortisol in human) is defined as a physiological marker of stress (Kim and Diamond, 2002). Until recently, the impact of stress has been mostly studied on the hippocampus, a key target of stress hormones. Exposure to stress affects structural and synaptic plasticity in the hippocampus, and the increased expression of glucocorticoid receptors in this structure has been linked to learning and memory deficits (Magarinos et al., 1996; Garcia, 2001). In addition to the hippocampus, the PFC is also a region that can be up-regulated by environmental stimulation, and its role in stress, particularly its medial part, is well documented. A number of studies have reported the particular vulnerability of the dopamine mesocortical system, and acute stress is also known to induce a higher glutamate efflux in the PFC. Although a key initial event in stress is long-term changes in multiple neurotransmitter systems, particularly the release of catecholamines (McEwen, 2000; Vermetten and Bremner, 2002), it is only recently that this region has been considered as an important player in the regulation of circulating glucocorticoids by its direct connection with the hypothalamus (Dioro et al., 1993; Sullivan and Gratton, 2002). Chronic stress has also been reported to impair spatial working memory (Mizoguchi et al., 2000; also see Chapter 7 of this volume) and induce atrophy in distal dendrites of cortical neurons (Trentani et al., 2002).

We recently investigated whether an acute stress could modify the characteristics of hippocampal–prefrontal LTP induced *in vivo* in the rat. Behavioral stress protocol was based on Balfour and Reid (1979). Rats placed on an elevated and unsteady platform during 30 min showed behavioral ("freezing" behavior) as well as endocrine signs of stress. Indeed, we could measure a significant and dramatic increase in plasma corticosterone levels at the end of the 30 min period of stress when compared to nonstressed rats. Then, animals were anesthetized and immediately placed in the stereotaxic frame. When tetanic stimulation was applied in the ventral hippocampus within 180 min after the end of the stress period, LTP in the PFC was completely blocked during the 120 min post-tetanus recording. We thus found that a mild acute stress (platform stress) causes a remarkable and long-lasting inhibition of LTP in the frontal cortex evoked by stimulation of hippocampal outflow (Rocher et al., 2002; Fig. 5).

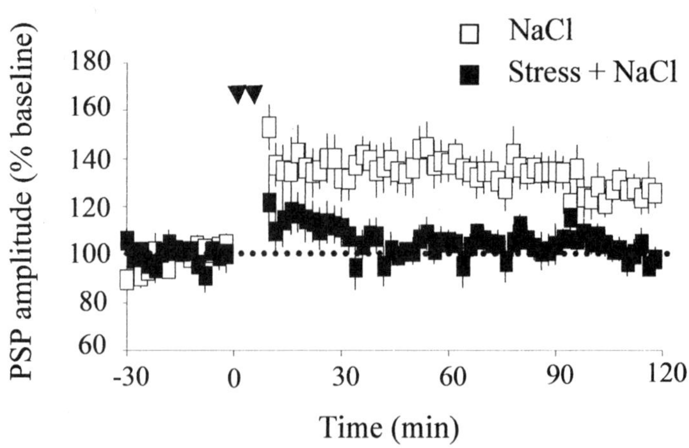

Figure 5: Long-lasting impairment of hippocampal–prefrontal LTP in rats previously exposed to an acute behavioral stress.

These results extend to the PFC the well-known inhibitory effect of stress on LTP in the hippocampus first demonstrated by Foy et al. (1987). Thus, the hippocampal–frontal circuitry, which is important for spatial and temporal contexts, is particularly sensitive to stress. As shown in the hippocampus, stress and glucocorticoids may inhibit LTP by favoring LTD (Xu et al., 1997, 1998), and this impairment in synaptic plasticity may be responsible for the deleterious effect of stress on memory. The PFC could also be a target for glucocorticoids involved in the stress response (Wellman, 2001).

The underlying mechanisms for the effects of stress on synaptic plasticity in the PFC will be explored, taking into consideration changes in the level of corticosteroids receptors, different neurotransmitters, and their receptors. A particular attention is paid to the glutamate and dopamine systems, since the mesocortical dopamine system is particularly vulnerable to stress and since prefrontal LTP is strongly dependent on dopamine tone.

6. GENERAL CONCLUSION

The potential to modify prefrontal synapses by hippocampal stimulation and the significant contribution of the mesoprefrontal dopamine circuit in these changes illustrate how synaptic plasticity, at least in the PFC, may be differentially regulated according to the animal's or human's behavioral state. Mesoprefrontal dopamine activity is known to affect behavioral performance in both humans and animals on tasks dependent on PFC function that implicate the planning of behaviors in appropriate sequences. The hippocampal–PFC communication can be considered as an important network for the transfer of spatial information (context) that is used to execute prospective strategies for action. The conclusion emerging from these data is that the hippocampus, the PFC, and the mesoprefrontal dopamine system are in a cooperative relationship with respect to working memory. Whether these cortical networks subserve different sub-functions in the overall cognitive operation when information needs to be rapidly processed or move into long-term stores need to be investigated in future studies. A further clarification of the relationships between these structures with the contribution of dopamine neurons could help the identification of specific defects in schizophrenia. Indeed, schizophrenic patients performing poorly on working memory have a significant alteration in prefrontal D_1 receptors, and from functional imaging, it has been postulated a missing link between the hippocampal formation and the PFC, i.e. a disrupted functional integration among these brain regions (Fletcher, 1998).

What might be the mechanistic basis for the dopamine and D_1 receptor-dependent enhancement on hippocampal–prefrontal LTP? Based on electrophysiological and biochemical data, one possibility would be a synergistic action of glutamatergic and dopaminergic inputs leading to a postsynaptic convergence on regulatory pathways at the PKA level. Several substrates of PKA, such as ionotropic glutamate receptors, the phosphoprotein DARPP-32, and the transcription factor CREB, could be considered as good candidates. Another possibility that may account for the control of synaptic strength is the D_1 receptor-stimulated Ca^{2+} release. One of the recently cloned proteins, calcyon, which confers this ability to D_1 receptor by potentiating a crosstalk between Gs- and Gq-coupled receptors, is abundant in the PFC. The function of calcyon in synaptic plasticity should be clarified in future studies.

With recent findings, we have provided evidence that exposure to stress dramatically impairs hippocampal–prefrontal LTP. Whereas these effects are well-known in the hippocampus, its effects on synaptic plasticity in the PFC had never been explored. Stress is known as a vulnerability factor for several psychiatric disorders such as depression. Prolonged and repeated depression is associated with atrophy in the hippocampus and the PFC. In major depressive disorder, imaging studies have consistently reported decreases in metabolic activity in multiple areas of the PFC and a subsequent return to baseline metabolism level after antidepressant treatment. These data have provided support for a model of limbic-cortical dysregulation for depression proposed by Mayberg in 1997. We have preliminary data showing that indeed a certain class of antidepressants restores prefrontal LTP impaired by prior acute stress. Beneficial effects on neuronal plasticity, defined as a reversal of the effects of stress in this paradigm, can be considered as a further indication that the hippocampal–prefrontal circuitry is important in depression.

REFERENCES

Artola A, Brocher S, Singer W (1990) Different voltage-dependent thresholds for inducing long-term depression and long-term potentiation in slices of the visual cortex. Nature 347:69-72.

Barbas H, Blatt GJ (1995) Topographically specific hippocampal projections target functionally distinct prefrontal areas in the rhesus monkey. Hippocampus 5:511-533.

Balfour DJ, Reid A (1979) Effects of betamethasone on the stimulation of corticosterone secretion in rats. Arch Int Pharmacodyn Ther 237:67-74.

Blank T, Nijholt I, Teichert U, Kugler H, Behrsing H, Fienberg A, Greengard P, Spiess J (1997) The phosphoprotein DARPP-32 mediates cAMP-dependent potentiation of striatal N-methyl-D-aspartate responses. Proc Natl Acad Sci USA 94:14859-14864.

Blitzer RD, Connor JH, Brown GP, Wong T, Shenolikar S, Iyengar R, Landau EM (1998) Gating of CaMKII by cAMP-regulated protein phosphatase activity during LTP. Science 280:1940-1942.

Blond O, Crepel F, Otani S (2002) Long-term potentiation in rat prefrontal slices facilitated by phased application of dopamine. Eur J Pharmacol 438:115-116.

Bolshakov VY, Golan H, Kandel ER, Siegelbaum SA (1997) Recruitment of new sites of synaptic transmission during the cAMP-dependent late phase of LTP at CA3-CA1 synapses in the hippocampus. Neuron 19:635-651.

Brozoski TJ, Brown RM, Rosvold HE, Goldman-Rakic PS (1979) Cognitive deficits caused by regional depletion of dopamine in prefrontal cortex of rhesus monkey. Science 205:929-932.

Burette F, Jay TM, Laroche S (1997) Reversal of LTP in the hippocampal afferent fiber system to the prefrontal cortex in vivo with low-frequency patterns of stimulation that do not produce LTD. J Neurophysiol 78:1155-1160.

Burette F, Jay TM, Laroche S (2000) Synaptic depression of the hippocampal to prefrontal cortex pathway during a non spatial working memory task. Curr Psychol Lett 1:9-23.

Carmichael ST, Price JL (1995) Limbic connections of the orbital and medial prefrontal cortex in macaque monkeys. J Comp Neurol 363:615-641.

Carr DB, Sesack SR (1996) Hippocampal afferents to the rat prefrontal cortex: synaptic targets and relation to dopamine terminals. J Comp Neurol 369:1-15.

Davies CH, Starkey SJ, Pozza MF, Collingridge GL (1991) GABA autoreceptors regulate the induction of LTP. Nature 349:609-611.

Diorio D, Viau V, Meaney MJ (1993) The role of the medial prefrontal cortex (cingulate gyrus) in the regulation of hypothalamic-pituitary-adrenal responses to stress. J Neurosci 13:3839-3847.

Escobar ML, Chao V, Bermudez-Rattoni F (1998) In vivo long-term potentiation in the insular cortex: NMDA receptor dependence. Brain Res 779:314-319.

Ferino F, Thierry AM, Glowinski J (1987) Anatomical and electrophysiological evidence for a direct projection from Ammon's horn to the medial prefrontal cortex in the rat. Exp Brain Res 65:421-426.

Fletcher P (1998) The missing link: a failure of fronto-hippocampal integration in schizophrenia. Nat Neurosci 1:266-267.

Floresco SB, Seamans JK, Phillips AG (1997) Selective roles for hippocampal, prefrontal cortical, and ventral striatal circuits in radial-arm maze tasks with or without a delay. J Neurosci 17:1880-1890.

Foy MR, Stanton ME, Levine S, Thompson RF (1987) Behavioral stress impairs long-term potentiation in rodent hippocampus. Behav Neural Biol 48:138-149.

Frey U, Matthies H, Reymann KG, Matthies H (1991) The effect of dopaminergic D1 receptor blockade during tetanization on the expression of long-term potentiation in the rat CA1 region in vitro. Neurosci Lett 129:111-114.

Froc DJ, Chapman CA, Trepel C, Racine RJ (2000) Long-term depression and depotentiation in the sensorimotor cortex of the freely moving rat. J Neurosci 20:438-445.

Gabbott P, Headlam A, Busby S (2002) Morphological evidence that CA1 hippocampal afferents monosynaptically innervate PV-containing neurons and NADPH-diaphorase reactive cells in the medial prefrontal cortex (Areas 25/32) of the rat. Brain Res 946:314-322.

Garcia R (2001) Stress, hippocampal plasticity, and spatial learning. Synapse 40:180-183.

Garris PA, Collins LB, Jones SR, Wightman RM (1993) Evoked extracellular dopamine in vivo in the medial prefrontal cortex. J Neurochem 61:637-647.

Gigg J, Tan AM, Finch DM (1994) Glutamatergic hippocampal formation projections to prefrontal cortex in the rat are regulated by GABAergic inhibition and show convergence with glutamatergic projections from the limbic thalamus. Hippocampus 4:189-198.

Goldman-Rakic PS, Selemon LD, Schwartz ML (1984) Dual pathways connecting the dorsolateral prefrontal cortex with the hippocampal formation and parahippocampal cortex in the rhesus monkey. Neuroscience 12:719-743.

Groenewegen HJ, Uylings HB (2000) The prefrontal cortex and the integration of sensory, limbic and autonomic information. In: Progress in Brain Research, vol 126 (Uylings HBM, Van Eden CG, De Bruin JPC, Feestra MGP, and Pennartz CMA, eds), pp 3-28, Elsevier, Amsterdam.

Gurden H, Tassin JP, Jay TM (1999) Integrity of the mesocortical DA system is necessary for complete expression of in vivo hippocampal-prefrontal cortex long-term potentiation. Neuroscience 94:1019-1027.

Gurden H, Takita M, Jay TM (2000) Essential role of D1 but not D2 receptors in the NMDA receptor-dependent long-term potentiation at hippocampal-prefrontal cortex synapses in vivo. J Neurosci 20:RC106 (1-5).

Hemmings HC Jr, Greengard P, Tung HY, Cohen P (1984) DARPP-32, a dopamine-regulated neuronal phosphoprotein, is a potent inhibitor of protein phosphatase-1. Nature 310:503-505.

Herry C, Vouimba RM, Garcia R (1999) Plasticity in the mediodorsal thalamo-prefrontal cortical transmission in behaving mice. J Neurophysiol 82:2827-2832.

Impey S, Mark M, Villacres EC, Poser S, Chavkin C, Storm DR (1996) Induction of CRE-mediated gene expression by stimuli that generate long lasting LTP in area CA1 of the hippocampus. Neuron 16:973-982.

Jay TM, Glowinski J, Thierry AM (1989) Selectivity of the hippocampal projection to the prelimbic area of the prefrontal cortex in the rat. Brain Res 505:337-340.

Jay TM, Witter MP (1991) Distribution of hippocampal CA1 and subicular efferents in the prefrontal cortex of the rat studied by means of anterograde transport of Phaseolus vulgaris-leucoagglutinin. J Comp Neurol 313:574-586.

Jay TM, Thierry AM, Wiklund L, Glowinski J (1992) Excitatory amino acid pathway from the hippocampus to the prefrontal cortex. Contribution of AMPA receptors in hippocampo-prefrontal cortex transmission. Eur J Neurosci 4:1285-1295.

Jay TM, Burette F, Laroche S (1995a) NMDA receptor-dependent long-term potentiation in the hippocampal afferent fibre system to the prefrontal cortex in the rat. Eur J Neurosci 7:247-250.

Jay TM, Glowinski J, Thierry AM (1995b) Inhibition of hippocampo-prefrontal cortex excitatory responses by the mesocortical DA system. Neuroreport 6:1845-1848.

Jay TM, Burette F, Laroche S (1996a) Plasticity of the hippocampal-prefrontal cortex synapses. J Physiol (Paris) 90:361-366.

Jay TM, Burette F, Laroche S (1996b) Dopaminergic modulation of long-term potentiation in the hippocampal-prefrontal cortex pathway. Soc Neurosci Abstr 22:322.

Jay TM, Gurden H, Yamaguchi T (1998) Rapid increase in PKA activity during long-term potentiation in the hippocampal afferent fibre system to the prefrontal cortex in vivo. Eur J Neurosci 10:3302-3306.

Kerr JN, Wickens JR (2001) Dopamine D-1/D-5 receptor activation is required for long-term potentiation in the rat neostriatum in vitro. J Neurophysiol 85:117-124.

Kim JJ, Diamond DM (2002) The stressed hippocampus, synaptic plasticity and lost memories. Nat Rev Neurosci 3:453-462.

Laroche S, Jay TM, Thierry AM (1990) Long-term potentiation in the prefrontal cortex following stimulation of the hippocampal CA1/subicular region. Neurosci Lett 114:184-190.

Law-Tho D, Desce JM, Crepel F (1995) Dopamine favours the emergence of long-term depression versus long-term potentiation in slices of rat prefrontal cortex. Neurosci Lett 188:125-128.

Lee HK, Barbarosie M, Kameyama K, Bear MF, Huganir RL (2000) Regulation of distinct AMPA receptor phosphorylation sites during bidirectional synaptic plasticity. Nature 405:955-959.

Leonard AS, Hell JW (1997) Cyclic AMP-dependent protein kinase and protein kinase C phosphorylate N-methyl-D-aspartate receptors at different sites. J Biol Chem 272:12107-12115.

Lewis BL, O'Donnell P (2000) Ventral tegmental area afferents to the prefrontal cortex maintain membrane potential 'up' states in pyramidal neurons via D(1) dopamine receptors. Cereb Cortex 10:1168-1175.

Lezcano N, Mrzljak L, Eubanks S, Levenson R, Goldman-Rakic P, Bergson C (2000) Dual signaling regulated by calcyon, a D1 dopamine receptor interacting protein. Science 287:1660-1664.

Lisman JE, Zhabotinsky AM (2001) A model of synaptic memory: a CaMKII/PP1 switch that potentiates transmission by organizing an AMPA receptor anchoring assembly. Neuron 31:191-201.

Magarinos AM, McEwen BS, Flugge G, Fuchs E (1996) Chronic psychosocial stress causes apical dendritic atrophy of hippocampal CA3 pyramidal neurons in subordinate tree shrews. J Neurosci 16:3534-3540.

Mayberg HS (1997) Limbic-cortical dysregulation: a proposed model of depression. J Neuropsychiatry Clin Neurosci 9:471-481.

McEwen BS (2000) The neurobiology of stress: from serendipity to clinical relevance. Brain Res 886:172-189.

Mizoguchi K, Yuzurihara M, Ishige A, Sasaki H, Chui DH, Tabira T (2000) Chronic stress induces impairment of spatial working memory because of prefrontal dopaminergic dysfunction. J Neurosci 15:1568-1574.

Moore AN, Waxham MN, Dash PK (1996) Neuronal activity increases the phosphorylation of the transcription factor cAMP response element-binding protein (CREB) in rat hippocampus and cortex. J Biol Chem 271:14214-14220.

Mulder AB, Arts MP, Lopes da Silva FH (1997) Short- and long-term plasticity of the hippocampus to nucleus accumbens and prefrontal cortex in the rat, in vivo. Eur J Neurosci 9:1603-1611.

Öngür D, Price JL (2000) The organization of networks within the orbital and medial prefrontal cortex of rats, monkeys and humans. Cereb Cortex 10:206-219.

Otani S, Blond O, Desce JM, Crepel F (1998) Dopamine facilitates long-term depression of glutamatergic transmission in rat prefrontal cortex. Neuroscience 85:669-676.

Reynolds JN, Wickens JR (2000) Substantia nigra dopamine regulates synaptic plasticity and membrane potential fluctuations in the rat neostriatum in vivo. Neuroscience 99:199-203.

Roberson ED, Sweatt JD (1996) Transient activation of cyclic AMP-dependent protein kinase during hippocampal long-term potentiation. J Biol Chem 271:30436-30441.

Rocher C, Spedding M, Munoz C, Jay TM (2002) Stress-induced alteration on cortical synaptic plasticity can be prevented by an antidepressant: tianeptine. FENS: 154.14, 3rd Forum of Eur Neurosci, Paris, France.

Sawaguchi T, Goldman-Rakic PS (1994) The role of D1-dopamine receptor in working memory: local injections of dopamine antagonists into the prefrontal cortex of rhesus monkeys performing an oculomotor delayed-response task. J Neurophysiol 71:515-528.

Schultz W (2002) Getting formal with dopamine and reward. Neuron 36:241-263.

Seamans JK, Floresco SB, Phillips AG (1998) D1 receptor modulation of hippocampal-prefrontal cortical circuits integrating spatial memory with executive functions in the rat. J Neurosci 18:1613-1621.

Sesack SR, Deutch AY, Roth RH, Bunney BS (1989) Topographical organization of the efferent projections of the medial prefrontal cortex in the rat: an anterograde tract-tracing study with Phaseolus vulgaris leucoagglutinin. J Comp Neurol 290:213-242.

Sesack SR, Snyder CL, Lewis DA (1995) Axon terminals immunolabeled for dopamine or tyrosine hydroxylase synapse on GABA-immunoreactive dendrites in rat and monkey cortex. J Comp Neurol 363:264-280.

Simon H, Scatton B, Le Moal M (1980) Dopaminergic A10 neurons are involved in cognitive functions. Nature 286:150-151.

Snyder GL, Fienberg AA, Huganir RL, Greengard P (1998) A dopamine/D1 receptor/protein kinase A/dopamine- and cAMP-regulated phosphoprotein (Mr 32 kDa)/protein phosphatase-1 pathway regulates dephosphorylation of the NMDA receptor. J Neurosci 18:10297-10303.

Snyder GL, Allen PB, Fienberg AA, Valle CG, Huganir RL, Nairn AC, Greengard P (2000) Regulation of phosphorylation of the GluR1 AMPA receptor in the neostriatum by dopamine and psychostimulants in vivo. J Neurosci 20:4480-4488.

Sullivan RM, Gratton A (2002) Prefrontal cortical regulation of hypothalamic-pituitary-adrenal function in the rat and implications for psychopathology: side matters. Psychoneuroendocrinol 27:99-114.

Surmeier DJ, Bargas J, Hemmings HC Jr, Nairn AC, Greengard P (1995) Modulation of calcium currents by a D1 dopaminergic protein kinase/phosphatase cascade in rat neostriatal neurons. Neuron 14:385-397.

Swanson-Park JL, Coussens CM, Mason-Parker SE, Raymond CR, Hargreaves EL, Dragunow M, Cohen AS, Abraham WC (1999) A double dissociation within the hippocampus of dopamine D1/D5 receptor and beta-adrenergic receptor contributions to the persistence of long-term potentiation. Neuroscience 92:485-497.

Takita M, Izaki Y, Jay TM, Kaneko H, Suzuki SS (1999) Induction of stable long-term depression in vivo in the hippocampal-prefrontal cortex pathway. Eur J Neurosci 11:4145-4148.

Thierry AM, Gioanni Y, Degenetais E, Glowinski J (2000) Hippocampo-prefrontal cortex pathway: anatomical and electrophysiological characteristics. Hippocampus 10:411-419.

Trentani A, Kuipers SD, Ter Horst GJ, Den Boer JA (2002) Selective chronic stress-induced in vivo ERK1/2 hyperphosphorylation in medial prefrontocortical dendrites: implications for stress-related cortical pathology? Eur J Neurosci 15:1681-1689.

Van Eden CG, Hoorneman EM, Buijs RM, Matthijssen MA, Geffard M, Uylings HB (1987) Immunocytochemical localization of dopamine in the prefrontal cortex of the rat at the light and electron microscopical level. Neuroscience 22:849-62.

Vermetten E, Bremner JD (2002) Circuits and systems in stress. I. Preclinical studies. Depress Anxiety 15:126-147.

Verwer RW, Meijer RJ, Van Uum HF, Witter MP (1997) Collateral projections from the rat hippocampal formation to the lateral and medial prefrontal cortex. Hippocampus 7:397-402.

Vickery RM, Morris SH, Bindman LJ (1997) Metabotropic glutamate receptors are involved in long-term potentiation in isolated slices of rat medial frontal cortex. J Neurophysiol 78:3039-3046.

Wellman CL (2001) Dendritic reorganization in pyramidal neurons in medial prefrontal cortex after chronic corticosterone administration. J Neurobiol 49:245-253.

Williams GV, Goldman-Rakic PS (1995) Modulation of memory fields by dopamine D1 receptors in prefrontal cortex. Nature 376:572-575.

Witter MP, Groenewegen HJ, Lopes da Silva FH, Lohman AH (1989) Functional organization of the extrinsic and intrinsic circuitry of the parahippocampal region. Prog Neurobiol 33:161-253.

Xu L, Anwyl R, Rowan MJ (1997) Behavioural stress facilitates the induction of long-term depression in the hippocampus. Nature 387:497-500.

Xu L, Holscher C, Anwyl R, Rowan MJ (1998) Glucocorticoid receptor and protein/RNA synthesis-dependent mechanisms underlie the control of synaptic plasticity by stress. Proc Natl Acad Sci USA 95:3204-3208.

Chapter 6

CHANGES OF NEURONAL ACTIVITY IN THE PREFRONTAL CORTEX RELATED TO THE EXPRESSION AND EXTINCTION OF CONDITIONED FEAR RESPONSES

Cyril Herry[1] and René Garcia[2]
[1]*Neurosciences Cognitives CNRS UMR5106, Université de Bordeaux I, 33405 Talence, France; e-mail: herry9@yahoo.com*
[2]*Neurobiologie Comportementale, Université de Nice-Sophia Antipolis, 06108 Nice, France; e-mail: rene.garcia@unice.fr*

Keywords: Fear conditioning and extinction, post-traumatic stress disorder, regional blood flow, animal models, neuronal inhibition and activation, synaptic plasticity, medial prefrontal cortex.

Abstract: One of the fundamental roles of the prefrontal cortex is to inhibit inappropriate responses, as indicated by studies showing that lesions of this structure can result in perseverative behaviors. However, analyses of the involvement of prefrontal neurons in inhibition of conditioned fear responses, during extinction, have led to contradictory observations. Recent electrophysiological studies suggest that prefrontal neuronal activity does not interfere with the expression of conditioned fear before extinction, but may strongly contribute to modulate the post-extinction expression of fear responses. Here we will discuss all of these studies, along with some possible mechanisms of interactions between the prefrontal cortex and the amygdala (long-term consolidation of extinction) and the hippocampus (modulation of the expression of extinction). The implications of these interactions for pathophysiology and therapy of post-traumatic stress disorder and relapse will also be discussed.

1. INTRODUCTION

One of the symptoms which characterizes large lesions of the medial prefrontal cortex is the development of a greater degree of perseveration, i.e.

a failure or inability to inhibit inappropriate responses (Eichenbaum et al., 1983; Dunnett et al., 1999; Gemmel and O'Mara, 1999; Hauser, 1999; Dias and Aggleton, 2000). However, since this observation mostly derived from studies of non-fear related responses, a fundamental question that it raises is whether the same cortical area is also implicated in the inhibition of conditioned fear responses when a conditioned stimulus (CS) is repetitively presented without the aversive unconditioned stimulus (US). This is of particular interest in the context of the behavioral therapy of post-traumatic stress disorder (PTSD). Indeed, PTSD develops as a result of fear conditioning (aversive CS-US association). In addition, treatments that involve exposure to fearful stimuli (extinction procedure) are effective in most PTSD patients (Rothbaum and Schwartz, 2002). However, a failure (Bremner et al., 1996) or a reduction (Peri et al., 2000) of extinction of conditioned fear responses in PTSD patients and the relapse following extinction of PTSD symptoms (Tarrier et al., 1999) continue to remain challenging. Accordingly, a dysfunction in prefrontal inhibitory mechanisms may lead either to this resistance to extinction of fear or to post-therapy symptom reactivation. Studies to test this possibility have been conducted using two main approaches (i.e. lesion and electrophysiological studies) with the animal models of exposure therapy, in which fear responses (e.g. freezing) to a tone CS (previously paired with footshock US) gradually decline over CS-alone presentations. The use of recent advances in neuroimaging technology has also provided additional clues. In particular, this approach has revealed that PTSD patients exhibit a reduced volume of the hippocampus (Bremner et al., 1995; Stein et al. 1997). Moreover, exposure to traumatic CS produces, in these patients, increased activity in the amygdala (Shin et al., 1997; Liberzon et al., 1999) and reduced activity in the medial prefrontal cortex (Bremner et al., 1999). These three structures (hippocampus, amygdala, and prefrontal cortex) being functionally interconnected, interactions between them may constitute key components controlling the direction of plasticity of prefrontal neuronal activity during the expression of either traumatic or extinction memory. We will focus our discussion on this aspect.

2. PLASTICITY RELATED TO TRAUMATIC MEMORY

Abnormal traumatic recall and fear responding (e.g. increased heart rate and blood pressure) can occur in PTSD patients in the absence of CS (e.g. intrusive memories, recurrent dreams, and "flashbacks" of the traumatic event). This suggests the potential existence of abnormalities in the circuits implicated in the emotional regulation of memory (e.g. the prefrontal cortex and the amygdala). Experimentally, traumatic memory is activated and expressed during exposure of PTSD patients to a provocative stimulus (e.g. traumatic pictures and sounds in combat veterans; Prins et al., 1995), or when an aversive CS is presented to a previously conditioned individual (both animals and humans). In this latter case, animals (Gerwitz et al., 1997; Morrow et al., 1999; Quirk et al., 2000) or humans (Bechera et al., 1999) with lesions of the medial prefrontal cortex have been found to express normal traumatic memory. The only exception concerns lesions located in the more dorsal part of the medial prefrontal cortex, for which potentiation of conditioned fear behavior (freezing responses) has been reported both in rats (Morgan and LeDoux, 1995) and mice (Vouimba et al., 2000). Although these results are far from uniform, it has been repeatedly suggested that a class of prefrontal neurons modulates fear responding via their inhibitory connections with amygdalar neurons that are involved in the expression of traumatic memory (LeDoux, 1993, Devinsky et al., 1995; Vouimba et al., 2000; Vermetten and Bremner, 2002). Consequently, damage to this class of prefrontal neurons produces potentiation of certain emotional responses (but also see Holson, 1986; Jaskiw and Weinberger, 1992; Frysztak and Neafsey, 1994). However, this potentiation remains difficult to explain, especially in the light of data from neuroimaging and electrophysiological studies reporting the absence of prefrontal neuronal activity effects on expression of traumatic memory (particularly before any acquisition of CS-no US association).

Firstly, in humans, Bremner and colleagues have reported consistent neuroimaging data demonstrating changes in blood flow in the medial prefrontal cortex (Brodmann's area 25) in PTSD patients. In particular, this group (Bremner et al., 1997, 1999) showed that exposure to traumatic conditioned stimuli (e.g. war sounds such as helicopter sounds, explosions, and machine gun fire) results in a decreased blood flow bilaterally in the medial prefrontal cortex in these patients (Fig. 1). This observation was also confirmed in a more recent study using war sounds as a provocation of PTSD symptoms (Fernandez et al., 2001). In general, it is admitted that neuronal firing (or synaptic efficacy) and blood flow increments or decrements are tightly paired (Kety and Schmidt, 1945). Hence, a decreased blood flow in bilateral medial prefrontal cortex may correspond to a decreased neuronal firing or synaptic efficacy. If decreased prefrontal activity (Brodmann's area 25) results in reduced inhibitory effects on amygdalar neurons involved in the expression of traumatic memory, increased prefrontal activity in the same area during similar provocation of PTSD symptoms should alter the expression of fear responses. However, two other neuroimaging studies, using helicopter sounds, explosions, and machine gun fire as provocative stimuli, have provided contrasting data. In one of these studies, no differences in prefrontal activation (Brodmann's

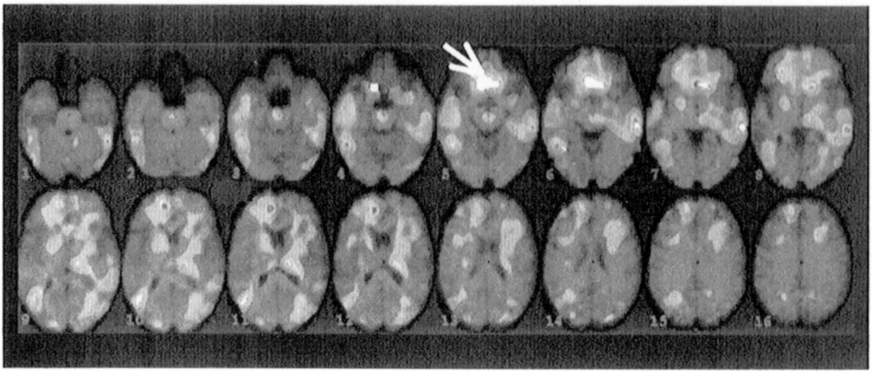

Figure 1: Scanning of the brain of a patient with combat-related PTSD during exposure to traumatic CS (combat-related sounds). The CS triggered expression of traumatic memory (e.g. increased heart rate and blood pressure), which was associated with decreased blood flow in the medial prefrontal cortex (arrow). From Bremner et al. (1999) with permission.

area 25) were found between combat veterans with and without PTSD or noncombattant controls (Liberzon et al., 1999). In the other study, it was observed that these stimuli produce an increased, instead of decreased, activity in the same frontal area (Zubieta et al., 1999). However, in addition to these different patterns of changes in neuronal activity in the medial prefrontal cortex, all of these studies reveal similar war sounds-associated distresses (scored via psychophysiological measures, such as increased skin conductance and heart rate). Therefore, similar symptom provocation paradigms can induce contrasting effects on neuronal activity in the medial prefrontal cortex, indicating a dissociation between the evoked direction of activity (decrease or increase or even no change) and the expression of traumatic memory. Additional studies are still needed to better understand this divergence in humans.

Secondly, animal studies examining the effects of aversive CS on prefrontal neuronal activity have also led to contradictory results concerning the medial prefrontal cortex (lesions of which produced either no change or potentiation of freezing behavior, as seen above). Garcia et al. (1999) and Milad and Quirk (2002) measured changes in spontaneous activity in the medial prefrontal cortex during re-exposure of animals (mice and rats, respectively) to a tone initially paired with foot shock. In the first study, animals were conditioned with "light–tone" presentations signaling the omission of shock (conditioned inhibition procedure) or without this procedure ("light–tone" unpaired, the tone being always reinforced). During conditioning tests (i.e. expression of traumatic memory), the "light–tone" paired group displayed less freezing than the "light–tone" unpaired group, indicating the acquisition of conditioned inhibition. Analyses of changes in multi-unit activity in the dorso-medial prefrontal cortex showed a greater decrease in the activity in the "light–tone" paired group than in the other group, with a strong correlation found between the magnitude of expression of traumatic memory (freezing behavior) and the decrease in prefrontal activity. However, in the second study (Milad and Quirk, 2002), the authors reported the absence of changes in unit firing rate in the same area during the expression of freezing behavior to the tone CS. This divergence is probably due to different experimental conditions. Moreover, prefrontal cells displaying CS-evoked decreased activity are not easily encountered, and once encountered (see Fig. 2, next page), they rapidly switch from depression to normal activity (R. Garcia, unpublished observation). Despite this divergence between the observations of Garcia et al. (1999) and Milad and Quirk (2002), these findings yield at least one common point of convergence. Indeed, here also, as with the neuroimaging studies, one can conclude that levels of neuronal activity in the medial prefrontal cortex do not affect the magnitude of conditioned fear responses. Consequently, during post-conditioning CS presentation, prefrontal neuronal activity does not profoundly alter the activity of amygdalar neurons involved in the expression of traumatic memory expression.

This conclusion is also supported by synaptic plasticity studies. In particular, these studies have shown that glutamatergic synapses in the medial prefrontal cortex exhibit changes in the efficacy (depression or potentiation) following either a learning task or a train of electrical stimulation (high or low-frequency stimulation). High-frequency stimulation generally induces long-term potentiation (LTP), whereas low-frequency stimulation generally produces long-term depression (LTD) in behaving mice (Herry et al., 1999; Herry and Garcia, 2002). However, neither the direction of synaptic plasticity (LTP or LTD, as well as learning-induced potentiation or depression) was found to significantly alter the magnitude of freezing behavior (Herry et al., 1999; Herry and Garcia, 2002). For example, mice that received high-frequency stimulation before being re-exposed to the tone CS exhibited LTP that was still present during CS presentation, which, however, produced freezing levels similar to that expressed by the mice that did not receive tetanus (Fig. 3). Therefore, what the direction of the changes in plasticity of prefrontal neuronal activity signifies during the expression of traumatic memory remains unclear. However, because the prefrontal neurons play a key role in various cognitive functions, it is possible that the direction of changes in plasticity in this structure during CS-alone presentations may be related to processing of cognitive information such as the occurrence or the absence of danger (Herry et al., 1999). This may be mediated via its interactions with the amygdala.

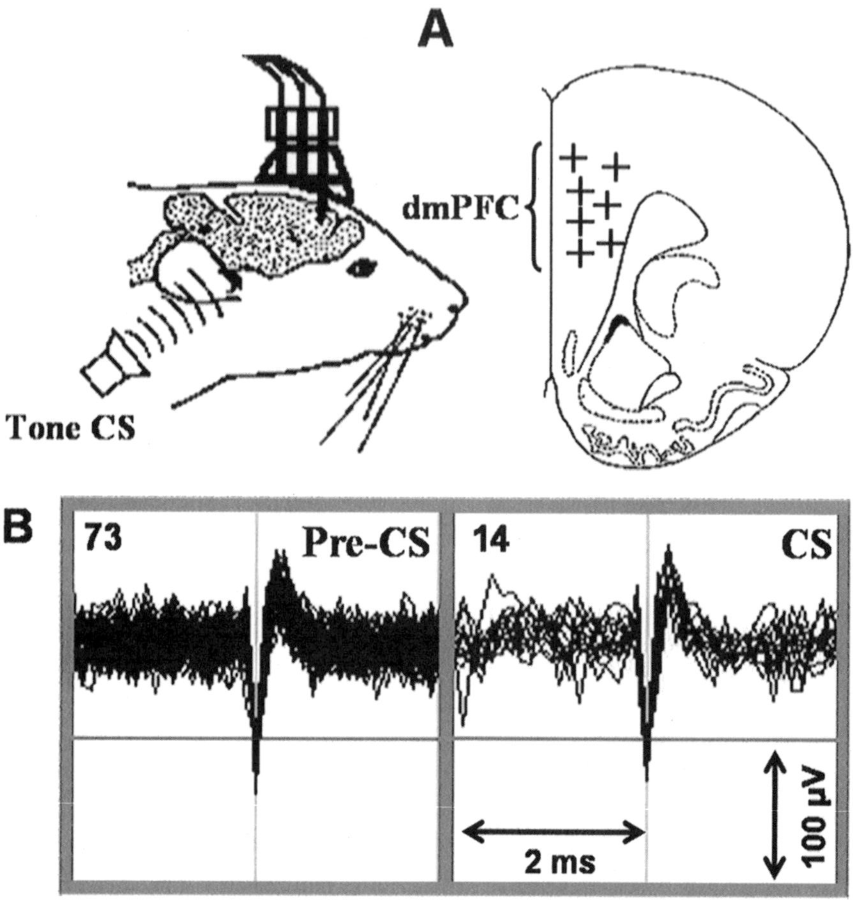

Figure 2: Example of a cell, located (in A) in the dorsal part of the medial prefrontal cortex (dmPFC), showing (in B) decreased spontaneous spike number during re-exposure of a mouse to a tone CS, which was initially paired with footshock. Recordings were made 20 sec before the CS presentation (Pre-CS: 73 spikes) and during CS presentation (20 sec: 14 spikes).

Currently, we partially understand the role of the amygdala in the induction of changes in neuronal activity within the medial prefrontal cortex during the expression of traumatic memory. Neurons in the amygdala, which are involved in the expression of this type of memory, also seem to inhibit spontaneous activity in the medial prefrontal cortex in the presence of an aversive CS (Garcia et al., 1999). Our hypothesis is that, for certain subjects (humans and animals), following conditioning, CS-alone presentations rapidly trigger mechanisms, which still remain undetermined (and are in the heart of the debate on the relationship between fear responding and prefrontal neuronal plasticity), freeing prefrontal neurons from the traumatic CS-related activation of the amygdala. Moreover, this disconnection does not affect expression of traumatic memory, which remains under the control of neurons within the amygdala. However, the rapid return to baseline activity within the medial prefrontal cortex during fear responding may correspond to the processing of information such as that the CS is not followed by the US.

3. PLASTICITY RELATED TO EXTINCTION

In a recent human study, Bechara et al. (1999) used the association between monochrome color slides as CS and a startling loud and obnoxious sound as US to examine fear conditioning and extinction of fear responding (scored via changes in skin conductance responses) in patients with ventro-

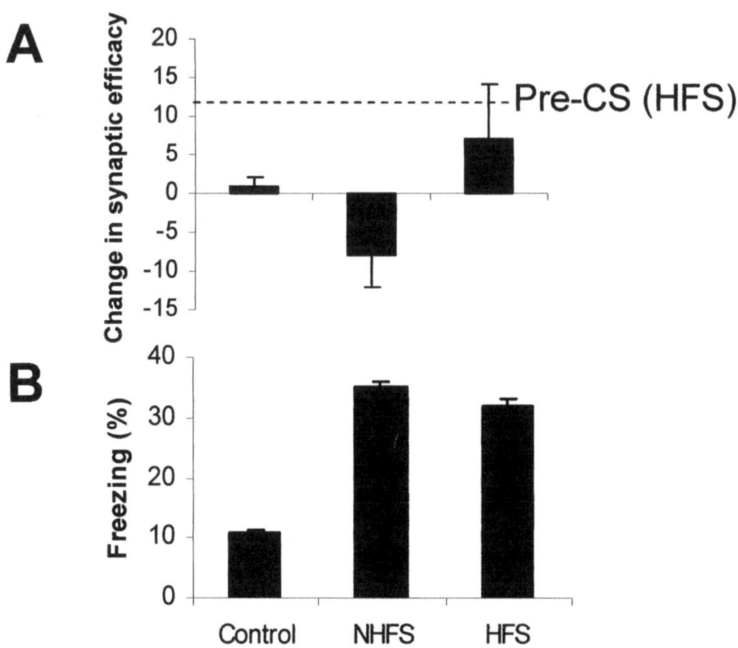

Figure 3: Mean percent changes in synaptic efficacy (measured by changes in the amplitude of prefrontal response) induced by thalamic stimulation in mice. A. During re-exposure to a tone CS, initially paired with footshock. Thalamic high-frequency stimulation (HFS) induced potentiation of synaptic efficacy (Pre-CS), the level of which was reduced (but remained above the baseline) during re-exposure to the CS. B. Despite this potentiation, the expression of traumatic memory (assessed by freezing) was similar to those of conditioned mice that did not receive HFS (NHFS). Adopted from Herry and Garcia (2002) with permission.

medial prefrontal cortex lesions. All patients acquired CS-associated skin conductance responses that were similar in magnitude to the responses displayed by control subjects. These responses progressively extinguished, with a rate similar to the control group, during the phase of repeated CS-alone presentations (Fig. 4). This finding reveals that neurons within the medial prefrontal cortex are not required for the acquisition of extinction.

Animal studies, both in mice (Vouimba et al., 2000) and rats (Gerwitz et al., 1997; Quirk et al., 2000), have also led to an identical conclusion. In these studies, an explicit CS (a light or a tone) was paired with footshock US, and prefrontal lesions (either dorsal or ventral area of the medial prefrontal cortex) were made before of after this fear conditioning. Neither lesion location was found to disrupt the rate of extinction. There are,

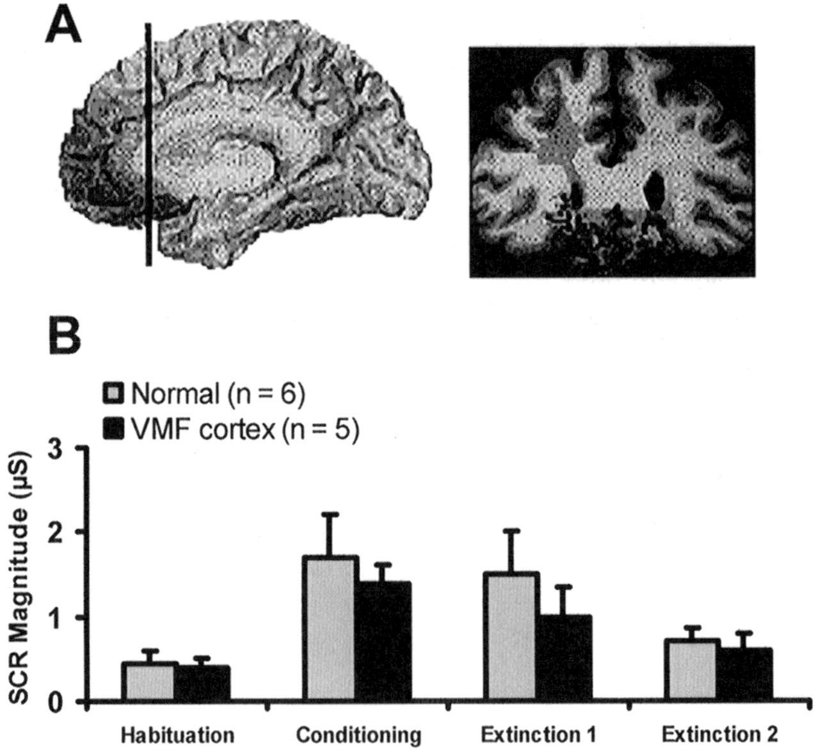

Figure 4: Magnitude of skin conductance responses (SCR) during habituation, CS-US association (conditioning), and CS-no US learning (extinction 1 and 2). The SCR generated by control and prefrontal patients (lesion location in VMF cortex shown in A) during the conditioning phase were significantly higher than those generated during the habituation or extinction phase (shown in B; all p values < 0.001). The two groups did not differ from each other. Adopted from Bechara et al. (1999) with permission.

however, two exceptions showing retardation of extinction following either electrolytic (Morgan and LeDoux, 1995) or 6-hydroxydopamine (Morrow et al., 1999) lesions of the medial prefrontal cortex. However, this phenomenon was observed in the second study only with a high US intensity (0.8 mA versus 0.4 mA).

Although most of these studies have clearly shown that prefrontal lesions do not interfere with extinction of fear responding, analyses of neuronal activity within the medial prefrontal cortex have indicated the occurrence of specific plasticity related to extinction. In humans, two studies in which subjects were scanned before and after treatment for PTSD indicate such changes. In the first study, Levin et al. (1999) observed, in one PTSD patient, that treatment by eye movement desensitization and reprocessing is associated with increased activity in two areas: the anterior cingulate gyrus and the left frontal lobe. Note that the patient was also on an antidepressant treatment (with a selective serotonergic reuptake inhibitor, SSRI) throughout the study. In the second study, Fernandez et al. (2001) found in one subject that pharmacological treatment (with a SSRI)-inducing extinction of behavioral activity in response to trauma reminders was associated with a conversion from depression to potentiation of neuronal activity in the medial prefrontal cortex. Although requiring replication, these two human studies suggest that successful treatments for PTSD are associated with potentiation of neuronal activity within the medial prefrontal cortex. Similar changes in plasticity were also observed in the original animal study (Herry et al., 1999). This study showed that extinction of a freezing response during repeated presentations of a tone CS, previously paired with footshock, initially suppressed the CS-induced depression in prefrontal synaptic efficacy within the medial prefrontal cortex, with further presentations resulting in LTP-like changes.

Contrary to lesion findings, the above data on neuronal activity plasticity suggest a possible role of the medial prefrontal cortex in the acquisition of extinction. Another recent study in which an acceleration of extinction was observed by pairing infralimbic tetanic stimulation with CS-alone presentations also supports this view (Milad and Quirk, 2002). However, other recent analyses of prefrontal synaptic plasticity (Herry and Garcia, 2002) are in agreement with lesion studies, rejecting any implication of prefrontal neuronal activity in fear inhibition during extinction. In this latter study, behavioral data indicate that all mice completely extinguished their freezing response toward a tone CS following 16 trials of CS-alone presentation. However, examination of individual changes in synaptic efficacy within the medial prefrontal cortex revealed two sub-groups of mice. One sub-group displayed maintenance of depression in response to

CS, whereas disappearance of this depression (with total restoration of baseline levels) characterized the other group. Despite this electrophysiological difference, the two sub-groups similarly extinguished their fear responding (Fig. 5). Likewise, mice receiving tetanic thalamic stimulation before extinction or thalamic low-frequency stimulation during extinction developed LTP and LTD, respectively, that did not affect the within-session rate of extinction (Herry and Garcia, 2002).

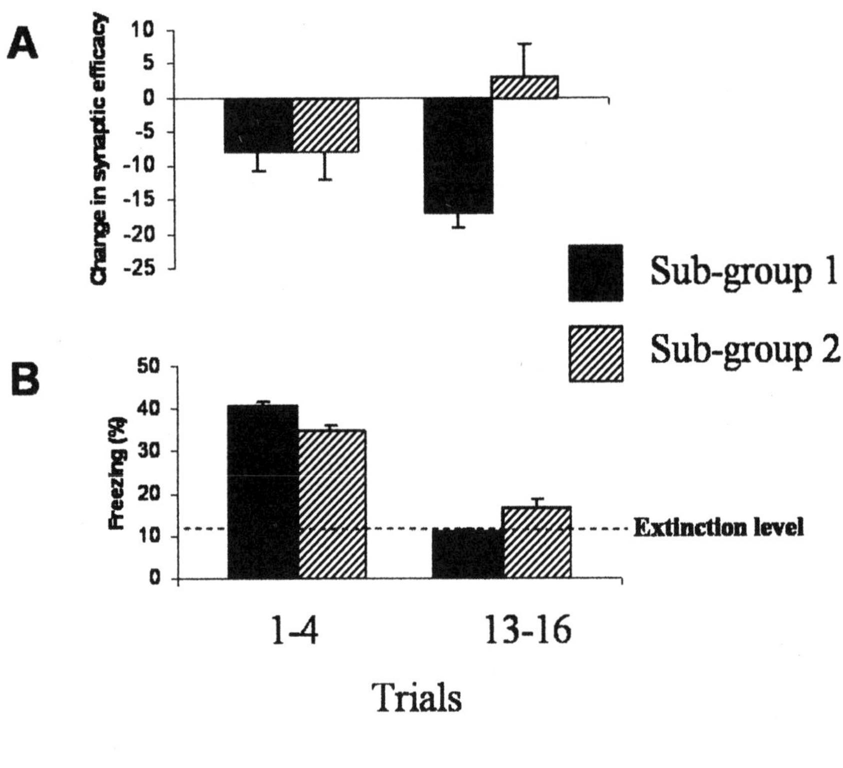

Figure 5: Mean percentage of changes in synaptic efficacy during re-exposure of mice to a tone CS (in A), initially paired with footshock. CS-alone presentations induced decreased synaptic efficacy (trials 1-4), which, during trials 13-16, was still present in half of animals (sub-group 1), while the other half exhibited a return to baseline synaptic efficacy (sub-group 2). However, mice in both sub-groups exhibited similar extinction of freezing behavior (in B). Adopted from Herry and Garcia (2002) with permission.

4. PLASTICITY RELATED TO THE EXPRESSION OF POST-EXTINCTION MEMORIES

It is now well recognized that conditioning and extinction result from the formation of distinct CS-related memories that are separately accessible after the extinction period (Bouton, 1993; Rescorla, 2001; Garcia, 2002a). If prefrontal neuronal plasticity is not involved in extinction learning, as shown above with lesion and electrophysiological studies, it may nevertheless favor, according to its direction (potentiation or depression), the post-extinction expression of one or the other form of CS-related memories (Garcia, 2002a).

First, Quirk et al. (2000) showed that 24 hours following the complete extinction of fear responding to a tone CS (previously paired with footshock US and then repeatedly presented alone), re-exposure to the CS preferentially activates the expression of traumatic memory in rats with

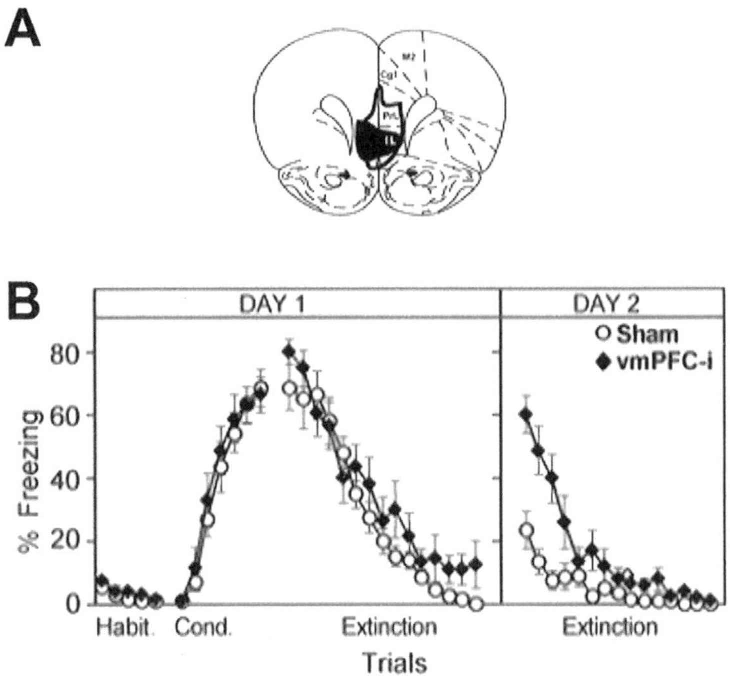

Figure 6: Effect of lesions of medial prefrontal cortex (**A**: lesion location) on consolidation or expression of extinction memory in rats. Mean percentage of freezing behavior (**B**) during habituation, conditioning, extinction and post-extinction test. Animals with prefrontal lesions exhibited high levels of fear return (Extinction-Day 2: the first 3 trials). Adopted from Quirk et al. (1999) with permission.

prefrontal lesions (Fig. 6). In contrast, the CS preferentially activates the expression of extinction memory in control rats. Therefore, neurons within the medial prefrontal cortex may be required either for memory consolidation of extinction or for the expression of this memory.

Second, a week following the extinction of learned fear, mice can express, in the presence of a tone CS, either traumatic memory or extinction memory (Herry and Garcia, 2002). Likewise, rats also express either CS-related traumatic memory or CS-related extinction memory, when tested 24 hours after the extinction session (Milad and Quirk, 2002). Analyses of synaptic plasticity within the medial prefrontal cortex in the first study (in mice) indicated that the expression of traumatic memory was associated with a depression of prefrontal synaptic efficacy. In contrast, the expression of extinction memory was accompanied by a potentiation of prefrontal synaptic efficacy (Fig. 7). Most notably, each mouse exhibiting post-extinction expression of traumatic memory displayed a resistance to the conversion of "traumatic" synaptic plasticity to baseline levels or to a potentiation during the acquisition of the extinction, while mice exhibiting post-extinction expression of extinction memory displayed a suppression of traumatic plasticity during the extinction (see also Fig. 5). Furthermore, mice receiving tetanic thalamic stimulation before extinction developed LTP that simulated or facilitated post-extinction expression of extinction memory, while induction of prefrontal LTD was associated with reactivation of conditioned freezing (Fig. 7).

The role of prefrontal neuronal plasticity in the expression or consolidation of extinction memory has also been studied using recordings of spontaneous single-unit activity (Milad and Quirk, 2002). Recordings were performed during conditioning (tone CS–footshock US pairings), extinction (CS-alone presentations) and post-extinction CS re-exposure, known to activate either expression of traumatic memory (characterized by CS-associated freezing) or expression of extinction memory (characterized by the absence of CS-related freezing). Neither conditioning nor extinction elicited any changes in single-unit firing rate in the medial prefrontal cortex. However, during post-extinction CS re-exposure, neurons in the more ventral portion of the medial prefrontal cortex displayed either no change in firing rate or an increased firing rate. Expression of traumatic memory was associated with the absence of plasticity in prefrontal neuronal activity, whereas expression of extinction memory was accompanied by the potentiation of neuronal activity in this area. Moreover, tetanic stimulation of this prefrontal area (that induces potentiation of synaptic efficacy) during extinction was associated with an inhibition of post-extinction expression of traumatic memory.

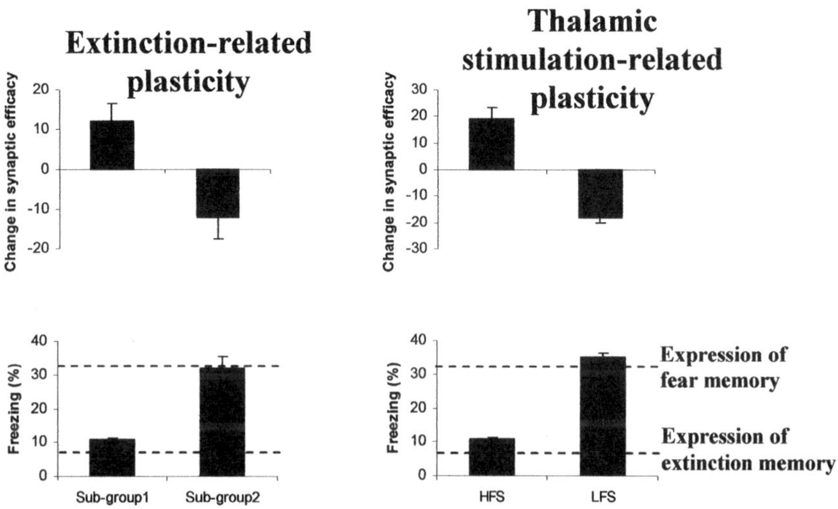

Figure 7: (Left) Extinction-related plasticity. Post-extinction CS-alone presentation (a week after extinction learning) can produce either increases of synaptic efficacy, associated with the expression of extinction (sub-group 1), or decreases of synaptic efficacy, associated with the expression of fear (sub-group 2), in medial prefrontal cortex. (Right) Thalamic stimulation mimics these effects. Following extinction, HFS-induced LTP inhibits expression of fear memory, while LFS-induced LTD produces post-extinction fear return. Adopted from Herry and Garcia (2002) with permission.

5. FUNCTIONAL CIRCUITS

Taken together, the above data show that the direction of the plasticity in neuronal activity in the medial prefrontal cortex may profoundly interact with memory consolidation of extinction and/or expression of this memory. The amygdalar-prefrontal loop has been recently implicated in this interaction (Garcia, 2002a, b).

Before any CS–no US learning, re-exposure to the CS-alone activates neurons in the amygdala that activate, in turn, circuits involved in the expression of traumatic memory (LeDoux, 2000). In addition, these amygdalar neurons produce, directly and/or indirectly, an inhibition of a class of pyramidal cells in the medial prefrontal cortex (Garcia et al., 1999), which is also characterized by a decrease in synaptic efficacy between these cells (Herry et al., 1999). As discussed above, this decrease may reflect the processing of the high certainty of the impending US. If the CS continues to be presented without the US, the modulatory effect of the amygdalar-prefrontal connection may either cease or be maintained. In the former case, synaptic efficacy in the medial prefrontal cortex returns to a baseline value (Herry and Garcia, 2002) or may even become higher than baseline levels (Herry et al., 1999). These changes in synaptic efficacy have no effect on the expression of traumatic memory before complete acquisition of CS–no US association, characterized by a complete extinction of conditioned fear responses. Although prefrontal cells can inhibit amygdalar neurons (Rosenkranz and Grace, 1999, 2001, 2002), the medial prefrontal cortex gains control over the amygdala only when amygdalar neurons lose their CS-evoked activity (for more details, see Garcia, 2002b), through repeated presentations of the CS without the US (Medina et al., 2002). In this case, post-extinction encounters with the CS may elicit the following prefrontal-amygdalar sequence: potentiation of prefrontal neuronal activity (Herry and Garcia, 2002; Milad and Quirk, 2002), which then blocks activation of amygdalar neurons involved in the expression of traumatic memory. Behaviorally, this corresponds to the expression of extinction memory. Cognitively, the potentiation of prefrontal neuronal activity may correspond to processing of the lack of the US. However, in the case of the maintenance of prefrontal depression during CS–no US learning, post-extinction presentation of the CS reactivates amygdalar neurons involved in the expression of traumatic memory. Cognitively, the depression of prefrontal synaptic efficacy may correspond to "irrational" processing of high certainty of the impending US despite initial learning of the CS–no US association.

There are several reasons to believe that changes in synaptic efficacy in the hippocampus and in hippocampal outputs to the medial prefrontal cortex are also involved in the prefrontal modulatory effect on post-extinction selection of expression of CS-related memory (either traumatic or extinction memory). First, lesion studies show that the hippocampus is involved in learning about the context in which trauma occurs (Kim and Fanselow, 1992; Phillips and LeDoux, 1992). Second, the hippocampus is not only involved in this contextual representation function, but hippocampal synaptic efficacy is also altered during re-exposure to the traumatic environment (Garcia et al., 1998). Most remarkable is that both hippocampal inputs (fimbria-fornix system) and outputs to the lateral septum displayed "traumatic" synaptic plasticity that outlasts extinction of the expression of traumatic memory (Garcia and Jaffard, 1996; Garcia et al., 1998). Third, anatomical and electrophysiological studies show that hippocampal neurons project to the medial prefrontal cortex (Condé et al., 1995; Jay and Witter, 1991; Jay et al., 1992). This hippocampal-prefrontal pathway is also known to display synaptic plasticity (Laroche et al, 2000). It is therefore possible that this pathway may display specific "traumatic" synaptic plasticity that does not produce resistance to extinction of conditioned fear responses but may even oppose to the consolidation or expression of extinction.

Does suppression of "traumatic" synaptic plasticity (a return to baseline levels; Herry and Garcia, 2002) or LTP (Herry and Garcia, 2002; but see also Milad and Quirk, 2002) within the medial prefrontal cortex associated with complete extinction of fear responding prevent further activation of expression of traumatic memory? Unfortunately, there is no study so far, which directly addresses this issue. Behaviorally, we know that if CS–US pairings and repeated CS-alone presentations take place in the same context, further presentations of the CS in a different environment preferentially activate the expression of traumatic memory (Frohardt et al., 2000; Corcoran and Maren, 2001). This phenomenon is known as renewal (Bouton and King, 1983). In contrast, if the CS is further presented in the extinction environment, the expression of extinction memory can be preferentially activated (Corcoran and Maren, 2001; Herry and Garcia, 2002). In the context of the circuits described above, and taking into account the role of the hippocampus in learning about environments, one can speculate that during post-extinction re-exposure to the conditioning context, the hippocampus may strongly inhibit the development of potentiation within the prefrontal cortex. First, behavioral studies combined with lesion approach have shown that muscimol infusion into the hippocampus disrupts the context-specific expression of extinction (Corcoran and Maren, 2001). More specifically, reversible inactivation of the hippocampus has no effect on the expression of traumatic memory in non-extinguished rats but blocked the expression of this memory in a context where extinguished rats should exhibit the renewal phenomenon (i.e. in a context different to the extinction environment). Second, hippocampal stimulation is known to produce both excitatory and inhibitory responses within the medial prefrontal cortex (Laroche et al., 1990; Thierry et al., 2000). The final balance between excitatory and inhibitory effects may control the direction of changes in plasticity in the medial prefrontal cortex as a function of environmental information (Garcia, 2002a). The "selected" direction modulates, in turn, amygdalar neurons involved in the expression of CS-related memories with, as a final result, the expression of "traumatic" memory or extinction memory or mixed expression.

6. CLINICAL IMPLICATIONS

Electrophysiological (Begic et al., 2001) and neuroimaging (Bremner et al., 1999; Pitman et al., 2001) data support the concept of an alteration of hippocampal functioning in relation to PTSD. Functional brain imaging data also argue for the involvement of the amygdala and the medial prefrontal cortex in the mechanisms underlying the expression of PTSD symptoms. Whereas neuronal activity increases in the amygdala during symptom provocation (Shin et al., 1997), the medial prefrontal cortex exhibits, on the contrary, decreased neuronal activity (Bremner et al., 1999). Although prefrontal data, both from animal and human studies, have yielded contradictory conclusions, more recent electrophysiological (Herry and Garcia, 2002) and neuroimaging (Fernandez et al., 2001) studies deserve, however, a little more consideration as potential tools for predicting treatment dropout. Indeed, following a complete elimination of PTSD-symptoms via an exposure therapy, follow-up data indicate that up to 40 % of treated individuals still display the original affective changes (Tarrier et al., 1999). If, as shown by Fernandez et al. (2001), extinction of PTSD symptoms is associated with a disappearance of depression in prefrontal neuronal activity, and if this plasticity signals low risk of symptom return, as shown in mice (Herry and Garcia, 2002), then post-treatment neuroimaging analyses might predict the long-term outcome of the treatment. This prediction would be defined by the restoration of amygdalar neuronal activity (Levin et al., 1999), and a restoration or potentiation of hippocampal and prefrontal neuronal activities (note: hippocampal neuronal activity is reduced in PTSD patients during symptom provocation; Bremner et al., 1999).

However, maintenance of depression of neuronal activity in the hippocampus and the medial prefrontal cortex, despite restoration of amygdalar activity and complete extinction of PTSD symptoms, would be predictive of treatment dropout. Hence, brain-mapping methods associated with PTSD treatment will not only provide insights into the basic mechanisms of this disorder but also improve diagnostics of potential relapses.

7. CONCLUSION

It is now well documented that extinction does not erase initial memory of conditioning but is an active learning process that can independently recruit mechanisms of long-term memory. Since the conditioning memory is also long-lasting, during post-extinction CS presentation, these two CS-related memories can compete in term of the expression. In addition, activation of each type of memory seems to occur via the amygdala. What are the factors or mechanisms which preferentially activate the expression of one or the other type of memory during further exposure to the CS? We propose that the plasticity in neuronal activity within the medial prefrontal cortex may be crucially involved in this selection. Plasticity within this structure depends, at least, on the plasticity in its hippocampal and amygdalar inputs. In humans, this plasticity may also be involved in "irrational" processing of insecurity (facilitating PTSD symptom return) or in mechanisms by which conditioned traumatic materials acquired the property of safety (inhibiting expression of traumatic memory, such as avoidance). Identifying the factors that contribute to the development of beneficial plasticity within the medial prefrontal cortex will enhance our understanding on the role of this structure in the extinction of conditioned fear responses.

REFERENCES

Bechara A, Damasio H, Damasio AR, Lee GP (1999) Different contributions of the human amygdala and ventromedial prefrontal cortex to decision-making. J Neurosci 19:5473-5481.

Begic D, Hotujac L, Jokic-Begic N (2001) Electroencephalographic comparison of veterans with combat-related post-traumatic stress disorder and healthy subjects. Int J Psychophysiol 40:167-172.

Bouton ME (1993) Context, time, and memory retrivial in the interference paradigms of Pavlovian learning. Psychol Bull 114:80-99.

Bouton ME, King DA (1983) Contextual control of the extinction of conditioned fear: tests for the associative value of the context. J Exp Psychol Anim Behav Process 9:248-265.

Bremner JD, Randall P, Scott TM, Bronen RA, Seibyl JP, Southwick SM, Delaney RC, McCarthy G, Charney DS, Innis RB (1995) MRI-based measurement of hippocampal volume in patients with combat-related posttraumatic stress disorder. Am J Psychiatry 152:973-981.

Bremner JD, Krystal JH, Charney DS, Southwick SM (1996) Neural Mechanisms in dissociative amnesia for childhood abuse: relevance to the current controversy surrounding the "false memory syndrome". Am J Psychiatry 153 (7 Suppl):71-82.

Bremner JD, Innis RB, Ng CK, Staib LH, Salomon RM, Bronen RA, Duncan J, Southwick SM, Krystal JH, Rich D, Zubal G, Dey H, Soufer R, Charney DS (1997) Positron emission tomography measurement of cerebral metabolic correlates of yohimbine administration in combat-related posttraumatic stress disorder. Arch Gen Psychiatry 54:246-254.

Bremner JD, Staib LH, Kaloupek D, Southwick SM, Soufer R, Charney DS (1999) Neural correlates of exposure to traumatic pictures and sound in Vietnam combat veterans with and without posttraumatic stress disorder: a positron emission tomography study. Biol Psychiatry 45:806-816.

Condé F, Maire-Lepoivre E, Audinat E, Crepel F (1995) Afferent connections of the medial frontal cortex of the rat. II. Cortical and subcortical afferents. J Comp Neurol 352:567-593.

Corcoran KA, Maren S (2001) Hippocampal inactivation disrupts contextual retrieval of fear memory after extinction. J Neurosci 21:1720-1726.

Devinsky O, Morrell MJ, Vogt BA (1995) Contributions of anterior cingulate cortex to behaviour. Brain 118:279-306.
Dias R, Aggleton JP (2000) Effects of selective excitotoxic prefrontal lesions on acquisition of nonmatching- and matching-to-place in the T-maze in the rat: differential involvement of the prelimbic-infralimbic and anterior cingulate cortices in providing behavioural flexibility. Eur J Neurosci 12:4457-4466.
Dunnett SB, Nathwani F, Brasted PJ (1999) Medial prefrontal and neostriatal lesions disrupt performance in an operant delayed alternation task in rats. Behav Brain Res 106:13-28.
Eichenbaum H, Clegg RA, Feeley A (1983) Reexamination of functional subdivisions of the rodent prefrontal cortex. Exp Neurol 79:434-451.
Fernandez M, Pissiota A, Frans O, von Knorring L, Fischer H, Fredrikson M (2001) Brain function in a patient with torture related post-traumatic stress disorder before and after fluoxetine treatment: a positron emission tomography provocation study. Neurosci Lett 297:101-104.
Frohardt RJ, Guarraci FA, Bouton ME (2000) The effects of neurotoxic hippocampal lesions on two effects of context after fear extinction. Behav Neurosci 114:227-240.
Frysztak RJ, Neafsey EJ (1994) The effect of medial frontal cortex lesions on cardiovascular conditioned emotional responses in the rat. Brain Res 643:181-193.
Garcia R (2002a) Postextinction of Conditioned Fear: Between Two CS-Related Memories. Learn Mem 9:361-363.
Garcia R (2002b) Stress, synaptic plasticity, and psychopathology. Rev Neurosci 13:195-208.
Garcia R, Jaffard R (1996) Changes in synaptic excitability in the lateral septum associated with contextual and auditory fear conditioning in mice. Eur J Neurosci 8:809-815.
Garcia R, Tocco G, Baudry M, Thompson RF (1998) Exposure to a conditioned aversive environment interferes with long-term potentiation induction in the fimbria-CA3 pathway. Neuroscience 82:139-145.
Garcia R, Vouimba RM, Baudry M, Thompson RF (1999) The amygdala modulates prefrontal cortex activity relative to conditioned fear. Nature 402:294-296.
Gemmell C, O'Mara SM (1999) Medial prefrontal cortex lesions cause deficits in a variable-goal location task but not in object exploration. Behav Neurosci 113:465-474.
Gewirtz JC, Falls WA, Davis M (1997) Normal conditioned inhibition and extinction of freezing and fear-potentiated startle following electrolytic lesions of medical prefrontal cortex in rats. Behav Neurosci 111:712-726.

Hauser MD (1999) Perseveration, inhibition and the prefrontal cortex: a new look. Curr Opin Neurobiol 9:214-222.

Herry C, Garcia R (2002) Prefrontal cortex long-term potentiation, but not long-term depression, is associated with the maintenance of extinction of learned fear in mice. J Neurosci 22:577-583.

Herry C, Vouimba RM, Garcia R (1999) Plasticity in the mediodorsal thalamo-prefrontal cortical transmission in behaving mice. J Neurophysiol 82:2827-2832.

Holson RR (1986) Mesial prefrontal cortical lesions and timidity in rats. I. Reactivity to aversive stimuli. Physiol Behav 37:221-230.

Jaskiw GE, Weinberger DR (1992) Ibotenic acid lesions of medial prefrontal cortex augment swim-stress-induced locomotion. Pharmacol Biochem Behav 41:607-609.

Jay TM, Witter MP (1991) Distribution of hippocampal CA1 and subicular efferents in the prefrontal cortex of the rat studied by means of anterograde transport of Phaseolus vulgaris-leucoagglutinin. J Comp Neurol 313:574-586.

Jay TM, Thierry AM, Wiklund L, Glowinski J (1992) Excitatory Amino Acid Pathway from the Hippocampus to the Prefrontal Cortex. Contribution of AMPA Receptors in Hippocampo-prefrontal Cortex Transmission. Eur J Neurosci 4:1285-1295.

Kety SS, Schmidt CF (1945) The determination of cerebral blood flow in man by the use of nitrous oxide in low concentrations. Am J Physiol 143:53-66.

Kim JJ, Fanselow MS (1992) Modality-specific retrograde amnesia of fear. Science 256:675-677.

Laroche S, Davis S, Jay TM (2000) Plasticity at hippocampal to prefrontal cortex synapses: dual roles in working memory and consolidation. Hippocampus 10:438-446.

Laroche S, Jay TM, Thierry AM (1990) Long-term potentiation in the prefrontal cortex following stimulation of the hippocampal CA1/subicular region. Neurosci Lett 114:184-190.

LeDoux JE (1993) Emotional memory systems in the brain. Behav Brain Res 58: 69-79.

LeDoux JE (2000) Emotion circuits in the brain. Annu Rev Neurosci 23:155-184.

Levin P, Lazrove S, van der Kolk B (1999) What psychological testing and neuroimaging tell us about the treatment of Posttraumatic Stress Disorder by Eye Movement Desensitization and Reprocessing. J Anxiety Disord 13:159-172.

Liberzon I, Taylor SF, Amdur R, Jung TD, Chamberlain KR, Minoshima S, Koeppe RA, Fig LM (1999) Brain activation in PTSD in response to trauma-related stimuli. Biol Psychiatry 45:817-826.

Medina JF, Christopher Repa J, Mauk MD, LeDoux JE (2002) Parallels between cerebellum- and amygdala-dependent conditioning. Nat Rev Neurosci 3:122-131.

Milad MR, Quirk GJ (2002) Neurons in medial prefrontal cortex signal memory for fear extinction. Nature 420:70-74.

Morgan MA, LeDoux JE (1995) Differential contribution of dorsal and ventral medial prefrontal cortex to the acquisition and extinction of conditioned fear in rats. Behav Neurosci 109:681-688.

Morrow BA, Elsworth JD, Rasmusson AM, Roth RH (1999) The role of mesoprefrontal dopamine neurons in the acquisition and expression of conditioned fear in the rat. Neuroscience 92:553-564.

Peri T, Ben-Shakhar G, Orr SP, Shalev AY (2000) Psychophysiologic assessment of aversive conditioning in posttraumatic stress disorder. Biol Psychiatry 47:512-519.

Phillips RG, LeDoux JE (1992) Differential contribution of amygdala and hippocampus to cued and contextual fear conditioning. Behav Neurosci 106:274-285.

Pitman RK, Shin LM, Rauch SL (2001) Investigating the pathogenesis of posttraumatic stress disorder with neuroimaging. J Clin Psychiatry 62 (Suppl 17):47-54.

Prins A, Kaloupek DG, Keane TM (1995) Psychophysiological evidence for autonomic arousal and startle in traumatized adult populations. In: Neurobiological and Clinical Consequences of Stress: From Normal Adaptation to PTSD (Friedman MJ, Charney DS, and Deutch AY, eds), pp 291-314, Raven Press, New York.

Quirk GJ, Russo GK, Barron JL, Lebron K (2000) The role of ventromedial prefrontal cortex in the recovery of extinguished fear. J Neurosci 20:6225-6231.

Rescorla RA (2001) Retraining of extinguished Pavlovian stimuli. J Exp Psychol Anim Behav Process 27:115-124.

Rosenkranz JA, Grace AA (1999) Modulation of basolateral amygdala neuronal firing and afferent drive by dopamine receptor activation in vivo. J Neurosci 19:11027-11039.

Rosenkranz JA, Grace AA (2001) Dopamine attenuates prefrontal cortical suppression of sensory inputs to the basolateral amygdala of rats. J Neurosci 21:4090-4103.

Rosenkranz JA, Grace AA (2002) Cellular mechanisms of infralimbic and prelimbic prefrontal cortical inhibition and dopaminergic modulation of basolateral amygdala neurons in vivo. J Neurosci 22:324-337.

Rothbaum BO, Schwartz AC (2002) Exposure therapy for posttraumatic stress disorder. Am J Psychother 56:59-75.

Shin LM, Kosslyn SM, McNally RJ, Alpert NM, Thompson WL, Rauch SL, Macklin ML, Pitman RK (1997) Visual imagery and perception in posttraumatic stress disorder. A positron emission tomographic investigation. Arch Gen Psychiatry 54:233-241.

Stein MB, Koverola C, Hanna C, Torchia MG, McClarty B (1997) Hippocampal volume in women victimized by childhood sexual abuse. Psychol Med 27:951-959.

Tarrier N, Sommerfield C, Pilgrim H, Humphreys L (1999) Cognitive therapy or imaginal exposure in the treatment of post-traumatic stress disorder. Twelve-month follow-up. Br J Psychiatry 175:571-575.

Thierry AM, Gioanni Y, Degenetais E, Glowinski J (2000) Hippocampo-prefrontal cortex pathway: anatomical and electrophysiological characteristics. Hippocampus 10:411-419.

Vermetten E, Bremner JD (2002) Circuits and systems in stress. II. Applications to neurobiology and treatment in posttraumatic stress disorder. Depress Anxiety 16:14-38.

Vouimba RM, Garcia R, Baudry M, Thompson RF (2000) Potentiation of conditioned freezing following dorsomedial prefrontal cortex lesions does not interfere with fear reduction in mice. Behav Neurosci 114:720-724.

Zubieta JK, Chinitz JA, Lombardi U, Fig LM, Cameron OG, Liberzon I (1999) Medial frontal cortex involvement in PTSD symptoms: a SPECT study. J Psychiatr Res 33:259-264.

Chapter 7

STRESS AND PREFRONTAL CORTICAL DYSFUNCTION IN THE RAT

Kazushige Mizoguchi
Pharmacology Department, Central Research Laboratories, Tsumura and Company, 3586 Yoshiwara, Ami-machi, Inashiki-gun, Ibaraki 300-1192, Japan

Keywords: Stress, prefrontal cortex, hippocampus, dopamine, serotonin, acetylcholine, working memory, reference memory, delayed-alternation task, T-maze, rotarod, depression, rat.

Abstract: Although the mechanism responsible for cognitive deficits or a depressive state in stress-related neuropsychiatric disorders has not been fully elucidated, dopaminergic or serotonergic dysfunction in the prefrontal cortex (PFC) is thought to be involved. In rats, the mesoprefrontal dopaminergic system, in particular, is activated in response to acute stress, whereas chronic stress reduces dopaminergic transmission in the PFC, causing working memory impairment through a D_1 receptor mechanism. However, chronic stress does not affect reference memory, which is attributed to hyperactivity of hippocampal cholinergic transmission. In addition, chronic stress induces a depressive behavioral state, caused by a reduction in dopaminergic and serotonergic transmission in the PFC. These findings provide important information for the understanding of the mechanisms underlying PFC dysfunction in stress-related neuropsychiatric disorders. In this chapter, I mainly discuss the mechanisms of the chronic stress-induced PFC dysfunction in rats, with reference to our recent findings.

1. INTRODUCTION

Exposure to stress is known to precipitate or exacerbate many neuropsychiatric disorders such as depression, Parkinson's disease, schizophrenia, and others (Schwab and Zieper, 1965; Mazure, 1995). All these disorders involve prefrontal cortical (PFC) dysfunction that causes a working memory deficit (Mattes, 1980; Weinberger et al., 1986; Deutch,

1993; Fibiger, 1995). Several antidepressants increase dopamine (DA) levels in the PFC (Tanda et al., 1994), and raising the DA level in patients with Parkinson's disease with L-3,4-dihydroxyphenylalanine improves their working memory deficit (Lange et al., 1992). These findings suggest that a reduced dopaminergic transmission in the PFC is responsible for the working memory deficits in neuropsychiatric disorders. In addition, serotonergic dysfunction is thought to be associated with depressive states and suicidal behavior, and several serotonergic abnormalities have been found in the PFC of depressives. For example, decreases in the serotonin (5-hyroxytryptamine; 5-HT) concentration, the 5-HT transporter (Owens and Nemeroff, 1994; Austin et al., 2002), and 5-HT responsiveness (Mann et al., 1996) have been shown in depressives. Abnormal density (Yates et al., 1990; Zanardi et al., 2001) or mRNA editing (Gurevich et al., 2002) of 5-HT_1 or 5-HT_2 receptors have also been demonstrated. Moreover, serotonergic antidepressants such as selective 5-HT reuptake inhibitors are effective for the clinical relief of several depressive symptoms (Lane et al., 1995; Montgomery, 1996; Perry et al., 1996). These findings suggest that serotonergic dysfunction is also involved in the pathophysiology of neuropsychiatric disorders.

A large number of animal studies indicate that exposure to acute or chronic stress can alter the activity of neurotransmitter systems in the brain that affect behavior. In particular, the PFC shows vulnerability to acute stress (Abercrombie et al., 1989). For example, exposure to acute stress in monkeys or rats has been shown to produce working memory impairment, which can be blocked by agents that prevent the increase in DA turnover (Arnsten and Goldman-Rakic, 1986; Murphy et al., 1996b) or that antagonize DA receptors (Murphy et al., 1996a; Arnsten and Goldman-Rakic, 1998), indicating a hyperdopaminergic mechanism. These stress responses are compatible with the observation that overstimulation of D_1 receptors in the PFC impairs working memory performance (Zahrt et al., 1997). On the other hand, a reduction in PFC dopaminergic function or a blockade of DA receptors in the PFC of monkeys or rats impairs working memory performance (Brozoski et al., 1979; Simon et al., 1980; Bubser and Schmidt, 1990), indicating a hypodopaminergic mechanism. These findings have led to the hypothesis that there is a narrow range of optimal DA receptor stimulation for correct PFC function (Zahrt et al., 1997; Arnsten and Goldman-Rakic, 1998), which indicates an important role for DA modulation of the neural processes within the PFC in working memory performance. In addition, a number of studies have shown that serotonergic neurons in the PFC are also sensitive to acute stress. For example, 5-HT release in the PFC is greatly increased in response to acute stress, raising an implication in emotional behavior relating to anxiety or fear (Yoshioka et al., 1995). Nevertheless, the mechanisms of the PFC dysfunction, induced by chronic stress and thought to be involved in the pathogenesis of neuropsychiatric disorders, are not well understood. Recently, we found that chronic stress induces a reduction in dopaminergic and serotonergic transmission in the rat PFC, which can cause working memory impairment or a depressive state (Mizoguchi et al., 2000, 2002a,b). In this chapter, I mainly discuss the mechanisms of the chronic stress-induced PFC dysfunction in rats, referring to our recent findings.

2. EXPERIMENTAL METHODS

2.1 Stress Exposure

Exposure of rats to stress has often been used to investigate the pathogenesis of stress-related neuropsychiatric disorders including depression. Experimental methods used to induce such stress have included forced running stress (Hatotani et al., 1977; Kitayama et al., 1994), restraint (Cancela et al., 1991; Albonetti and Farabollini, 1993), learned helplessness (Danysz et al., 1988), or unpredictable stress (Biagini et al., 1993; Papp et al., 1994). The results described in this chapter were obtained using water immersion and restraint stress (Mizoguchi et al., 2000, 2001a,b, 2002a,b). Briefly, the rats were placed in a stress cage made of wire net, and immersed for 2 h to the level of the xiphoid process in a water bath maintained at 21°C. The rats were subjected to this form of stress once a day for 4 weeks (chronic stress). To avoid the acute influence of the last stress exposure, and to ensure the long-term consequences of the chronic stress, the rats were allowed a 10-day recovery period.

2.2 Evaluation of Working Memory

Delayed-alternation tasks are widely considered to be particularly sensitive in demonstrating working memory impairment after lesion of the PFC in all species of mammals (Markowitsch and Pritzel, 1977). In rats, this task, usually performed in a T-maze (Moran, 1993; Zahrt et al., 1997), is one of the most common methods for evaluating spatial working memory performance associated with the PFC (Van Haaren et al., 1985).

As shown in Figure 1, working memory performance in rats was examined using the T-maze apparatus (Mizoguchi et al., 2000). Briefly, the animals' food allowance was maintained at about 90% of the normal intake until the end of the T-maze test. The rats were initially habituated to the T-maze apparatus [dimensions: stem arm, 75 length (L) x 13 width (W) x 20 height (H) cm; two branch arms, 50 (L) x 13 (W) x 20 (H) cm each] for 4 days until

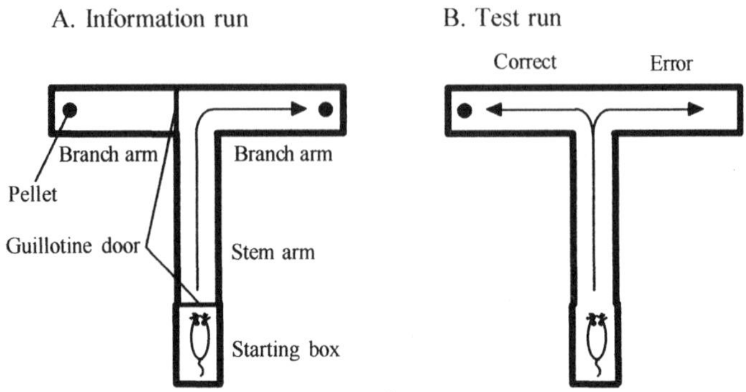

Figure 1: Schematic illustration of T-maze apparatus.

they were readily eating food pellets placed at the end of each branch arm. After habituation, the rats were trained for the delayed-alternation task. In the first trial (information run in Fig. 1), each rat was placed in the starting box of the stem arm, with the condition that one branch arm was blocked by a guillotine door, and the rat was rewarded for entering either branch arm. Subsequently, the rat was returned to the starting box, with the condition that both branch arms were open, and was rewarded (test run in Fig. 1). Thereafter, rats were only rewarded when they entered the branch arm that was not chosen previously (correct choice in test run, win-shift strategy). At the end of the training trial, the rats demonstrating a rate of >90% correct choices were selected, exposed to stress for 4 weeks, and allowed a 10-day recovery period. Then, following the information run, each rat was subjected to several delay times (0, 10, 30, and 60 sec), and was allowed a test run. Ten trials for each delay time were performed. The number of errors per test run was recorded.

2.3 Evaluation of Depressive Behavioral State

Although several methods, e.g. a forced swimming test (Porsolt et al., 1977) or a tail suspension test (Steru et al., 1985), are used to evaluate a depressive behavioral state of rats or mice, we selected the rotarod test, because it shows a higher sensitivity in evaluating the effects of serotonergic antidepressants such as trazodone, mianserin, and clomipramine than the forced swimming test (Morimoto and Kito, 1994). Furthermore, this test does not involve any habituation or adaptation to water, which could cause problems in measuring the duration of immobility as seen in the forced swimming test. The experimental procedure of the rotarod test has been described elsewhere (Dunhan and Miya, 1957; Ahmad and Nicholls, 1990; Mizoguchi et al., 2002a,b). Briefly, the time (sec) that the rats remained on a rotating rod (10 cm diameter; 7 rpm; Muromachi Kikai, Tokyo, Japan) was recorded automatically in each condition. In addition, muscle strength was evaluated by a traction test (Kuribara et al., 1977; Mizoguchi et al., 2002a,b), and spontaneous locomotor activity was measured over a period of 5 min using Animex apparatus (ANIMEX AUTO, MK-110; Muromachi Kikai; Mizoguchi et al., 2002a, b).

3. EFFECT OF STRESS ON PFC FUNCTION

In this section, I mainly describe the effects of chronic stress on the PFC function, with regard to working memory performance, the depressive state, and dopaminergic and serotonergic transmission.

3.1 Working Memory

In the delayed-alternation task (Fig. 2), chronic stress did not affect performance accuracy under the no delay condition, suggesting that chronic stress did not affect motivation, motor function, and previously acquired long-term memory for efficient rewarding in the task, i.e. reference memory.

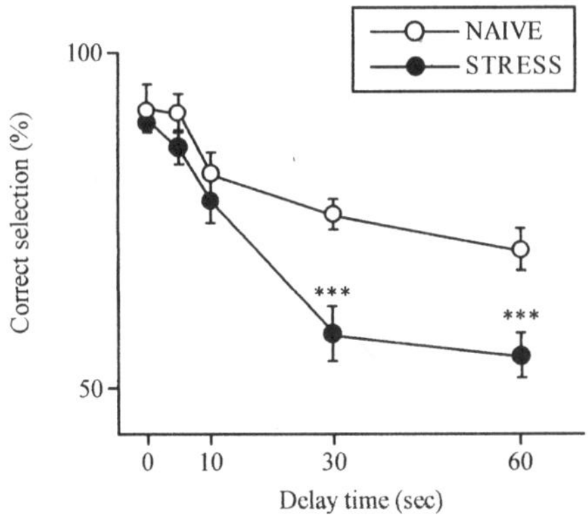

Figure 2: Effect of chronic stress on working memory performance in the delayed-alternation task with no delay (0 sec) or several delay conditions (5, 10, 30, and 60 sec). Each point is the mean ± S.E.M. of 10 rats per group. Asterisks indicate a significant difference from naive nonstressed rats, $***p < 0.001$.

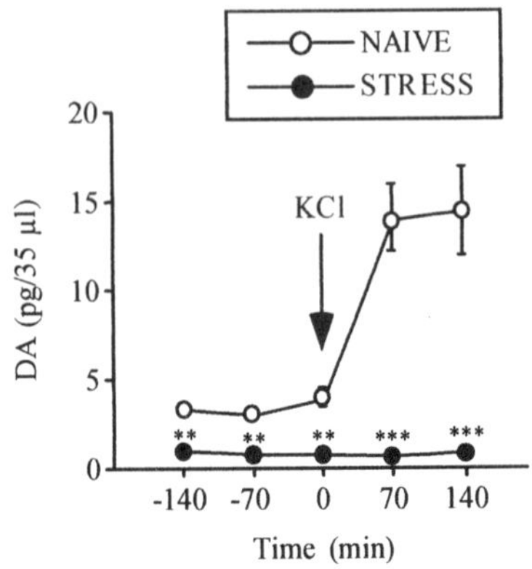

Figure 3: Effect of chronic stress on DA release in the PFC under a basal or KCl-stimulated condition. The basal DA release in chronically stressed rats was much lower than that in naive nonstressed rats. A remarkable increase in DA release was observed in naive nonstressed rats on the perfusion of high KCl (100 mM). In contrast, no KCl-induced increase in DA release was observed in chronically stressed rats. Each point is the mean ± S.E.M. of 8 or 9 rats per group. Asterisks indicate a significant difference from naive nonstressed rats, $^{**}p < 0.01$, $^{***}p < 0.001$. Data are derived from Mizoguchi et al. (2000) with permission.

However, chronic stress produced a marked decrease in performance accuracy accompanied by prolongation of the delay time. This indicates that chronic stress impairs the maintenance of a novel short-term memory, in other words, working memory, which is the term applied to the aspect of memory responsible for the recall of information immediately after it has been presented.

3.2 Dopaminergic Transmission in the PFC

It has been shown that DA has a beneficial influence on the spatial working memory function of the rat PFC (Simon, 1980; Bubser and Schmidt, 1990). In addition, dopamine release in the PFC shows vulnerability to acute stress (Abercrombie et al., 1989). Considering these findings, it is conceivable that chronic stress may affect dopaminergic transmission in the PFC. Indeed, a microdialysis study revealed that chronic stress greatly decreased the DA release in the PFC (Fig. 3). This finding supports the chronic stress-induced working memory impairment (Fig. 2).

3.3 Tissue Concentration of DA and its Metabolites

We examined the time course of changes in the tissue concentrations of DA and its metabolites, dihydroxyphenylacetic acid (DOPAC) and homovanillic acid (HVA), in the PFC during the chronic stress session and the recovery period (Table 1). This neurochemical study revealed that short-term stress (1 week) increased the concentrations of DOPAC and HVA, indicating an increase in DA turnover and thus supporting the concept that a hyperdopaminergic mechanism is responsible for acute stress-induced working memory impairment (Arnsten and Goldman-Rakic, 1998). However, these increases were not observed in the long-term stressed (4 weeks) rats, and the stressed and recovered rats showed a marked decrease in the concentrations of DA, DOPAC, and HVA, supporting the reduced DA release (Fig. 3). Thus, during the chronic stress session and the recovery period, the dopaminergic neurons innervating in the PFC are initially activated, but subsequently inactivated. Also, the chronic stress-induced reduction in DA release (Fig. 3) is thought to be related to a decrease in the DA stored at the synaptosomes. As regards the mechanism of this decrease, it seemed possible that a reduction in DA synthesis in the area projecting to the PFC, i.e. the ventral tegmental area (VTA), resulted in the decrease. However, chronic stress did not affect the number of DA-containing neurons, that was identified by a tyrosine hydroxylase (a rate-limiting enzyme in catecholamine biosynthesis)-immunohistochemistry, in the VTA (Mizoguchi et al., 2000), suggesting that the decrease in DA was not due to the reduction in DA synthesis in the originating area.

The factors that modulate the stress-induced dopaminergic dysfunction in the PFC are unknown. It is possible that some stress-sensitive neurotransmitters or hormones contribute to the dysfunction. For example, GABA (Hegarty and Vogel, 1995), norepinephrine (Gresch et al., 1993), and glutamate (Jedema and Moghaddam, 1994) can modulate the activity of

Table 1. Time course of changes in tissue concentrations of DA and its metabolites in the PFC during a 4-week stress session and a 10-day recovery period

		Stress period				
	Naive	1 W	2 W	3 W	4 W	4 W + 10 D
DA	1150.0 ± 113.5	1342.1 ± 168.8	1435.8 ± 118.5	719.7 ± 104.5*	585.3 ± 70.2***	696.0 ± 63.8**
DOPAC	80.2 ± 7.8	133.4 ± 11.8**	112.6 ± 10.4	105.3 ± 14.5	75.8 ± 11.9	55.7 ± 5.2***
HVA	57.5 ± 4.6	89.6 ± 8.6**	88.5 ± 7.5**	70.9 ± 9.1	56.8 ± 8.5	39.1 ± 5.2**

At week 1 of the stress period, both DOPAC and HVA levels were significantly increased. The HVA level was still increased at week 2. At weeks 3 and 4, the DA levels were significantly decreased, and the DOPAC and HVA levels returned to initial values. At day 10 of the recovery period, the levels of DA, DOPAC and HVA were significantly decreased. Each value is the mean ± S.E.M (nanograms per gram of tissue) of 9 rats per group. Asterisks indicate a significant difference from the value of naive nonstressed rats for each monoamine, $*p < 0.05$, $**p < 0.01$, $***p < 0.001$. W, week; D, day. Data are derived from Mizoguchi et al. (2000) with permission.

dopaminergic neurons during the stress response. Considering that the long-term stress period was required for the expression of stress-induced dopaminergic dysfunction, as indicated in Table 1, another factor with long-term effects, such as glucocorticoids, may be implicated in the dysfunction. It is well known that glucocorticoid secretion is potently activated by exposure to stress. Mesencephalic and mesoprefrontal dopaminergic neurons have glucocorticoid receptors (Härfstrand et al., 1986; Diorio et al., 1993), and the administration of glucocorticoids can modify DA metabolism (Versteeg et al., 1983; Rothschild et al., 1985) and increase the DA release in the PFC (Imperato et al., 1989). Conversely, suppression of endogenous glucocorticoids by adrenalectomy reduces dopaminergic transmission in the nucleus accumbens (Piazza et al., 1996). A similar reduction is also observed in the PFC (K. Mizoguchi, unpublished observations). Thus, glucocorticoids can positively regulate the dopaminergic activity in the PFC. Also, Sapolsky et al. (1984) have demonstrated that chronic stress downregulates glucocorticoid receptors in the brain. Furthermore, chronic stress attenuates the response to exogenous glucocorticoids on the glucocorticoid negative feedback (Haracz et al., 1988; Young et al., 1990; Mizoguchi et al., 2001a). Since the attenuated response to glucocorticoids is considered to reduce the actions of glucocorticoids at the feedback sites, including the PFC (Diorio et al., 1993), the reduction in glucocorticoid-induced actions in the PFC may be involved in the chronic stress-induced dopaminergic dysfunction. Indeed, attenuated feedback is one of the most consistent findings in patients with depression (Carroll et al., 1981; Kalin et al., 1982; Holsboer, 1983; Arana et al., 1985) and is thought to contribute to some of the symptoms of depression (Steckler et al., 1999).

3.4 Relationship between Working Memory Impairment and Reduced Dopaminergic Transmission

Although the types of DA receptors, involved in the working memory function in the PFC, remain to be determined, several studies have identified the importance of D_1 receptors (Sawaguchi and Goldman-Rakic, 1991; Seamans et al., 1995). As shown in Figure 4, the stress-induced working memory impairment was ameliorated by D_1 receptor stimulation in the PFC with the D_1 receptor agonist SKF 81297 in a dose-dependent manner, suggesting that this impairment is caused by reduced D_1 receptor stimulation. The reversal of the SKF 81297 response due to the D_1 receptor antagonist, SCH 23390, confirms the hypothesis of action at the D_1 receptor rather than nonspecific drug actions. These results are consistent with the observation that a low dose (e.g. 100 ng/kg) of SKF 81297 ameliorates spatial working memory impairment in aged monkeys with naturally occurring DA depletion (Arnsten et al., 1994; Cai and Arnsten, 1997). In addition, the doses of SKF 81297 that had a partial or sufficient ameliorative effect on the stress-induced working memory impairment (i.e. 1 or 10 ng, Fig. 4) were within the extent of the proper dose-response relationship of D_1 receptor agonist for the working memory performance (Zahrt et al., 1997).

D_1 receptors are located postsynaptically on cortical neurons (Tassin et al., 1978, 1982), and the decrease in the DA level in the PFC induced by electrolytic lesion upregulates their density in the PFC (Tassin et al., 1982). Conversely, long-term administration of the D_1 receptor agonist SKF 38393 downregulates their density in the PFC (Gambarana et al., 1995). In the chronically stressed rats, D_1 receptors in the PFC were upregulated (Mizoguchi et al., 2000), confirming the reduction in DA transmission at the receptor level, and the ameliorative effects of SKF 81297. Thus, the chronic stress-induced dopaminergic dysfunction appears to mainly occur at presynaptic sites of the dopaminergic neurons in the PFC.

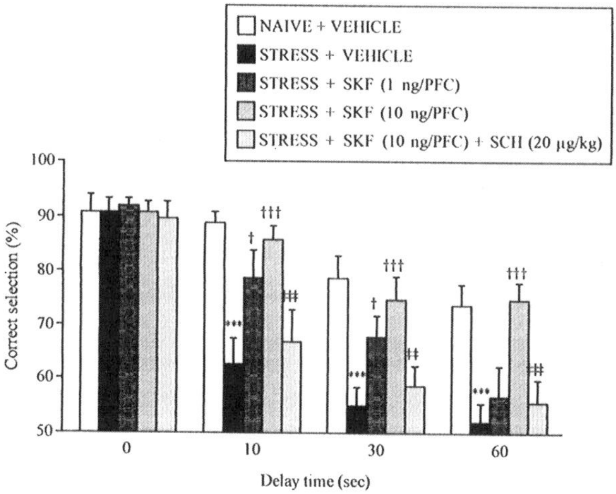

Figure 4: Effect of bilateral infusions of SKF 81297 (SKF) into the PFC on chronic stress-induced working memory impairment in the delayed-alternation task with no delay (0 sec) or several delay conditions (10, 30, and 60 sec). SKF 81297 was infused at day 10 of the recovery period. The task was performed immediately after the infusions. Intra-PFC infusion of 1 ng SKF 81297 in chronically stressed rats caused a significant amelioration of the impaired performance at 10 and 30 sec delay conditions, but not at 60 sec. At 10 ng, SKF 81297 sufficiently and significantly ameliorated the impaired performance in all delay conditions. These ameliorative effects of 10 ng SKF 81297 were reversed by pretreatment with 20 μg/kg (i.p.) SCH 23390 (SCH). Each column is the mean ± S.E.M. of 10 rats per group. Asterisks indicate a significant difference from naive nonstressed and vehicle (saline)-infused rats, $^{***}p < 0.001$; daggers, a significant difference from stressed and vehicle-infused rats, $^{\dagger}p < 0.05$, $^{\dagger\dagger\dagger}p < 0.001$; double daggers, a significant difference from stressed and SKF (10)-infused rats, $^{\ddagger\ddagger}p < 0.01$, $^{\ddagger\ddagger\ddagger}p < 0.001$. Data are derived from Mizoguchi et al. (2000) with permission.

Taken together, the results from the series of experiments indicate that chronic stress induces working memory impairment through a D_1 receptor-mediated hypodopaminergic mechanism in the PFC. Thus, acute stress impairs the working memory function through a hyperdopaminergic mechanism (Arnsten and Goldman-Rakic, 1998), whereas chronic stress impairs this function through a hypodopaminergic mechanism.

3.5 Cholinergic Transmission in the Hippocampus

Several reports have shown that the hippocampal cholinergic system is also involved in the formation and maintenance of short-term working memory, or retention and retrieval processes in long-term reference memory (Pope et al., 1987; Murai et al., 1995; Izquierdo et al., 1998). Therefore, we examined the effects of chronic stress on acetylcholine (ACh) release in the rat hippocampus. Consequently, chronic stress increased the hippocampal cholinergic transmission in response to stimuli (Mizoguchi et al., 2001b), suggesting that hippocampal cholinergic neurons become hypersensitive with chronic stress. Gonzalez and Pazos (1992) have also shown that chronic immobilization stress causes an increase in the density of muscarinic ACh receptors in the hippocampus. Thus, the hippocampal cholinergic system may be activated by chronic stress. As regards the significance of this cholinergic hyperactivity, it is possible that the hyperactivity occurs in compensation for the reduced dopaminergic function in the PFC. Alternatively, it may be involved in the maintenance of long-term reference memory that is not impaired by chronic stress (Fig. 2).

3.6 Depressive State

The rotarod is an established test for evaluating pharmacological actions of psychotropic agents such as skeletal muscle relaxants, anticonvulsants, and antidepressants in the central or peripheral nervous system (Dunhan and Miya, 1957). Morimoto and Kito (1994) have shown that this test is useful to evaluate the antidepressive effects of serotonergic and adrenergic antidepressants. As shown in Figure 5, chronic stress impaired the rotarod performance, concomitant with unchanged traction performance and locomotor activity, suggesting that the impaired rotarod performance is not due to muscle relaxation or motor dysfunction. As antidepressants increase the riding time on the rotating rod in normal rats (Morimoto and Kito, 1994), the impaired rotarod performance suggests a depressive behavioral state.

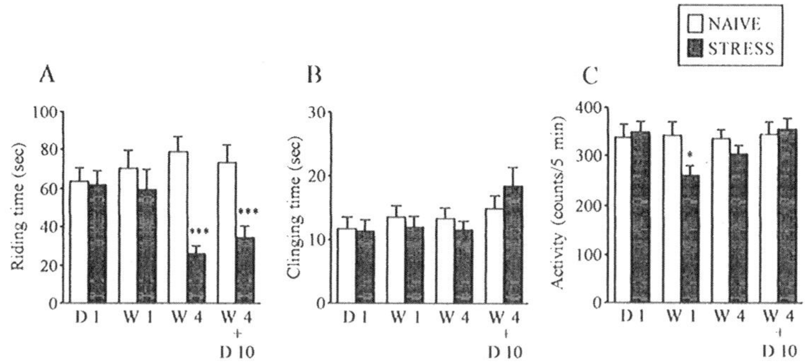

Figure 5: Time course of changes in behavioral performance during a 4-week stress session and a 10-day recovery period. A, rotarod performance; B, traction performance; C, locomotor activity. In the rotarod test, the riding time was not affected at day 1 or week 1. However, the riding time was significantly decreased at week 4 and the recovery. In the traction test, the clinging time did not change during the stress and recovery periods. In the locomotor test, the activity was significantly decreased at week 1, but was not affected at day 1, week 4, or the recovery. Each column is the mean ± S.E.M. of 10 or 16 rats per group. Asterisks indicate a significant difference from naive nonstressed rats in each time point, $*p < 0.05$, $***p < 0.001$. D, day; W, week. Data are derived from Mizoguchi et al. (2002a) with permission.

3.7 Serotonergic Transmission in the PFC

The dopaminergic and serotonergic systems in the PFC are thought to be involved in the depressive state. It has already been indicated that chronic stress reduces dopaminergic transmission in the PFC (Fig. 3). As shown in Figure 6, chronic stress also reduces serotonergic transmission in the PFC. These findings support the chronic stress-induced depressive state.

Regarding the mechanism of the serotonergic dysfunction, some stress-sensitive hormones, such as glucocorticoids, may be involved. For example, long-term treatment of rats with glucocorticoids has been shown to decrease 5-HT responsiveness (Karten et al., 1999) or the efficiency of 5-HT transport into synaptosomes (Slotkin et al., 1996). These mechanisms may be involved in serotonergic dysfunction.

Several reports have shown that dopaminergic and serotonergic neurons in the PFC are closely associated. For example, dopaminergic activity in the PFC can be positively regulated by 5-HT_1 receptor-stimulation (Wedzony et al., 1996; Sakaue et al., 2000) or by an increase in 5-HT levels via local application of a 5-HT reuptake inhibitor in the PFC (Tanda et al., 1994; Matsumoto et al., 1999). Since chronic stress reduces serotonergic transmission in the PFC (Fig. 6), it is possible that chronic stress-induced serotonergic dysfunction in the PFC causes the dopaminergic dysfunction.

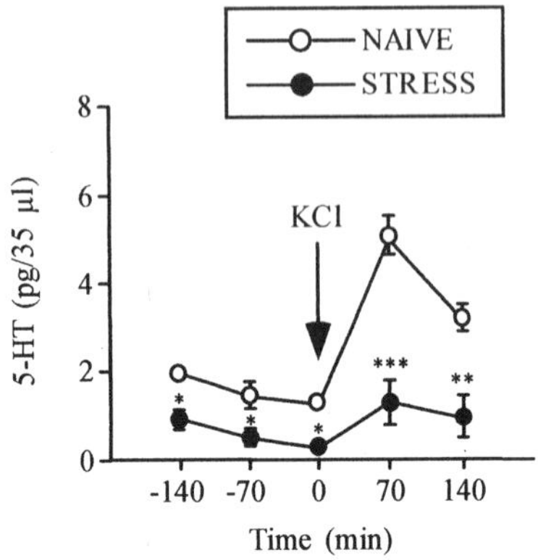

Figure 6: Effect of chronic stress on 5-HT release in the PFC under a basal or KCl-stimulated condition. The basal 5-HT release in chronically stressed rats was significantly lower than that in naive nonstressed rats. A remarkable increase in 5-HT release was observed in naive nonstressed rats on perfusion of high KCl (100 mM). However, this increase was significantly attenuated in chronically stressed rats. Each point is the mean ± S.E.M. of 8 or 9 rats per group. Asterisks indicate a significant difference from naive nonstressed rats, $*p < 0.05$, $**p < 0.01$, $***p < 0.001$. Data are derived from Mizoguchi et al. (2002a) with permission.

3.8 Relationship between Depressive State and Reduced Dopaminergic and Serotonergic Transmission

To clarify the involvement of the reduction in serotonergic and dopaminergic transmission in the PFC in the chronic stress-induced depressive state, the beneficial effects of the serotonergic antidepressant, trazodone and those of SKF 81297 were examined. As shown in Figure 7, the chronic stress-induced impairment of rotarod performance was ameliorated by trazodone in a dose-dependent manner. Since the traction performance and locomotor activity were not affected by trazodone, the ameliorative effect of trazodone does not appear to be caused by an effect on muscle strength or motor function. Considering that trazodone has a 5-HT reuptake inhibitory action (Clements-Jewery et al., 1980), it is postulated that chronic stress induces a depressive state caused by a reduction in serotonergic transmission in the PFC (Fig. 6). Since trazodone also has a 5-HT_2 receptor antagonistic action, dysfunction of 5-HT_2 receptors may be attributed to the depressive state. Indeed, chronic forced swim stress (Takao et al., 1995) or long-term treatment with adrenocorticotropic hormone or glucocorticoids (Kuroda et al., 1992, 1993) increases the density of 5-HT_2 receptors in the rat frontal cortex.

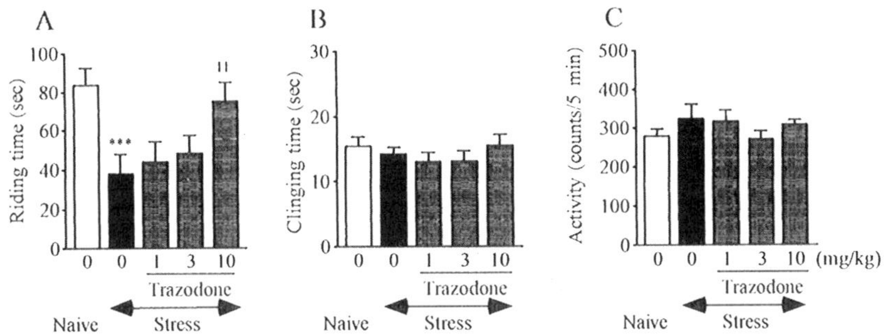

Figure 7: Effect of trazodone on chronic stress-induced behavioral deficit. A, rotarod performance; B, traction performance; C, locomotor activity. Trazodone was administered (p.o.) daily during a 10-day recovery period. The behavioral analyses were performed twenty-four hours after the last administration. In the rotarod test, 1 or 3 mg/kg trazodone did not ameliorate the chronic stress-induced decrease in riding time, but 10 mg/kg trazodone significantly ameliorated it. Neither chronic stress nor any dose of the drug tested affected the clinging time in the traction test or locomotor activity. Each column is the mean ± S.E.M. of 10 rats per group. Asterisks indicate a significant difference from naive nonstressed and vehicle (0 mg/kg)-treated rats, $^{***}p < 0.001$; daggers, a significant difference from stressed and vehicle-treated rats, $^{\dagger\dagger}p < 0.01$. Data are derived from Mizoguchi et al. (2002a) with permission.

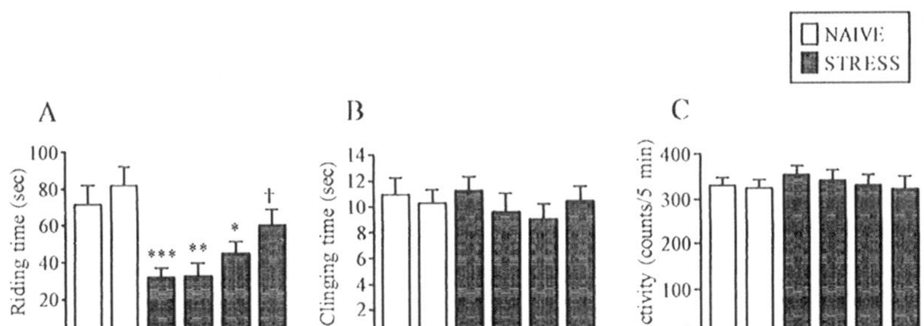

Figure 8: Effect of bilateral infusions of SKF 81297 into the PFC on chronic stress-induced behavioral deficit. A, rotarod performance; B, traction performance; C, locomotor activity. SKF 81297 was infused at day 10 of the recovery period. The behavioral analyses were performed immediately after the infusions. In the rotarod test, although intra-PFC infusion of 1 or 10 ng SKF 81297 did not significantly ameliorate the chronic stress-induced decrease in riding time, the extent of the significant difference in stressed rats receiving these doses, versus naive nonstressed and vehicle (0 ng/PFC)-infused rats, was decreased. At 100 ng, SKF 81297 significantly ameliorated the decrease. Neither chronic stress nor any dose of SKF 81297 tested affected the clinging time in the traction test or locomotor activity. Each column is the mean ± S.E.M. of 10 rats per group. Asterisks indicate a significant difference from naive nonstressed and vehicle-infused rats, $^{*}p < 0.05$, $^{**}p < 0.01$, $^{***}p < 0.001$; a dagger, a significant difference from stressed and vehicle-infused rats, $^{\dagger}p < 0.05$. Data are derived from Mizoguchi et al. (2002b) with permission.

The SKF 81297 infusion study (Fig. 8) revealed that the chronic stress-induced depressive state was ameliorated by intra-PFC infusions of SKF 81297 in a dose-dependent manner. Since traction performance and locomotor activity were not affected by SKF 81297 infusions, the ameliorative effect of SKF 81297 appears to be caused by an intra-PFC mechanism, rather than an effect on muscle strength or motor function. Considering that chronic stress reduces dopaminergic transmission in the PFC (Fig. 3), these results suggest that the chronic stress-induced depressive state is caused by a D_1 receptor-mediated hypodopaminergic mechanism in the PFC. This hypothesis is supported by a previous report showing that desensitization of D_1 receptors in the PFC produced a behavioral deficit in an animal model of depression (Gambarana et al., 1995).

Taken together, the results from the series of experiments indicate that chronic stress induces the depressive state via dopaminergic and serotonergic dysfunction in the PFC.

With regard to the dose-response relationship of SKF 81297 on the behavior of rats, intra-PFC infusion of 10 ng SKF 81297 ameliorated the chronic stress-induced working memory impairment (Fig. 4). In contrast, intra-PFC infusion of 100 ng SKF 81297 has been shown to impair working memory performance in rats (Zahrt et al., 1997). However, the dose of SKF 81297 which significantly ameliorated the stress-induced depressive state (i.e. 100 ng) was outside the range of the beneficial dose-response relationship of SKF 81297 for working memory performance. Furthermore, intra-PFC infusion of 100 ng SKF 81297 in the naive nonstressed rats did not affect the rotarod performance (Fig. 8). These findings suggest that the involvement of the dopaminergic system in the PFC differs in working memory performance and the depressive state. The contribution of the serotonergic system to the depressive state may explain this difference.

4. CLINICAL RELEVANCE

The response of the central nervous system to stress is often critical to an organism's adaptation to a stressful environment. In humans, however, an over-response to stress can be maladaptive, resulting in the expression or exacerbation of many neuropsychiatric disorders. Such disorders often exhibit a number of features that indicate abnormal functioning of the PFC (Mattes, 1980; Weinberger et al., 1986; Deutch, 1993; Fibiger, 1995). The influence of dopaminergic and serotonergic systems on PFC function in neuropsychiatric disorders remains unclear, but some observations support the idea that these systems play a role in the pathogenesis of several. For example, Dolan et al. (1994) have provided evidence that neuropsychological symptoms in depression, including cognitive deficits, are associated with profound hypometabolism, involving the medial PFC in particular. Similarly, it has been reported that both bipolar and unipolar depressives are characterized by decreases in cerebral blood flow and the rate of glucose metabolism in the PFC (Drevets et al., 1997). Furthermore, agents such as bupropion that enhance DA transmission have been successfully used as antidepressants (Calabrese and Markovitz, 1991). Various other serotonergic antidepressants, such as fluoxetine, clomipramine, and imipramine, also increase the release of DA as well as 5-HT in the rat PFC (Tanda et al., 1994), indicating that the PFC is a target site of antidepressants. These findings implicate a reduction in dopaminergic and serotonergic transmission in the PFC in the pathogenesis of depression. A similar association has been suggested in patients with Parkinson's disease who suffer from depression (Cummings, 1992; Deutch, 1993). Depression occurs in large populations of patients with Parkinson's disease, and such patients have greater frontal lobe dysfunction and a more frequent occurrence of reduced dopaminergic function than non-depressed patients with the same disease. In addition, negative or defect symptoms of schizophrenia, which include not only impaired working memory but also low volition, social withdrawal, and impaired insight and judgment, are suspected to be attributed to reduced dopaminergic transmission in the PFC (Knable and Weinberger, 1997). Thus, dopaminergic and serotonergic systems in the PFC are thought to play an important role in many neuropsychic activities, including working memory impairment and depressive states. Although other neuropsychic activities of chronically stressed rats have not been examined here, it is likely that the significance of dopaminergic and serotonergic dysfunction induced by chronic stress is not restricted to working memory impairment and the depressive state, but involves the disruption of other neuropsychic activities in stress-related neuropsychiatric disorders.

5. CONCLUDING REMARKS

In this chapter, I have described the evidence that exposure to chronic stress in rats is sufficient to produce PFC dysfunction. Thus, dopaminergic and serotonergic neurons in the PFC show vulnerability to chronic stress, which causes working memory impairment or a depressive state, whereas cholinergic neurons in the hippocampus show resistance to this stress, which may be involved in the maintenance of reference memory (Fig. 9). These findings will lead to a deeper understanding of the mechanisms underlying the pathogenesis of stress-related neuropsychiatric disorders.

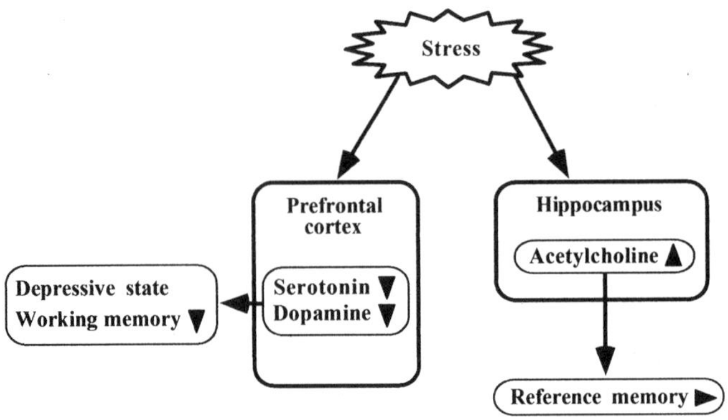

Figure 9: Mechanism of PFC hypofunction and hippocampal hyperfunction induced by chronic stress. Closed triangles indicate decrease, increase, impairment, or maintenance.

REFERENCES

Abercrombie ED, Keefe KA, di Frischia DS, Zigmond MJ (1989) Differential effect of stress on in vivo dopamine release in striatum, nucleus accumbens, and medial frontal cortex. J Neurochem 52:1655-1658.

Ahmad B, Nicholls PJ (1990) Development of tolerance to the CNS effects of aminoglutethimide in mice. Eur J Pharmacol 182:237-244.

Albonetti ME, Farabollini F (1993) Effect of single and repeated restraint on the social behavior of male rats. Physiol Behav 53:937-942.

Arana GW, Baldessarini RJ, Ornsteen M (1985) The dexamethasone suppression test for diagnosis and prognosis in psychiatry. Arch Gen Psychiatry 42:1193-1204.

Arnsten AFT, Goldman-Rakic PS (1986) Reversal of stress-induced delayed response deficits in rhesus monkeys by clonidine and naloxone. Soc Neurosci Abstr 12:1464.

Arnsten AFT, Goldman-Rakic PS (1998) Noise stress impairs prefrontal cortical cognitive function in monkeys. Arch Gen Psychiatry 55:362-368.

Arnsten ATF, Cai JX, Murphy BL, Goldman-Rakic PS (1994) Dopamine D_1 receptor mechanisms in the cognitive performance of young adult and aged monkeys. Psychopharmacology (Berl) 116:143-151.

Austin M, Whitehead R, Edgar C, Janosky J, Lewis D (2002) Localized decrease in serotonin transporter-immunoreactive axons in the prefrontal cortex of depressed subjects committing suicide. Neuroscience 114:807.

Biagini G, Pich EM, Carani C, Marrama P, Gustafsson JÅ, Fuxe K, Agnati LF (1993) Indol-pyruvic acid, a tryptophan ketoanalogue, antagonizes the endocrine but not the behavioral effects of repeated stress in a model of depression. Biol Psychiatry 33:712-719.

Brozoski TJ, Brown RM, Rosvold HE, Goldman PS (1979) Cognitive deficit caused by regional depletion of dopamine in prefrontal cortex of rhesus monkey. Science 205:929-932.

Bubser M, Schmidt WJ (1990) 6-Hydroxydopamine lesion of the rat prefrontal cortex increases locomotor activity, impairs acquisition of delayed alternation tasks, but does not affect uninterrupted tasks in the radial maze. Behav Brain Res 37:157-168.

Cai JX, Arnsten ATF (1997) Dose-dependent effects of the dopamine D_1 receptor agonists A77636 or SKF 81297 on spatial working memory in aged monkeys. J Pharmacol Exp Ther 282:1-7.

Calabrese JR, Markovitz PJ (1991) Treatment of depression. New pharmacologic approaches. Prim Care 18:421-433.

Cancela LM, Rossi S, Molina VA (1991) Effect of different restraint schedules on the immobility in the forced swim test: modulation by an opiate mechanism. Brain Res Bull 26:671-675.

Carroll BJ, Feinberg M, Greden JF, Tarika J, Albala AA, Haskett RF, James NM, Kronfol Z, Lohr N, Steiner M, de Vigne JP, Young E (1981) A specific laboratory test for the diagnosis of melancholia: standardization, validation and clinical utility. Arch Gen Psychiatry 38:15-22.

Clements-Jewery S, Robson PA, Chidley LJ (1980) Biochemical investigations into the mode of action of trazodone. Neuropharmacology 19:1165-1173.

Cummings JL (1992) Depression and Parkinson's disease: a review. Am J Psychiatry 149:443-454.

Danysz W, Plaznik A, Kostowski W, Malatynska E, Jarbe TU, Hiltunen AJ, Archer T (1988) Comparison of desipramine, amitriptyline, zimelidine and alaproclate in six animal models used to investigate antidepressant drugs. Pharmacol Toxicol 62:42-50.

Deutch AY (1993) Prefrontal cortical dopamine systems and the elaboration of functional corticostriatal circuits: implications for schizophrenia and Parkinson's disease. J Neural Transm 91:197-221.

Diorio D, Viau V, Meaney MJ (1993) The role of the medial prefrontal cortex (cingulate gyrus) in the regulation of hypothalamic-pituitary-adrenal responses to stress. J Neurosci 13:3839-3847.

Dolan RJ, Bench CJ, Brown RG, Scott LC, Frackowiak RS (1994) Neuropsychological dysfunction in depression: the relationship to regional cerebral blood flow. Psychol Med 24:849-857.

Drevets WC, Price JL, Simpson Jr JR, Todd RD, Reich T, Vannier M, Raichle ME (1997) Subgenual prefrontal cortex abnormalities in mood disorders. Nature 386:824-827.

Dunhan NW, Miya TS (1957) A note on a simple apparatus for detecting neurological deficit in rats and mice. J Am Pharm Ass 46:208-209.

Fibiger HC (1995) Neurobiology of depression: focus on dopamine. Adv Biochem Psychopharmacol 49:1-17.

Gambarana C, Ghiglieri O, Graziella de Montis M (1995) Desensitization of the D_1 dopamine receptors in rats reproduces a model of escape deficit reverted by imipramine, fluoxetine and clomipramine. Prog Neuropsychopharmacol Biol Psychiatry 19:741-755.

Gonzalez AM, Pazos A (1992) Modification of muscarinic acetylcholine receptors in the rat brain following chronic immobilization stress: an autoradiographic study. Eur J Pharmacol 223:25-31.

Gresch PJ, Sved AF, Zigmond MJ, Finlay JM (1993) Local application of desipramine increases basal and stress-induced dopamine efflux in medial prefrontal cortex. Soc Neurosci Abstr 19:1828.

Gurevich I, Tamir H, Arango V, Dwork AJ, Mann JJ, Schmauss C (2002) Altered editing of serotonin 2C receptor pre-mRNA in the prefrontal cortex of depressed suicide victims. Neuron 34:349-356.

Haracz JL, Minor TR, Wilkins JN, Zimmermann EG (1988) Learned helplessness: an experimental model of the DST in rats. Biol Psychiatry 23:388-396.

Härfstrand A, Fuxe K, Cintra A, Agnati LF, Zini I, Wikstrom AC, Okret S, Yu ZY, Goldstein M, Steinbusch H, Verhofstad A, Gustafsson JÅ (1986) Glucocorticoid receptor immunoreactivity in monoaminergic neurons of rat brain. Proc Natl Acad Sci USA 83:9779-9783.

Hatotani N, Nomura J, Yamaguchi T, Kitayama I (1977) Clinical and experimental studies on the pathogenesis of depression. Psychoneuroendocrinology 2:115-130.

Hegarty AA, Vogel WH (1995) The effect of acute and chronic diazepam treatment on stress-induced changes in cortical dopamine in the rat. Pharmacol Biochem Behav 52:771-778.

Holsboer F (1983) The dexamethasone suppression test in depressed patients: clinical and biochemical aspects. J Steroid Biochem 19:251-257.

Imperato A, Puglisi-Allegra S, Casolini P, Zocchi A, Angelucci L (1989) Stress-induced enhancement of dopamine and acetylcholine release in limbic structure: role of corticosterone. Eur J Pharmacol 165:337-339.

Izquierdo I, Izquierdo LA, Barros DM, Mello e Souza T, de Souza MM, Quevedo J, Rodrigues C, Sant'Anna MK, Madruga M, Medina JH (1998) Differential involvement of cortical receptor mechanisms in working, short-term and long-term memory. Behav Pharmacol 9:421-427.

Jedema HP, Moghaddam B (1994) Glutametargic control of dopamine release during stress in the rat prefrontal cortex. J Neurochem 63:785-788.

Kalin NH, Weiler SJ, Shelton SE (1982) Plasma ACTH and cortisol concentration before and after dexamethasone. Psychiatry Res 7:87-92.

Karten YJ, Nair SM, van Essen L, Sibug R, Joels M (1999) Long-term exposure to high corticosterone levels attenuates serotonin responses in rat hippocampal CA1 neurons. Proc Natl Acad Sci USA 96:13456-13461.

Kitayama I, Nakamura S, Yaga T, Murase S, Nomura J, Kayahara T, Nakao K (1994) Degeneration of locus coeruleus axons in stress-induced depression model. Brain Res Bull 35:573-580.

Knable MB, Weinberger DR (1997) Dopamine, the prefrontal cortex and schizophrenia. J Psychopharmacol 11:123-131.

Kuribara H, Higuchi Y, Tadokoro S (1977) Effects of central depressants on rota-rod and traction performance in mice. Jpn J Pharmacol 27:117-126.

Kuroda Y, Mikuni M, Nomura N, Takahashi K (1993) Differential effect of subchronic dexamethasone treatment on serotonin-2 and ß-adrenergic receptors in the rat cerebral cortex and hippocampus. Neurosci Lett 155:195-198.

Kuroda Y, Mikuni M, Ogawa T, Takahashi K (1992) Effect of ACTH, adrenalectomy and the combination treatment on the density of 5-HT_2 receptor binding site in neocortex of rat forebrain and 5-HT_2 receptor-mediated wet-dog shake behaviors. Psychopharmacology (Berl) 108:27-32.

Lane R, Baldwin D, Preskorn S (1995) The SSRIs: advantages and differences. J Psychopharmacol 92 (suppl):163-178.

Lange KW, Robbins TW, Marsden CD, James M, Owen AM, Paul GM (1992) L-dopa withdrawal in Parkinson's disease selectively impairs cognitive performance in tests sensitive to frontal lobe dysfunction. Psychopharmacology (Berl) 107:394-404.

Mann JJ, Malone KM, Diehl DJ, Perel J, Cooper TB, Mintun MA (1996) Demonstration in vivo of reduced serotonin responsivity in the brain of untreated depressed patients. Am J Psychiatry 153:174-182.

Markowitsch HJ, Pritzel M (1977) Comparative analysis of prefrontal learning functions in rats, cats, and monkeys. Psychol Bull 84:817-837.

Matsumoto M, Togashi H, Mori K, Ueno K, Miyamoto A, Yoshioka M (1999) Characterization of endogenous serotonin-mediated regulation of dopamine release in the rat prefrontal cortex. Eur J Pharmacol 383:39-48.

Mattes JA (1980) The role of frontal lobe dysfunction in childhood hyperkinetics. Comp Psychiatry 21:358-369.

Mazure CM (1995) Does stress cause psychiatric illness? In: Progress in Psychiatry (Spiegel D, ed), pp 270-298, American Psychiatric Press, Washington, DC.

Mizoguchi K, Yuzurihara M, Ishige A, Sasaki H, Chui DH, Tabira T (2000) Chronic stress induces impairment of spatial working memory because of prefrontal dopaminergic dysfunction. J Neurosci 20:1568-1574.

Mizoguchi K, Yuzurihara M, Ishige A, Sasaki H, Chui DH, Tabira T (2001a) Chronic stress differentially regulates glucocorticoid negative feedback response in rats. Psychoneuroendocrinology 26:443-459.

Mizoguchi K, Yuzurihara M, Ishige A, Sasaki H, Tabira T (2001b) Effect of chronic stress on cholinergic transmission in rat hippocampus. Brain Res 915:108-111.

Mizoguchi K, Yuzurihara M, Ishige A, Sasaki H, Tabira T (2002a) Chronic stress impairs rotarod performance in rats: implications for depressive state. Pharmacol Biochem Behav 71:79-84.

Mizoguchi K, Yuzurihara M, Ishige A, Sasaki H, Tabira T (2002b) Dopamine-receptor stimulation in the prefrontal cortex ameliorates stress-induced rotarod impairment. Pharmacol Biochem Behav 72:723-728.

Montgomery SA (1996) Selective serotonin re-uptake inhibitors in long-term treatment of depression. In: Selective serotonin re-uptake inhibitors, 2nd edition: advances in basic research and clinical practice (Feighner JP, Boyer WF, eds), pp 35-62, John Wiley & Sons, Chichester.

Moran PM (1993) Differential effects of scopolamine and mecamylamine on working and reference memory in the rats. Pharmacol Biochem Behav 45:533-538.

Morimoto S, Kito G (1994) Rotarod method in young rats and the antidepressive effect: is the rotarod method capable of evaluating antidepressive effects? Folia Pharmacol Jpn 104:39-49.

Murai S, Saito H, Masuda Y, Odashima J, Itoh T (1995) AF64A disrupts retrieval processes in long-term memory of mice. Neuroreport 6:349-352.

Murphy BL, Arnsten ATF, Goldman-Rakic PS, Roth RH (1996a) Increased dopamine turnover in the prefrontal cortex impairs spatial working memory performance in rats and monkeys. Proc Natl Acad Sci USA 96:1325-1329.

Murphy BL, Arnsten ATF, Jentsch JD, Roth RH (1996b) Dopamine and spatial memory in rats and monkeys: pharmacological reversal of stress-induced impairment. J Neurosci 16:7768-7775.

Owens MJ, Nemeroff CB (1994) Role of serotonin in the pathophysiology of depression: focus on the serotonin transporter. Clin Chem 40:288-295.

Papp M, Klimek V, Willner P (1994) Effects of imipramine on serotonergic and beta-adrenergic receptor binding in a realistic animal model of depression. Psychopharmacology (Berl) 114:309-314.

Perry PJ, Alexander B, Liskow B (1996) Selective serotonin reuptake inhibitors. In: Psychotropic drug handbook, 7th edition, pp 167-178, American Psychiatric Press, Washington, DC.

Piazza PV, Barrot M, Rougé-Pont F, Marinelli M, Maccari S, Abrous N, Simon H, le Moal M (1996) Suppression of glucocorticoid secretion and antipsychotic drugs have similar effects on the mesolimbic dopaminergic transmission. Proc Natl Acad Sci USA 93:15445-15450.

Pope CN, Ho BT, Wright AA (1987) Neurochemical and behavioral effects of N-ethyl-acetylcholine aziridinium chloride in mice. Pharmacol Biochem Behav 26:365-371.

Porsolt RD, Le Pichon M, Jalfre M (1977) Depression: a new animal model sensitive to antidepressant treatments. Nature 266:730-732.

Rothschild AJ, Langlais PJ, Schatzberg AF, Miller MM, Saloman MS, Lerbinger JE, Cole JO, Bird ED (1985) The effect of a single acute dose of dexamethasone on monoamine and metabolite levels in the rat brain. Life Sci 36:2491-2505.

Sakaue M, Somboonthum P, Nishihara B, Koyama Y, Hashimoto H, Baba A, Matsuda T (2000) Postsynaptic 5-hydroxytryptamine(1A) receptor activation increases in vivo dopamine release in rat prefrontal cortex. Br J Pharmacol 129:1028-1034.

Sapolsky RM, Krey LC, McEwen BS (1984) Stress down-regulates corticosterone receptors in a site-specific manner in the brain. Endocrinology 114:287-292.

Sawaguchi T, Goldman-Rakic PS (1991) D_1 dopamine receptors in prefrontal cortex: involvement in working memory. Science 251:947-950.

Schwab RS, Zieper I (1965) Effects of mood, motivation, stress, and alertness on the performance in Parkinson's disease. Psychiat Neurol (Basel) 150:345-357.

Seamans JK, Floresco SB, Phillips AG (1995) Selective impairment on a delay radial arm task following local administration of a selective D_1, but not a D_2, antagonist into the prefrontal cortex. Soc Neurosci Abstr 21:1942.

Simon H, Scatton B, le Moal M (1980) Dopaminergic A-10 neurons are involved in cognitive functions. Nature 286:150-151.

Slotkin TA, McCook EC, Ritchie JC, Seidler FJ (1996) Do glucocorticoids contribute to the abnormalities in serotonin transporter expression and function seen in depression? An animal model. Biol Psychiatry 40:576-584.

Steckler T, Holsboer F, Reul JM (1999) Glucocorticoids and depression. Baillière's Best Pract Res Clin Endocrinol Metab 13:597-614.

Steru L, Chermat R, Thierry B, Simon P (1985) The tail suspension test: a new method for screening antidepressants in mice. Psychopharmacology (Berl) 85:367-370.

Takao K, Nagatani T, Kitamura Y, Kawasaki K, Hayakawa H, Yamawaki S (1995) Chronic forced swim stress of rats increases frontal cortical 5-HT_2

receptors and the wet-dog shakes they mediate, but not frontal cortical beta-adrenoceptors. Eur J Pharmacol 294:721-726.

Tanda G, Carboni E, Frau R, Di Chiara G (1994) Increase of extracellular dopamine in the prefrontal cortex: a trait of drugs with antidepressant potential? Psychopharmacology (Berl) 115:285-288.

Tassin JP, Bockaert J, Blanc G, Stinus L, Thierry AM, Lavielle S, Prémont J, Glowinski J (1978) Topographical distribution of dopaminergic innervation and dopaminergic receptors of the anterior cerebral cortex of the rats. Brain Res 154:241-251.

Tassin JP, Simon H, Hervé D, Blanc G, le Moal M, Glowinski J, Bockaert J (1982) Non-dopaminergic fibers may regulate dopamine-sensitive adenylate cyclase in the prefrontal cortex and the nucleus accumbens. Nature 259:696-698.

Van Haaren F, De Bruin JP, Heinsbroek RP, Van de Poll NE (1985) Delayed spatial response alternation: effects of delay-interval duration and lesions of the medial prefrontal cortex on response accuracy of male and female Wistar rats. Behav Brain Res 18:41-49.

Versteeg DHG, van Zoest I, de Kloet ER (1983) Acute changes in dopamine metabolism in the medial basal hypothalamus following adrenalectomy. Experientia 40:112-114.

Wedzony K, Mackowiak M, Fijal K, Golembiowska K (1996) Ipsapirone enhances the dopamine outflow via 5-HT_{1A} receptors in the rat prefrontal cortex. Eur J Pharmacol 305:73-78.

Weinberger DR, Berman KF, Zec RF (1986) Physiologic dysfunction of dorsolateral prefrontal cortex in schizophrenia, I: regional cerebral blood flow evidence. Arch Gen Psychiatry 43:114-124.

Yates M, Leake A, Candy JM, Fairbairn AF, McKeith IG, Ferrier IN (1990) 5HT_2 receptor changes in major depression. Biol Psychiatry 27:489-496.

Yoshioka M, Matsumoto M, Togashi H, Saito H (1995) Effects of conditioned fear stress on 5-HT release in the rat prefrontal cortex. Pharmacol Biochem Behav 51:515-519.

Young EA, Akana S, Dallman MF (1990) Decreased sensitivity to glucocorticoid fast feedback in chronically stressed rats. Neuroendocrinology 51:536-542.

Zahrt J, Taylor JR, Mathew RG, Arnsten AFT (1997) Supranormal stimulation of D_1 dopamine receptors in the rodent prefrontal cortex impairs spatial working memory performance. J Neurosci 17:8528-8535.

Zanardi R, Artigas F, Moresco R, Colombo C, Messa C, Gobbo C, Smeraldi E, Fazio F (2001) Increased 5-hydroxytryptamine-2 receptor binding in the frontal cortex of depressed patients responding to paroxetine treatment: a positron emission tomography scan study. J Clin Psychopharmacol 21:53-58.

Chapter 8

STRATEGY SWITCHING AND THE RAT PREFRONTAL CORTEX

Matthijs G. P. Feenstra and Jan P. C. de Bruin
Netherlands Institute for Brain Research, Amsterdam, The Netherlands

Keywords: Flexibility, reversal learning, instrumental learning, extinction, dopamine, noradrenaline, D_1-receptor.

Abstract: The prefrontal cortex (PFC) is important for cognitive flexibility - the PFC appears to be involved whenever a novel strategy has to be adopted or a switch from an old to a new strategy is needed. A serial reversal (Acquisition-Reversal-Extinction) task was used to study when and how the rat PFC is involved in various phases of this task. Each phase requires switching of behavioral strategies. Experimental data are summarized that suggest involvement of the medial PFC in the formation of new action–outcome relationships during the first stages of the serial reversal phase. This corresponds to a similar involvement of the dopaminergic innervation of the medial PFC. No involvement was detected during the discrimination phase, the later reversal phase, or the extinction phase, corroborating the view that PFC is actively involved in the formation of representations of goals and the means to achieve them (Miller, 2000). The ARE-task is taken as a basis to review the possible role of the PFC and its dopaminergic innervation in the various phases and the underlying processes of formation and adaptation of goal-directed actions.

1. INTRODUCTION – PFC FUNCTION

In the introduction of one of the oldest experimental studies on the prefrontal cortex (PFC), four functions are stated to '...have been assigned to the frontal lobes - movement, inhibition, attention and association' (Franz, 1907). After reporting on his experiments in cats and monkeys, Franz concludes that '...the frontal lobes are concerned in normal and daily associational processes and that through them we are enabled to form habits and, in general, to learn.' A central position in associational processes and learning is still (or again) considered essential in the PFC function. Later reviewers stressed the importance of PFC in inhibition and short-term memory (Mishkin, 1964; Goldman-Rakic, 1987; see also Roberts, 1998), but Fuster (1997) argued that these functions, together with motor set, underlie the 'superordinate prefrontal function of temporally organizing goal-directed behavior'. The PFC is now thought to form and actively maintain representations of 'goals and the means to achieve them' (Miller, 2000) and to learn and use rules or strategies (Wise et al, 1996). Thus, in the end, it provides the means for cognitive control and flexibility of behavior (Miller and Cohen, 2001). This is relevant in view of the outcome of the pioneer imaging studies performed twenty years ago, which indicated that it "participates in any form of structured brain work a subject can do when awake" (Roland, 1984). All these reviews were based primarily on human and non-human primate data. However, rodent PFC has similar anatomical (Kolb, 1984; Uylings and van Eden, 1990; Groenewegen and Uylings, 2000) and functional (Kolb, 1984, 1990; de Bruin, 1994; Kesner, 2000, 2002; Brown and Bowman, 2002) characteristics and provides excellent opportunities to study PFC function in the flexibility of behavior on the molecular, cellular, and circuit levels (see also Chapters 1and 2). Flexibility and cognitive control of goal-directed behavior require many components; i.e. cue retrieval, maintaining and shifting of attentional set, working memory, inhibition and excitation of responses, outcome monitoring, *etc.* In the present chapter, we will review the evidence that PFC is involved in these functions and suggest that catecholamine afferents may be important for these functions.

2. PFC OF THE RAT

The rat PFC consists of a medial and a lateral part, which are in the pregenual frontal pole connected by an orbital part (Fig. 1). Medial subareas are IL (infralimbic), PL (prelimbic), AC (anterior congulate), and Fr2 (frontal 2). Lateral subareas are AIv and AId (agranular insular ventral and dorsal). Orbital subareas are MO, VO, and VLO (medial, ventral, and ventrolateral orbital). Comprehensive descriptions of morphology and anatomy of the rat PFC have been provided by Kolb (1984), Uylings and van Eden (1990), Groenewegen and Berendse (1994), and Groenewegen and Uylings (2000). The PFC is part of parallel corticostriatal thalamic circuits (Groenewegen and Berendse, 1994; Groenewegen and Uylings, 2000). Through this organization, the PFC may control striatal output, but is itself also under striatal control (Fig. 2). Both are under strong modulatory control of monoamine afferents. The PFC receives a relatively dense innervation of

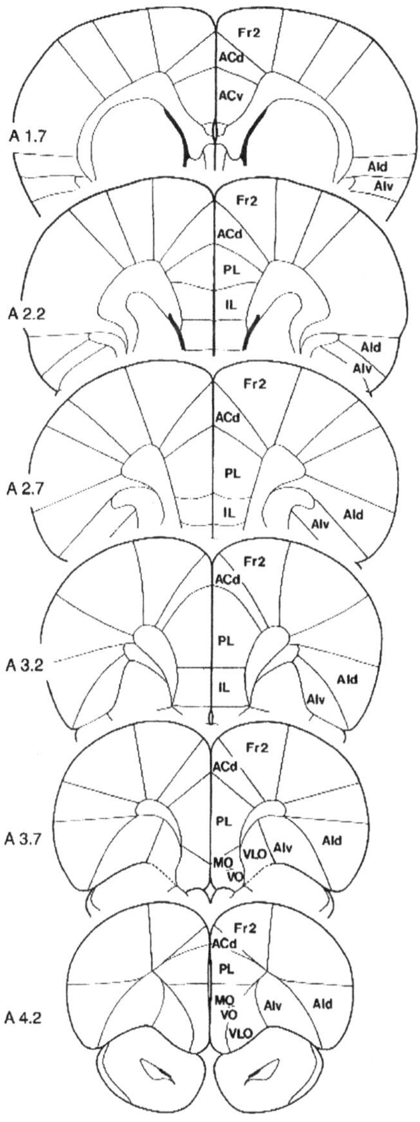

Figure 1: Delineation of the prefrontal cortex of the rat. Adopted from De Brabander (1992).

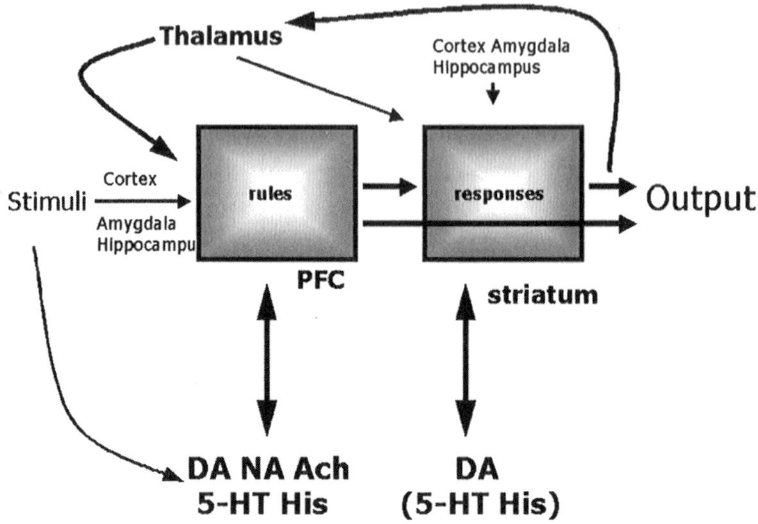

Figure 2: Schematic representation of the central position of corticostriatal pathway within the PFC circuit. Subcortical limbic and monoaminergic inputs to the corticostriatal pathway are also depicted. The hypothesized roles of PFC and striatum are shown in the blocks.

all amines that have been grouped together as (arousal) transmitters of the reticular activating system (Marocco et al., 1994). Among them are the catecholamines, dopamine (DA) and noradrenaline (NA), the other monoamines serotonin (5-HT) and histamine (His), and acetylcholine (ACh). Common characteristics are the reciprocal connections with the PFC, the innervation of wide areas of the brain from a few nuclei, and the fact that activation of *in vivo* release is observed as part of an arousal response after presentation of salient novel, appetitive, or aversive stimuli (Feenstra, 2000). Striatal areas are predominantly under DAergic control.

3. COGNITIVE FLEXIBILITY

Based on the wide variety of functions that the PFC has been claimed to mediate or support, more unifying concepts were recently introduced, e.g. executive functions, cognitive control, and flexibility of behavior. This reflects the idea that the PFC has a supervisory role in goal-directed behavior, and is activated when goals are set, actions are selected, and outcomes are evaluated, but not activated anymore when goals are familiar, tasks are becoming routines, and rules can be applied more or less automatically (e.g. Passingham, 1993). The term cognitive flexibility appears to do justice to the key issue, i.e. maintaining background control and taking action when needed. Many behavioral tasks have been used to study PFC function in rodents and other species with transient or permanent brain lesions. Most of them lack tight experimental control that is needed, particularly in combination with *in vivo* electrophysiological or neurochemical analyses. De Bruin et al. (2000) set up an instrumental task in the Skinner box to study flexibility of behavior. In this task, new components were introduced in phases (Acquisition, Reversal, Extinction; hence termed as "ARE-task"), and tight experimental control is possible. It can therefore be easily combined (time-locked) with ongoing invasive measurements. The flexibility of the task itself allows small alterations and adaptations when wanted or needed.

4. THE ARE-TASK: SERIAL REVERSAL LEARNING

The ARE task consists of four phases as shown in Table 1. In the shaping phase, the rat learns that pressing the lever results in the presentation of a reward pellet. In this phase, only one of the two levers randomly comes out into the box and has to be pressed within a certain time, before it is retracted. This is the essential rule of the task, and may be called the motor rule (cf. Miller and Cohen, 2001). The next step is the introduction of a FR3 (fixed ratio 3) rule, i.e. the rat has to press the lever three times before the reward is presented. This may be called the response ratio rule. In the second phase (one lever acquisition), both levers come out and a discrimination has to be learned. The discrimination is in principle a spatial one, i.e. pressing either the left or the right lever is rewarded (discrimination rule). In contrast to most spatial discriminations, however, the reward is not delivered with spatial selectivity, but in a central food receptacle, exactly between the two levers. The third phase introduces a reversal of action-outcome contingencies, i.e. the previously rewarded lever is not rewarded anymore, and the previously neutral one is now the reinforced lever (rule reversal). Reversals can be introduced repeatedly, each time when the previous one has been learned. A series of reversals urges the rat to adopt a flexible way of responding. The final phase is the extinction phase, in which both levers still come out, but none is rewarded. The old rules will not be destroyed, but remain stored in the 'background' (Bouton, 1994; Garcia, 2002). Transient inactivation of PFC areas by local injection of lidocaine was used to test which phase was dependent on ongoing neuronal activity in that area.

Discrimination learning was rapid, and optimal performance was reached within one or two (64-trial) sessions (Fig. 3). Inactivation of medial or lateral PFC during the discrimination phase did not affect response accuracy.

Table 1: Behavioral training - ARE (acquisition, reversal, extinction)

1. SHAPING (learning that lever pressing leads to a food pellet delivery)		
stimulus **(light, L or R)**	→ response **(lever press, L or R)**	→ reward **(pellet)**
2. DISCRIMINATION		
(learning that pressing one lever (i.e. the left one) leads to a food pellet delivery)		
a. stimulus **(light, L+R)**	→ response **(lever press, L)**	→ reward **(pellet)**
b. stimulus (light, L+R)	→ response (lever press, R)	→ no reward
3. First REVERSAL		
(learning that pressing the other lever (i.e. the right one) leads to a food pellet delivery)		
a. stimulus (light, L+R)	→ response (lever press, L)	→ no reward
b. stimulus **(light, L+R)**	→ response **(lever press, R)**	→ reward **(pellet)**
4. EXTINCTION		
(learning that lever pressing no longer leads to a food pellet delivery)		
stimulus (light, L+R)	→ response (lever press, L or R)	→ no reward

Reversal learning took considerably more time. The number of incorrect responses (pressing the previously rewarded lever) was high in the beginning and decreased as the number of correct responses increased. After the first session of 64 trials, rats reached approximately 50% correct responses (Fig. 3). Inactivation of the medial, but not the lateral, PFC during the first reversal impaired switching to the other lever (de Bruin et al., 2000). These rats showed perseveration, in that they pressed the wrong lever for a longer time. Performance reached close to 100% correct responses at the end of the second session. Subsequent reversals showed the same pattern, but switching occurred more and more rapidly, and optimal levels of responding were already reached in the first session. Inactivation of the medial or lateral PFC no longer impaired switching from the third reversal on. Imposing serial reversals for several weeks led to a new strategy, in which rats switched levers from the first response on (they adopt a day-to-day alternation rule) (van der Plasse and Feenstra, unpublished observations).

In extinction sessions, responding was high in the beginning, but decreased within the session to low values. With subsequent extinction sessions, responding reached very low values more rapidly, although a relatively high level of responding was always present in the first block of trials. Inactivation of the medial PFC did not affect responding, but inactivation of the lateral PFC retarded extinction in the third session (de Bruin et al., 2000).

These results suggest that medial PFC is needed for switching responses, but only when the 'reversal rule' has to be learned. Once this was done and

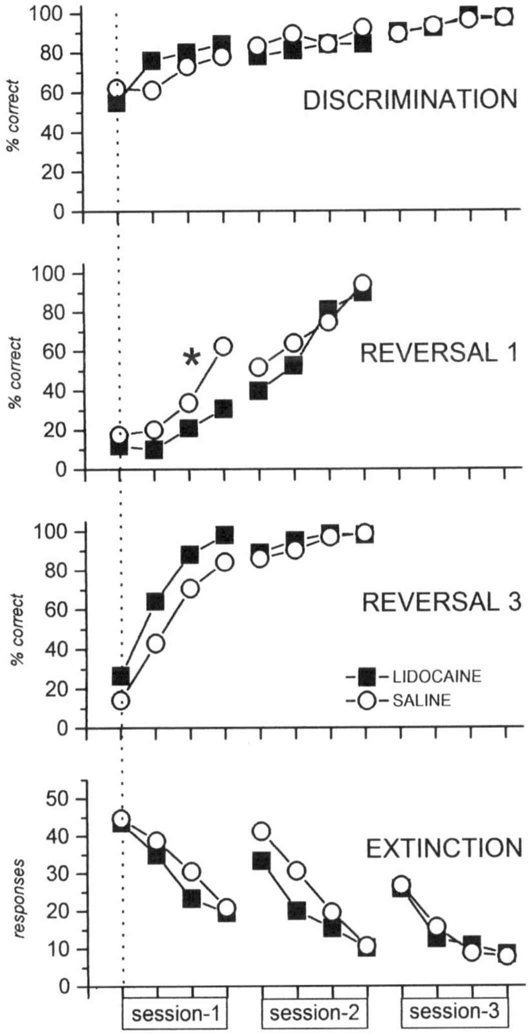

Figure 3: Effects of lidocaine or saline infusion into the medial PFC on behavior in the various phases of the ARE-task. The infusion was given before session 1 as indicated by the dotted line. Each session consists of 64 trials which are shown in blocks of 16 trials. Data are presented as % correct or total number of responses during extinction. * Significant difference between groups.

the new rule was (probably) consolidated, PFC inactivation did not lead to impaired reversal performance. Thus, medial PFC was only needed for the formation (rule learning), but apparently not for retrieval or for the application of this new rule.

During extinction, the results suggest that neither medial nor lateral PFC is needed for decreasing behavioral responses when they are not reinforced anymore. Lateral PFC may be involved in retrieval or the application of the 'extinction rule'.

5. THE ARE-TASK: PFC AND CATECHOLAMINES

The PFC and the related cortico-striatal circuit receive a strong modulatory input of monoamines (Fig. 2), and it has been shown that dopamine (DA) and noradrenaline (NA) are important for prefrontal functions (Arnsten, 1998). As DA is also strongly involved in reinforcement learning (Salamone and Correa, 2002) and in the selection of responses or strategies (Cools, 1980; Robbins, 1991), the effects of blocking DA receptors during the various phases of the ARE task was studied next. Local injection of the D_1-selective antagonist SCH23390 into the medial PFC did not affect discrimination learning, but retarded switching between levers during the first reversal. With regard to the effects of blocking D_1 receptors on lever press activity, it is interesting that during reversal learning, the total number of responses was decreased. However, this manipulation had no effect on the number of lever press responses during the discrimination phase of the task or during a separate task designed to lead to high frequencies of lever pressing. Injections of SCH23390 into the lateral PFC always led to a decreased number of responses. Local injection of the D_1-antagonist before the third reversal session did not affect performance. Responses during extinction sessions were also not affected by D_1-receptor blockade.

A complementary approach to study involvement of monoamines is to measure transmitter efflux during task performance. Microdialysis measurements of DA and NA efflux in the medial PFC present a reliable estimate of neuronal release (Feenstra, 2000). Using this approach, a study was performed in which rats that had learned the discrimination were tested during either another session of the discrimination phase (control group) or a first reversal session (reversal group) (van der Meulen et al., 2002). The results showed that task performance is always associated with an increase in DA efflux of about 50%. NA efflux increased to a lower extent, and this increase did not always reach a statistical significance. The increase of DA

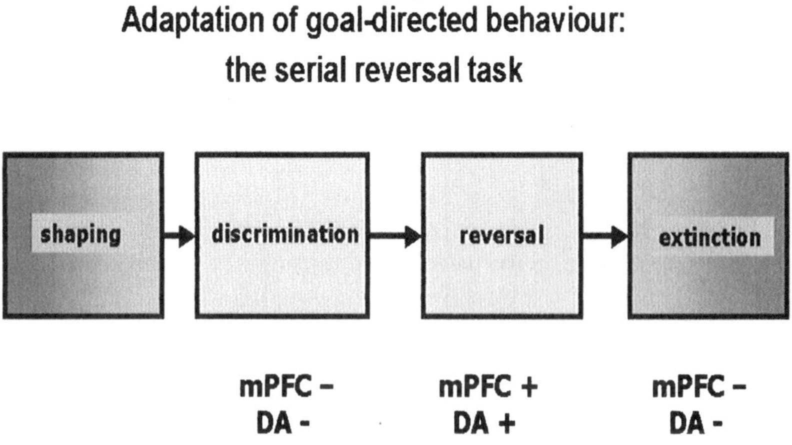

Figure 4: Summary of the results obtained with local injections of lidocaine and the D_1-antagonist SCH23390 into the medial PFC and with measurements of DA efflux in the medial PFC. Both treatments only affect performance during the first reversals, the only phase during which DA efflux increased more than control performance. Involvement is indicated by a + sign, no involvement by a – sign.

efflux, but not NA efflux, during reversal learning of the reversal group was larger compared to the control group. This is a remarkable result, as exposure to novelty as well as to unconditioned stimuli in general often leads to similar increases in prefrontal DA and NA efflux (Feenstra et al, 2000). Interestingly, when subjected to a reversal, the initial increase of DA was similar to that of control animals. Along with the increase in correct responses and reward acquisition, DA efflux in the PFC increased, suggesting that this increase coincided with the phase in the reversal learning when the rats started to expect the new response to be followed by a reward, i.e. when they began to form a representation of the outcome of their actions. Recent measurements during a *third reversal* and during extinction showed that DA efflux did not differ from controls in the reversal, but was only minimally activated during extinction (lower than controls) (van der Meulen et al, in preparation).

A summary of these results is presented in Figure 4. The correlation between the involvement of medial PFC and that of the DA afferents during the ARE task phases is striking. Given the close relation between PFC and striatal subregions and their modulation by DA, we need a further study in which the inactivation of the striatal subareas, the blockade of their DA receptors, and the measurement of DA efflux are carried out. Moreover, the observation that the formation of a new action-outcome association in the reversal phase depends on the PFC and DA suggests that the formation of the original action-outcome association during the shaping phase might be also dependent on PFC and DA (see below).

6. COMPONENTS OF THE ARE-TASK

In this section, we will review some aspects of what is known about the PFC and DA/NA involvement in a number of separate processes that may be considered to be part of cognitive flexibility as tested in the ARE task.

In every new phase, novelty detection takes place. Then, acquisition of a conditioned response is the essential step in goal-directed action. After every new phase, the newly formed associations will be consolidated, followed by retrieval of this information in later sessions or phases. Action monitoring and error detection are on-going processes and may be followed by subsequent corrective actions like inhibition. Finally, extinction or strategy switching is needed to adapt behavior to the altered conditions. It is beyond the scope of this review to discuss all these processes in detail, but we will review some that were recently studied in relation to the functions of the PFC of the rat and may be relevant to the findings obtained using the ARE-task.

6.1 Novelty Detection

Medial PFC is involved in the reactions to novelty in rats (Holson and Walker, 1986; Dias and Honey, 2002), as it is in human (Daffner et al., 2000). This is not only apparent from lesion studies, but from the studies with novelty exposure, which activates PFC neuronal activity (Handa et al, 1993) and causes DA and NA effluxes in the PFC (Feenstra et al., 1995, 2000). The DA responses remain when the stimulus is no longer novel but still salient (e.g. a reward): in this case, the response may shift to the first event that predicts the reward (Hollerman et al., 2000). NA is more specifically involved in novelty, as habituation of the NA responses develops rapidly (Vankov et al., 1995) and as behavioral habituation and novelty seeking may depend on NA (Mason and Fibiger, 1977; Sara et al., 1995). Based on these data, one may infer that novelty in general, including task novelty (Barceló et al., 2002), results in the activation of PFC-related circuits. Our findings using the ARE-task suggest, however, that not all task alterations require PFC involvement for adapting the behavior.

6.2 Acquisition of Conditioned Responses

The PFC is generally thought to be uninvolved in the acquisition of classical, Pavlovian conditioning or in operant learning. Numerous authors report that rats with PFC lesions were not impaired in learning a wide variety of associations or discriminations, e.g. fear conditioning (Quirk et al., 2000), discrimination learning in active avoidance (Li and Shao, 1998) or odor-reward association (Schoenbaum et al., 2002), acquisition of spatial or visual-cued version of cheeseboard or cross-arm tasks (Ragozzino et al., 1999a,b), and operant discrimination learning (de Bruin et al, 2000).

In view of a number of recent data obtained with various experimental techniques, this view is challenged. Baldwin et al. (2000) observed impaired learning of lever-pressing after local injections of a NMDA-antagonist into medial PFC, while Izaki et al. (2000) found that medial PFC lesions had a similar effect. Both groups indicated a special role for DA as well: local injections of a D_1-antagonist potentiated the effects of the NMDA-antagonist (Baldwin et al., 2002), and DA was reported to be selectively activated during learning of the lever-press response (Izaki et al., 1998). This initial (shaping) phase generally precedes all other experimental tasks and is often not included in the test for lesion effects or, at least, is not included in report. Inclusion of the initial phase may be very important in relation to PFC function, because 'motor set' has been suggested to be one of the main properties of the PFC and because the 'motor rule' is the first rule to be learned in this sequence of task phases (see above). Moreover, in the studies on task-related neuronal activity, Mulder et al. (2000) described neurons in the medial PFC that develop sustained task-related activity during the acquisition phase, indicating that a neuronal substrate may be available in the PFC. It is well-known that DA is not only activated by appetitive stimuli but also by aversive stimuli (Bertolucci-D'Angio et al., 1990; Feenstra, 2000), and Stark et al. (1999, 2000) showed that DA efflux in the medial PFC of the gerbils performing an aversively motivated shuttle-box task is selectively activated during learning, i.e. during strategy formation. These examples suggest the involvement of the PFC in making operant or 'planned' responses in which the response may be thought of as separated from the outcome, unlike food-searching tasks where the response directly leads to the outcome.

Although at the moment, no further information is available regarding the involvement of the PFC or DA in the shaping phase of the ARE-task (Fig. 4), the activation of PFC- and DA-dependent mechanisms is not only in line with the ideas put forward by Franz (1907), but also with the recent theories by Passingham (1993, 1998) that new actions are supported by the PFC but that upon practice, brain activation is shifted to other cortical and cerebellar areas (Shadmehr and Holcomb, 1997). As this is what we observe in the course of serial reversals, a similar mechanism may be expected during the shaping phase.

6.3 Consolidation

PFC involvement in the consolidation phase of association formation has been found in appetitively motivated odor discrimination. Using c-*fos* as a marker, Tronel and Sara (2002a) reported neuronal activation in medial PFC during consolidation. In addition, impaired consolidation was observed after local blockade of NMDA-receptors in the same area (Tronel and Sara, 2002b). Consolidation of appetitive instrumental learning was reported to take place in the core subarea of the nucleus accumbens (ventral striatum) (Hernandez et al., 2002). As other studies indicated that a corticostriatal network mediates instrumental learning, an additional prefrontal contribution to consolidation might be expected (Baldwin et al., 2000). Aversively motivated learning has been used frequently in studies of memory consolidation, and in general, consolidation of this type of memory does not appear to depend on PFC areas (Ambrogi Lorenzini et al., 1999). Memory for inhibitory avoidance learning, however, was dependent on the precentral PFC area (FR2 at level A3.7, Fig. 1) (Mello e Souza et al., 2000).

Memory consolidation is under a strong modulatory influence of emotional processes, and the effects of both catecholamines have been described in many learning paradigms (McGaugh, 2000).

Consolidation has not been studied as a separate process in the ARE-task. However, some evidence is available for a role during extinction (see below).

6.4 Extinction

Response inhibition has often been studied in extinction trials, where a predicted presentation of a reinforcer is omitted. This can occur in a classical conditioning paradigm, where the CS is not followed by the US anymore, or in an operant paradigm, where the action is not followed by the expected outcome. Inhibition assessed in extinction trials is, however, different from the inhibition in e.g. Go-No Go paradigms. In the latter case, responding or not responding has different consequences, and active control of behavior is called for, while in extinction there are no consequences, and adaptation might be expected to be more non-committal. Interestingly, however, PFC has been suggested to be involved in extinction of conditioned fear. In rats with lesions in the dorsomedial or ventromedial PFC, but not ventrolateral PFC, Morgan et al. (1993) and Morgan and LeDoux (1995, 1999) reported enhanced freezing responses when the rats were re-exposed to an explicit CS 24 h after the acquisition of the conditioned fear. It may not be the immediate expression of extinction, i.e. the acute inhibition of the behavioral reaction, through which the PFC controls the behavior. Rather, the PFC may be involved in the consolidation of this new association. This was shown by Quirk et al. (2000), who identified the infralimbic area as the site that is necessary for the consolidation of extinction of fear conditioning. Consolidation of extinction was observed only when LTP-like changes (spontaneous or artificially induced) were observed in PFC neurons (Herry and Garcia, 2002). In addition, stimulation of the infralimbic area at 24 h after the acquisition even accelerates the extinction (Millad and Quirk, 2002). These results suggest that PFC involvement is important during consolidation of a new association and in retrieval of this information. Studies that specifically relate prefrontal catecholamine activation to extinction are sparse, as extinction is difficult to separate from retrieval. However, Morrow et al. (1999) reported that catecholamine lesions in the medial PFC impaired extinction without affecting acquisition.

6.5 Retrieval

Re-exposure to a cue that was learned to have predictive properties in a behavioral situation leads to retrieval of the previously acquired information, maintenance of an active representation of that information, and, upon reinforcement, whether it is appetitive or aversive, to reconsolidation of the association (Sara, 2000a). Presentation of such a reminder cue may facilitate behavioral reaction and performance (Sara, 2000b). Using a paradigm in which presentation of the retrieval cue is separated from the task in which the information is to be used, Gisquet-Verrier et al. (1989) showed a similar reminder effect of cue presentation. Importantly, Botreau et al. (2001) observed that the medial PFC is required for using the information. The involvement of various arousal systems in the retrieval process may be suggested, since it was improved by the activation of NA-systems (Sara and Devauges, 1989). It is not yet known whether monoamines act in the PFC to support retrieval, or whether the PFC and monoamines control retrieval mechanisms taking place somewhere else in the brain.

A specific retrieval cue was not used in the ARE task (except for the shaping phase) where up to the present no inactivation studies have been carried out. However, retrieval of previously stored task-relevant information may be an important function of the PFC, as a working memory task learned before a PFC-lesion was more affected than the same task acquired after the lesion (Broersen, 2000; see also Becker et al., 1981). It is possible that the late impairment after inactivation of the lateral PFC would fit with these findings (see above), but they obviously need more experimental support.

7. THE PFC AND ITS CATECHOLAMINE AFFERENTS IN STRATEGY SWITCHING AND REVERSAL LEARNING

7.1 Rule Switching and Attentional Set-Shifting

In primates, (dorso)-lateral PFC damage has been associated strongly with deficits in shifting attentional set between, as compared to within, dimensions (Owen et al., 1991; Dias et al., 1996). Tests involving intra- and extra-dimensional shifts using various dimensions have been described (Shepp and Eimas, 1964; Oswald et al., 2001), but only the task recently developed by Birrell and Brown (2000) has been tested in relation to the PFC function in rats. Their results suggest a similar dependence of extradimensional shifts on PFC activity (the medial PFC) as in primates, whereas the emotional shift (or reversal, Dias et al., 1996) depends on the orbital PFC. Neurons in the latter area code the current incentive value of a cue (Schoenbaum and Setlow, 2001). In the terminology of Kesner (2002), they support rules based on reward value or affect. The tripartite memory system of Kesner (2000, 2002) appears to give the most versatile system to probe and explain PFC-dependent functions. Thus, a common finding has been that the tasks which involve a switch between rules depend on the medial PFC in the rat. De Bruin et al. (1994) showed that rats with PFC lesions were not impaired in learning a spatial version of the water maze, but had problems in switching from a spatial rule to a rule based on visual cues. Similar deficits in rule (or strategy) switching behavior were reported by Seamans et al. (1995), Joel et al. (1997), and Ragozzino et al. (1999a,b). As Gisquet-Verrier et al. (2000) point out, also other findings, at first sight unrelated, may be explained by assuming a rule-switching function for the medial PFC. Rats performing a working memory task were impaired only by imposing longer delays when they were allowed to master the task first with the shortest delay, before increasingly longer ones were tested (Delatour and Gisquet-Verrier, 1999). Using random delays from the start did not result in impairment. Also, findings by Granon and Poucet (2000) support a PFC function in rule switching, and these authors stress that medial PFC is "important for the learning of contingency rules that are not in the animal's natural behavioral repertoire or that counteract previously learned strategies". We feel that this presents a key issue to compare the wide variety of tasks that have been used in relation to PFC function; in particular, food-search tasks and instrumental tasks. The fact that reversals (or emotional shifts) sometimes depend on medial PFC (e.g. de Bruin et al.,

2000) and other times on orbital PFC (e.g. Brown and Bowman, 2001) may be explained following this theory.

7.2 Reversal

Many different learning situations have been called "reversal learning". We define it as the process in which a discriminative stimulus or action loses its association with a reinforcer, and the other stimulus or action now acquires this association. The essential characteristics are the discrimination between two choices and the exchange of selective reinforcer association between these choices. In the terminology of Dias et al. (1996), a reversal is an "emotional shift". In contrast to the attentional shifts (intradimensional or extradimensional), reversals are concerned with active changing of an existing, well-learned association. A contingency that is opposite to the acquired one is presented: if stimulus A or action X has always been reinforced by reward presentation, this association now has to be inhibited, and the inhibitory association with B or Y has to be overcome. We present just a few examples here of the vast literature on reversal learning. Li and Shao (1998) taught rats a visual discrimination in a T-maze that allowed them to escape from a foot shock. After lesions in the prelimbic or infralimbic, but not anterior cingulated, area of the medial PFC, reversal learning was impaired, while acquisition of the discrimination was unaffected. Chachich and Powell (1998) studied eyeblink conditioning in rabbits using an auditory discriminative stimulus. Lesions in the medial PFC severely impaired reversal learning. Another example is an odor discrimination task, in which a Go response after a positive odor leads to a reward (Schoenbaum et al., 2002). Lesions in the orbital PFC lead to a strong impairment in learning the reversal, but in a series of reversals, only the acquisition of the first was affected, and later ones were in fact more rapidly learned. Brown and Bowman (2002) taught rats to find a reward by digging in a bowl and learning discrimination between two stimuli within one attentional set (odor, digging medium, or texture of the bowl). Animals with lesions in the orbital PFC were selectively impaired in the reversal phase, but not when new discrimination had to be learned (intra-dimensional shift) or when attention had to be shifted to another stimulus dimension (extra-dimensional shift). As presented above, de Bruin et al. (2000) used an instrumental task in Skinner box, where rats had to choose between two levers (left and right) only one of which was rewarded. Transient inactivation of medial PFC impaired reversal learning in this test. These examples suggest that reversal learning in a wide variety of behavioral paradigms depends on the integrity of one or more subareas of the PFC — both in classical conditioning and instrumental learning, with many different cues (visual, auditory, odor, tactile, and spatial), and with aversively or appetitively motivated responses. The combination of these and other findings strongly support the role of PFC in the flexibility of behavior.

Previous reports suggest a differential involvement of medial PFC on the basis of the difficulty of stimulus discrimination so that more difficult discriminations are dependent on the medial PFC (see Birrell and Brown, 2000). But, the few examples presented here do not support this possibility. None of the discriminations required in the medial PFC-dependent reversal tasks was particularly difficult (e.g. the discrimination learning in the instrumental task was the phase that was most rapidly mastered; Fig. 3). The conclusion could be that not only stimulus qualities are important, but also reinforcement value and strength and (the complexity of) the action sequence are critical (see above).

7.3 DA and NA in Flexibility of Behavior

Both DA and NA systems have been shown to be involved in reversal learning. Reversal of visual stimuli in a vigilance task in monkeys led to a shift of phasic activation of locus coeruleus (LC) neurons to the new target stimulus (Aston-Jones et al., 1997). This shift preceded the behavioral shift in responding to the newly rewarded stimulus, indicating that NA may play a role in reversal learning. In addition, a longer-lasting increase in the tonic firing rate of the NA neurons was observed. The combination of these can be expected to lead to increased NA release from NA terminals throughout the forebrain. Further indications that NA is important for cognitive flexibility can be found in the literature: in rats, an increase in central NA activity was associated with a facilitation of shifting attention, i.e. improved flexibility (Devauges and Sara, 1990). This finding and the important role of NA in novelty detection led Sara (1998) to propose that NA 'facilitates shifts in attention, information processing and memory', in other words, flexibility. Aston-Jones et al. (2000) extended this view by suggesting that flexibility of behavior is supported by tonic LC activity (related to scanning attention), whereas focused attention would be related to phasic LC activity. The finding that cognitive flexibility is improved when central β-adrenergic receptors are blocked (Beversdorf et al., 2002) indicates that an inverted U-shaped relationship between LC activity and the performance might exist (cf. Aston-Jones et al., 2000). Whether NA acts in the PFC to promote flexibility of behavior is not yet known. The role of NA in novelty processing and attention suggests that it is the case, but NA has not been elevated like DA in novel conditions during the ARE task. The apparent inconsistency of these results and those reported by Aston-Jones et al. (1997) may be explained by the fact that their monkeys were over-trained on the original stimulus for periods up to one year, so that this association was 'stamped in' much more than in our rats. NA is activated, however, when reinforcers (e.g. rewards) are presented unexpectedly and out of the context of an action-outcome sequence (Dalley et al., 2001; Feenstra et al., 2001). Unlike DA, NA does not appear to be involved with expectation or prediction of outcomes: it facilitates flexibility only under special conditions of increased arousal.

The idea that DA is involved in switching behavioral strategies was put forward by Cools (1980) and Oades (1985). While these theories predominantly concerned striatal DA, recently strong experimental support for a role of prefrontal DA was also obtained. As noted above (see section 5), blockade of D_1-receptors in the medial PFC impaired operant reversal learning (de Bruin et al., 2000), while an extra increase of prefrontal DA was observed during reversal learning (van der Meulen et al., 2002). Moreover, strategy switching in cross-maze tasks was also impaired when D_1-receptors were blocked (Ragozzino, 2002). This relation between medial PFC function and DA activity in flexibility resembles the previously described relation between DA activity and PFC function in working memory (Arnsten, 1998; Goldman-Rakic et al., 2000) and attention (Granon et al., 2000) and confirms the view of Le Moal and Simon (1991) that DA supports 'the integrative functions of the neuronal systems onto which they project'. For a more comprehensive discussion on the role of DA (and other monoamines) in PFC function, the reader is recommended to refer to the review by Robbins (2000).

8. CONCLUSION

Rat PFC supports a complex set of processes, which can be summarized as cognitive flexibility. Behavioral strategies or rules based on external and internal information and on relevant experience may be formed and stored depending on the complexity of the task set, and the PFC is needed whenever rules have to be adapted or whenever novel strategies have to be adopted. The DA innervation in the PFC is firmly linked to this PFC function in flexibility.

It is clear from the experimental studies that individual differences exist between rats in their ability to adapt to novel task demands. It is interesting that in man, such individual differences have recently been related to the differences in the activation of the PFC (Carlsson et al., 2000; Duncan et al., 2000; Gray et al., 2003; Cazalis et al., 2003). In rats, such studies have been restricted to the role of prefrontal DA in attention, where poor attention could be improved by increasing DA transmission (Granon et al., 2000). In man, a relation between DA function and cognition has been also suggested

(Ashby et al., 1999). Future studies might unravel neurobiological mechanisms in the PFC of rats that might be related to the differences in their individual capabilities to form, adapt, and switch rules or strategies in the behavior.

REFERENCES

Ambrogi Lorenzini CG, Baldi E, Bucherelli C, Sacchetti B, Tassoni G (1999) Neural topography and chronology of memory consolidation: a review of functional inactivation findings. Neurobiol Learn Mem 71: 1-18.

Arnsten AFT (1998) Catecholamine modulation of prefrontal cortical cognitive function. Trends Cogn Sci 2: 436-446.

Ashby FG, Isen AM, Turken U. (1999) A neuropsychological theory of positive affect and its influence on cognition. Psychol Rev 106: 529-550.

Aston-Jones G, Rajkowski J, Kubiak P (1997) Conditioned responses of monkey locus coeruleus neurons anticipate acquisition of discriminate behavior in a vigilance task. Neuroscience 80: 697-715.

Aston-Jones G, Rajkowski J, Cohen J (2000) Locus coeruleus and regulation of behavioral flexibility and attention. In: Cognition, Emotion and Autonomic Responses: The Integrative Role of The Prefrontal Cortex and Limbic Structures. Progress in Brain Research vol 126, (Uylings HBM, van Eden CG, de Bruin JPC, Feenstra MGP, and Pennartz CMA, eds), pp 165-182, Elsevier, Amsterdam.

Baldwin AE, Holahan MR, Sadeghian K, Kelley AE (2000) N-Methyl-D-aspartate receptor-dependent plasticity within a distributed corticostriatal network mediates appetitive instrumental learning. Behav Neurosci 114: 84-98.

Baldwin AE, Sadeghian K, Kelley AE (2002) Appetitive instrumental learning requires coincident activation of NMDA and dopamine D1 receptors within the medial prefrontal cortex. J Neurosci 22: 1063-1071.

Barceló F, Periañez JA, Knight RT (2002) Think differently: a brain orienting response to task novelty. NeuroReport 13: 1887-1892.

Becker JT, Olton DS, Anderson CA, Breitinger ERP (1981) Cognitive mapping in rats: the role of the hippocampal and frontal systems in retention and reversal. Behav Brain Res 3: 1-22.

Bertolucci-D'Angio M, Serrano A, Driscoll P, Scatton B (1990) Involvement of mesocorticolimbic dopaminergic systems in emotional states. In: The Prefrontal Cortex: Its Structure, Function and Pathology. Progress in Brain Research, vol 85 (Uylings H.B.M., van Eden C., de

Bruin JPC, Corner MA, and Feenstra MGP, eds), pp 405-417, Elsevier, Amsterdam.

Beversdorf DQ, White DM, Chever DC, Hughes JD, Bronstein RA (2002) Central β-adrenergic modulation of cognitive flexibility. NeuroReport 13: 2505-2507.

Birrell JM, Brown VJ (2000) Medial prefrontal cortex mediates perceptual attentional set shifting in the rat. J Neurosci 20: 4320-4324.

Botreau F, Chéruel F, El Massioui N, Gisquet-Verrier P (2001) Involvement of medial prefrontal cortex and of dorsal striatum in retrieval processes in the rat. Soc Neurosci Abst 27: 189.1.

Bouton ME (1994) Conditioning, remembering, and forgetting. J Exptl Psychol: Anim Behav Process 20: 219-231.

Broersen LM (2000) Attentional processes and learning and memory in rats: the prefrontal cortex and hippocampus compared. In: Cognition, Emotion and Autonomic Responses: The Integrative Role of The Prefrontal Cortex and Limbic Structures. Progress in Brain Research vol 126, (Uylings HBM, van Eden CG, de Bruin JPC, Feenstra MGP, and Pennartz CMA, eds), pp 79-94, Elsevier, Amsterdam.

Brown VJ, Bowman EM (2002) Rodent models of prefrontal cortical function. Trends Neurosci 25: 340-343.

Carlsson I, Wendt PE, Risberg J (2000) On the neurobiology of creativity. Differences in frontal activity between high and low creative subjects. Neuropsychologia 38: 873-885.

Cazalis F, Valabrègue R, Pélégrini-Isaac M, Asloun S, Robbins TW, Granon S (2003) Individual differences in prefrontal cortical activation on the Tower of London planning task: implication for effortful processing. Eur J Neurosci 17: 2219-2003.

Chachich M, Powell DA (1998) Both medial prefrontal and amygdala central nucleus lesions abolish heart rate classical conditioning, but only prefrontal lesions impair reversal of eyeblink differential conditioning. Neurosci Lett 257: 151-154.

Cools AR (1980) Role of neosatriatal dopaminergic activity in sequencing and selecting behavioural strategies: facilitation of processes involved in selecting the best strategy in a stressful situation. Behav Brain Res 1: 361-378.

Daffner KR, Mesulam MM, Holcomb PJ, Calvo V, Acar D, Chabrerie A, Kikinis R, Jolesz FA, Rentz DM, Scinto LFM (2000) Disruption of attention to novel events after frontal lobe injury in humans. J Neurol Neurosurg Psychiatry 68: 18-24.

Dalley JW, McGaughy J, O'Connell MT, Cardinal RN, Levita L, Robbins TW (2001) Distinct changes in cortical acetylcholine and noradrenaline efflux during contingent and noncontingent performance of a visual attentional task. J Neurosci 21: 4908-4914.

de Brabander JM (1992) Sparing of function following damage to the prefrontal cortex. Ph D Thesis, University of Amsterdam.

de Bruin JPC (1994) Evolution of prefrontal cortex: comparative aspects of its behavioral functions. In: Flexibility and Constraint in Behavioral Systems (Greenspan RJ and Kyraciou CP, eds), pp 185-192, Wiley.

de Bruin JPC, Sànchez-Santed F, Heinsbroek RPW, Donker A, Postmes P (1994) A behavioural analysis of rats with damage to the medial prefrontal cortex using the Morris water maze: evidence for behavioural flexibility, but not for impaired spatial navigation. Brain Res 652: 323-333.

de Bruin JPC, Feenstra MGP, Broersen LM, van Leeuwen M, Arens C, de Vries S, Joosten RNJMA (2000) Role of the prefrontal cortex of the rat in learning and decision making: effects of transient inactivation. In: Cognition, Emotion and Autonomic Responses: The Integrative Role of The Prefrontal Cortex and Limbic Structures. Progress in Brain Research vol 126, (Uylings HBM, van Eden CG, de Bruin JPC, Feenstra MGP, and Pennartz CMA, eds), pp 103-113, Elsevier, Amsterdam.

Delatour B, Gisquet-Verrier P (1999) Lesions of the prelimbic-infralimbic cortices in rats do not disrupt response selection processes but induce delay-dependent deficits: evidence for a role in working memory? Behav Neurosci 113: 941-955.

Devauges V, Sara SJ (1990) Activation of the noradrenergic system facilitates an attentional shift in the rat. Behav Brain Res 39: 19-28.

Dias R, Robbins TW, Roberts AC (1996) Dissociation in prefrontal cortex of affective and attentional shifts. Nature 380: 69-72.

Dias R, Honey RC (2002) Involvement of the rat medial prefrontal cortex in novelty detection. Behav Neurosci 116: 498-503.

Duncan J, Seitz R, Kolodny J, Bor D, Herzsog H, Ahmed A, Newell FN, Emslie H (2000) A neural basis for general intelligence. Science 289: 457-460.

Feenstra MGP (2000) Dopamine and noradrenaline release in the prefrontal cortex in relation to conditioned and unconditioned stress and reward. In: Cognition, Emotion and Autonomic Responses: The Integrative Role of The Prefrontal Cortex and Limbic Structures. Progress in Brain Research vol 126, (Uylings HBM, van Eden CG, de Bruin JPC, Feenstra MGP, and Pennartz CMA, eds), pp 133-163, Elsevier, Amsterdam.

Feenstra MGP, Botterblom MHA, van Uum HFM (1995) Novelty stress increases dopamine release in the rat medial prefrontal cortex in vivo: inhibition by diazepam. Neurosci Lett 189: 81-84.

Feenstra MGP, Botterblom MHA, Mastenbroek S (2000) Dopamine and noradrenaline release in the prefrontal cortex in the light and dark phase:

Effects of novelty and handling and comparison to the nucleus accumbens. Neuroscience 100: 741-748.

Feenstra MGP, IJff B, Lesman A, de Weijer B, Geurtsen A, van der Vliet J, Joosten R, de Bruin J (2001) Effects of reward density and lever pressing on dopamine and noradrenaline efflux in the prefrontal cortex during operant responding. Soc Neurosci Abstr 27: 189.11.

Franz SI (1907) On the functions of the cerebrum. The frontal lobes. Arch Psychol 2: 5-64.

Fuster JM (1997) The Prefrontal Cortex. Anatomy, Physiology and Neuropsychology of The Frontal Lobe. Lippincott-Raven Press, New York.

Garcia R (2002) Postextinction of conditioned fear: between two CS-related memories. Learn Mem 9: 361-363.

Gisquet-Verrier P, Dekeyne A, Alexinsky T (1989) Differential effects of several retrieval cues over time: Evidence for time-dependent reorganization of memory. Animal Learn Behav 17: 394-408.

Gisquet-Verrier P, Winocur G, Delatour B (2000) Functional dissociation between dorsal and ventral regions of the medial prefrontal cortex in rats. Psychobiol 28: 248-260.

Goldman-Rakic PS (1987) Circuitry of primate prefrontal cortex and regulation of behavior by representational memory. In: Handbook of Physiology, The Nervous System V (Plum F and Mountcastle V, eds), pp 373-417, American Physiological Society, Bethesda.

Goldman-Rakic PS, Muly III EC, Williams GV (2000) D_1 receptors in prefrontal cells and circuits. Brain Res Rev 31: 295-301.

Granon S, Poucet B (2000) Involvement of the rat prefrontal cortex in cognitive functions: a central role for the prelimbic area. Psychobiol 28: 229-237.

Granon S, Passetti F, Thomas KL, Dalley JW, Everitt BJ, Robbins TW (2000) Enhanced and impaired attentional performance after infusion of D1 dopaminergic receptor agents into rat prefrontal cortex. J Neurosci 20: 1208-1215.

Gray JR, Chabris CF, Braver TS (2003) Neural mechanisms of general fluid intelligence. Nature Neurosci 6: 316-322.

Groenewegen HJ, Berendse HW (1994) Anatomical relationships between the prefrontal cortex and the basal ganglia in the rat. In: Motor and Cognitive Functions of The Prefrontal Cortex (Thierry A-M, Glowinski J, Goldman-Rakic PS, and Christen Y, eds), pp 51-77, Springer Verlag, Berlin.

Groenewegen HJ, Uylings HBM (2000) The prefrontal cortex and the integration of sensory, limbic and autonomic function. In: Cognition, Emotion and Autonomic Responses: The Integrative Role of The

Prefrontal Cortex and Limbic Structures. Progress in Brain Research vol 126, (Uylings HBM, van Eden CG, de Bruin JPC, Feenstra MGP, and Pennartz CMA, eds), pp 3-28, Elsevier, Amsterdam.

Handa RJ, Nunley KM, Bollnow MR (1993) Induction of c-fos mRNA in the brain and anterior pituitary gland by a novel environment. NeuroReport 4: 1079-1082.

Hernandez PJ, Sadeghian K, Kelley AE (2002) Early consolidation of instrumental learning requires protein synthesis in the nucleus accumbens. Nat Neurosci 5: 1327-1331.

Herry C, Garcia R (2002) Prefrontal cortex long-term potentiation, but not long-term depresion, is associated with the maintenance of extinction of learned fear in mice. J Neurosci 22: 577-583.

Hollerman JR, Tremblay L, Schultz W (2000) Involvement of basal ganglia and orbitofrontal cortex in goal-directed behavior. In: Cognition, Emotion and Autonomic Responses: The Integrative Role of The Prefrontal Cortex and Limbic Structures. Progress in Brain Research vol 126, (Uylings HBM, van Eden CG, de Bruin JPC, Feenstra MGP, and Pennartz CMA, eds), pp 193-215, Elsevier, Amsterdam.

Holson RR, Walker C (1986) Medial prefrontal cortical lesions and timidity in rats. II. Reactivity to novel stimuli. Physiol Behav 37: 231-238.

Izaki Y, Hori K, Nomura M (1998) Dopamine and acetylcholine elevation on lever-press acquisition in rat prefrontal cortex. Neurosci Lett 258: 33-36.

Izaki Y, Hori K, Nomura M (2000) Disturbance of rat lever-press learning by hippocampo- prefrontal disconnection. Brain Res 860: 199-202.

Joel D, Weiner I, Feldon J (1997) Electrolytic lesions of the medial prefrontal cortex in rats disrupt performance on an analog of the Wisconsin Card Sorting Test, but do not disrupt latent inhibition: implications for animal models of schizophrenia. Behav Brain Res 85: 187-201.

Kesner RP (2000) Subregional analysis of mnemonic functions of the prefrontal cortex in the rat. Psychobiol 28: 219-228.

Kesner RP (2002) Memory neurobiology. Encyclopedia of the Human Brain vol.2, pp 783-796.

Kolb B (1984) Functions of the prefrontal cortex of the rat: a comparative review. Brain Res Rev 8: 65-98.

Kolb B (1990) Animal models for human PFC related disorders. In: The prefrontal Cortex: Its Structure, Function and Pathology. Progress in Brain Research, vol 85 (Uylings H.B.M., van Eden C., de Bruin JPC, Corner MA, and Feenstra MGP, eds), pp 501-519, Elsevier, Amsterdam.

Le Moal M, Simon H (1991) Mesocorticolimbic dopaminergic network: functional and regulatory roles. Physiol Rev 71: 155-234.

Li L, Shao J (1998) Restricted lesions to ventral prefrontal subareas block reversal learning but not visual discrimination learning in rats. Physiol Behav 65: 371-379.

Marocco RT, Witte EA, Davidson MC (1994) Arousal systems. Curr Opn Neurobiol 4: 166-170.

Mason ST, Fibiger HC (1977) Altered exploratory behaviour after 6-OHDA lesion to the dorsal noradrenergic bundle. Nature 269:704-705.

McGaugh JL (2000) Memory – a century of consolidation. Science 287: 248-251.

Mello e Souza T, Vianna MRM, Rodrigues C, Quevedo J, Moleta BA, Izquierdo I (2000) Involvement of the medial precentral prefrontal cortex in memory consolidation for inhibitory avoidance learning in rats. Pharmacol Biochem Behav 66: 615-622.

Millad MR, Quirk GJ (2002) Neurons in medial prefrontal cortex signal memory for fear extinction. Nature 420: 70-74.

Miller EK (2000) The prefrontal cortex and cognitive control. Nat Rev Neurosci 1: 59-65.

Miller EK, Cohen JD (2001) An integrative theory of prefrontal cortex function. Annu Rev Neurosci 24: 167-202.

Mishkin M (1964) Perseveration of central sets after frontal lesions in monkeys. In: The Frontal Granular Cortex and Behavior (Warren JM and Akert K, eds), pp 219-241, McGraw-Hill, New York.

Morgan MA, Romanski LM, LeDoux JE (1993) Extinction of emotional learning: Contribution of medial prefrontal cortex. Neurosci Lett 163, 109-113.

Morgan MA, LeDoux JE (1995) Differential contribution of dorsal and ventral medial prefrontal cortex to the acquisition and extinction of conditioned fear in rats. Behav Neurosci 109: 681-688.

Morgan MA, LeDoux JE (1999) Contribution of ventrolateral prefrontal cortex to the acquisition and extinction of conditioned fear in rats. Neurobiol Learn Mem 72: 244-251.

Morrow BA, Elsworth JD, Rasmusson AM, Roth RH (1999) The role of mesoprefrontal dopamine neurons in the acquisition and expression of conditioned fear in the rat. Neuroscience 92: 553-564.

Mulder AB., Nordquist RE, Orgut OB, Pennartz CMA (2000) Plasticity of neuronal firing in deep layers of the medial prefrontal cortex in rats engaged in operant conditioning. In: Cognition, Emotion and Autonomic Responses: The Integrative Role of The Prefrontal Cortex and Limbic Structures. Progress in Brain Research vol 126, (Uylings HBM, van Eden CG, de Bruin JPC, Feenstra MGP, and Pennartz CMA, eds), pp 287-301, Elsevier, Amsterdam.

Oades RD (1985) The role of noradrenaline in tuning and dopamine in switching between signals in the CNS. Neurosci Biobehav Rev 9: 261-282.

Oswald CJP, Yee BK, Rawlins JNP, Bannerman DB, Good M, Honey RC (2001) Involvement of the entorhinal cortex in a process of attentional modulation: evidence from a novel variant of an IDS/EDS procedure. Behav Neurosci 115: 841-849.

Owen AM, Roberts AC, Polkey CE, Sahakian BJ, Robbins TW (1991) Extra-dimensional versus intra-dimensional set shifting performance following frontal lobe excisions, temporal lobe ixcisions or amygdalo-hippocampectomy in man. Neuropsychologia 29: 993-1006.

Passingham R (1993) The Frontal Lobes and Voluntary Action. Oxford University Press, Oxford.

Passingham RE (1998) Attention to action. In: The Prefrontal Cortex. Executive and Cognitive Functions (Roberts AC, Robbins TW, and Weiskrantz L, eds), pp 131-143, Oxford University Press, Oxford.

Quirk GK, Russo GK, Barron JL, Lebron K (2000) The role of ventromedial prefrontal cortex in the recovery of extinguished fear. J Neurosci 20: 6225-6231.

Ragozzino ME (2002) The effects of dopamine D_1 receptor blockade in the prelimbic-infralimbic areas on behavioral flexibility. Learn Mem 9: 18-28.

Ragozzino ME, Detrick S, Kesner RP (1999a) Involvement of the prelimbic-infralimbic areas of the rodent prefrontal cortex in behavioral flexibility for place and response learning. J Neurosci 19: 4585-4594.

Ragozzino ME, Wilcox C, Raso M, Kesner RP (1999b) Involvement of rodent prefrontal cortex subregions in strategy switching. Behav Neurosci 113: 32-41.

Robbins TW (1991) Cognitive deficits in schizophrenia and Parkinson's disease: neural basis and the role of dopamine In: The Mesolimbic Dopamine System: from Motivation to Action (Willner P and Scheel-Krüger J, eds), pp 497-528, Wiley, Chicester.

Robbins TW (2000) From arousal to cognition: the integrative position of the prefrontal cortex. In: Cognition, Emotion and Autonomic Responses: The Integrative Role of The Prefrontal Cortex and Limbic Structures. Progress in Brain Research vol 126, (Uylings HBM, van Eden CG, de Bruin JPC, Feenstra MGP, and Pennartz CMA, eds), pp 469-480, Elsevier, Amsterdam.

Roberts AC (1998) Introduction. In: The Prefrontal Cortex. Executive and Cognitive Functions (Roberts AC, Robbins TW, and Weiskrantz L, eds), pp 1-7, Oxford University Press, Oxford.

Roland PE (1984) Metabolic measurements of the working frontal cortex in man. Trends Neurosci 7: 430-435.

Salamone JD, Correa M (2002) Motivational views of reinforcement: implications for understanding the behavioral functions of nucleus accumbens dopamine. Behav Brain Res, 137: 3-25.

Sara SJ (1998) Learning by neurones: role of attention, reinforcement and behaviour. C R Acad Sci Life Sci 321: 193-198.

Sara SJ (2000a) Strenghtening the shaky trace through retrieval. Nat Rev Neurosci 1: 212-213.

Sara SJ (2000b) Retrieval and reconsolidation: toward a neurobiology of remembering. Learn Mem 7: 73-84.

Sara SJ, Devauges V (1989) Priming stimulation of the locus coeruleus facilitates memory retrieval in the rat. Brain Res 438: 401-411.

Sara SJ, Dyon-Laurent C, Hervé A (1995) Novelty seeking behavior in the rat is dependent upon the integrity of the noradrenergic system. Cogn Brain Res 2: 181-187.

Schoenbaum G, Setlow B (2001) Integrating orbitofrontal cortex into prefrontal theory: common processing themes across species and subdivisions. Learn Mem 8: 134-147.

Schoenbaum G, Nugent SL, Saddoris MP, Setlow B (2002) Orbitofrontal lesions in rats impair reversal but not acquisition of go, no-go odor discriminations. NeuroReport 13: 885-890.

Seamans JK, Floresco SB, Phillips AG (1995) Functional differences between the prelimbic and anterior cingulate regions of the rat prefrontal cortex. Behav Neurosci 109: 1063-1073.

Shadmehr R, Holcomb HH (1997) Neural correlates of motor memory consolidation. Scinece 277: 821-825.

Shepp BE, Eimas PD (1964) Intradimensional and extradimensional shifts in the rat. J Comp Phsysiol Psychol 57: 357-361.

Stark H, Bischof A, Scheich H (1999) Increase of extracellular dopamine in prefrontal cortex of gerbils during acquisition of the avoidance strategy in the shutle-box. Neurosci Lett 264: 77-80.

Stark H, Bischof A, Wagner T, Scheich H (2000) Stages of avoidance strategy formation in gerbils are correlated with dopaminergictransmission activity. Eur J Pharmacol 405: 263-275.

Tronel S, Sara SJ (2002a) Mapping of olfactory memory circuits: region-specific c-fos activation after odor-reward associative learning or after its retrieval. Learn & Mem 9: 105-111.

Tronel S, Sara SJ (2002b) Temporal dynamics of memory consolidation into the prefrontal cortex: NMDA receptors in the early phase, beta adrenergic receptors in the late phase. Soc Neurosci Abstr 284.12.

Uylings HBM, van Eden CG (1990) Qualitative and quantitative comparison of the prefrontal cortes in the rat and in primates. In: The Prefrontal Cortex: Its Structure, Function and Pathology. Progress in Brain

Research, vol 85 (Uylings H.B.M., van Eden C., de Bruin JPC, Corner MA, and Feenstra MGP, eds), pp 31-62, Elsevier, Amsterdam.

Van der Meulen J, Joosten R, de Bruin J, Feenstra M (2002) Effects of reversal learning on dopamine efflux in the medial prefrontal cortex measured with microdialysis. FENS Abstr 1, A 209.14.

Vankov A, Hervé-Minvielle A, Sara SJ (1995) Response to novelty and its rapid habituation in locus coeruleus neurons in the freely exploring rat. Eur J Neurosci 7: 180-1187.

Wise SP, Murray EA, Gerfen CR (1996) The frontal cortex - basal ganglia system in primates. Crit Rev Neurobiol 10: 317-356.

Chapter 9

INFORMATION PROCESSING IN THE PRIMATE PREFRONTAL CORTEX

Shintaro Funahashi
Department of Cognitive and Behavioral Sciences, Graduate School of Human and Environment Studies, Kyoto University, Sakyo-ku, Kyoto 606-8501, Japan

Keywords: Delayed-response task, delay-period activity, population vector, sensory-motor transformation, working memory.

Abstract: Working memory is a mechanism for short-term active storage of information and for processing stored information. Working memory is an important concept to understand prefrontal cortical functions. Although evidences for temporary storage mechanisms of information have been accumulated, little is known about neuronal mechanisms for processing information. To understand how information is processed in the nervous system, we need to know what information single-neuron activity represents and how represented information by single-neuron activities changes along the temporal sequence of the trial. We used two kinds of oculomotor version of the delayed-response (ODR) tasks and examined what information prefrontal single-neuron activity represents. We found that all cue-period activity represented cue positions and that most of delay-period activity represented cue positions. However, most of oculomotor activity represented saccade directions. These results suggest that prefrontal neurons participate in sensory-motor transformation. To examine prefrontal participation in sensory-motor transformation, we analyzed single-neuron activities using a population vector analysis. As a result, we found that information represented by a population of PFC neurons changes gradually during the delay period from information for visual cue to that for saccade.

1. INTRODUCTION

The prefrontal cortex has been known to participate in higher cognitive functions such as thinking, reasoning, planning, and decision- making (Fuster, 1997; Miller and Cohen, 2001; Stuss and Knight, 2002). To perform these functions, the prefrontal cortex is required to monitor the external world continually, pay attention to important stimuli, input necessary information from the external world, retrieve related information from long-term memory, stored such information temporarily for manipulation, integration, and processing, and then output the information to brain areas where it will be used. Each of these processes is related to complex neuronal mechanisms within the prefrontal cortex and requires interactions with other cortical and subcortical structures. Therefore, it is not easy to understand how the prefrontal cortex participates in these processes.

However, working memory is an important concept for understanding these processes. Working memory is a mechanism for short-term active storage of information as well as for processing stored information (Baddeley, 1986; Miyake and Shah, 1999). Since working memory has been thought to participate significantly in higher cognitive functions, it is expected that neuronal mechanisms for working memory would provide an important clue to understand mechanisms for thinking, planning, and decision-making. Lesion studies, neurophysiological studies, neuro-psychological studies, and PET and fMRI studies have revealed that the dorsolateral prefrontal cortex (DLPFC) plays a significant role for working memory (Goldman-Rakic, 1987; Petrides, 1994; Smith and Jonides, 1999; Duncan and Owen, 2000; Funahashi, 2001; Collette and Van der Linden, 2002). In addition, neurophysiological studies revealed that tonic sustained activation observed during the delay period (delay-period activity) is a neuronal correlate of temporary active storage mechanism of information (Goldman-Rakic, 1987; Miller, 2000; Funahashi and Takeda, 2002a). Therefore, examining neuronal mechanisms for working memory in the prefrontal cortex would provide important information to understand neuronal mechanisms for higher cognitive functions.

Although evidences for the neuronal mechanism of the temporary storage of information have been accumulated, little is known about mechanisms for manipulating, integrating, and processing information. To understand how information is processed and manipulated in the prefrontal cortex, we first need to know what information prefrontal neurons represent while monkeys perform a cognitive task. We then need to know how information represented by neuronal activity changes along the time course of a single trial of the cognitive task. There are evidences that single particular information is encoded by a population of prefrontal neurons. For example, many prefrontal neurons respond to visual stimuli and have visual receptive fields (Mikami et al., 1982; Suzuki and Azuma, 1983; Funahashi et al., 1990). When a visual cue was presented at a particular position in the visual field, a number of prefrontal neurons were activated. However, the preferred direction, the strength of the spatial tuning, the magnitude and the temporal pattern of the visual response are different from neuron to neuron (Funahashi et al., 1990). In spite of these differences, the activity of all these neurons represents information regarding characteristics of the visual stimulus. Therefore, to understand how information is processed in the prefrontal cortex, we need to examine the information represented by a population of prefrontal activities and its temporal change depending upon the behavioral context of the task.

2. PREFRONTAL ACTIVITY RELATED TO WORKING MEMORY PROCESSES

2.1 Working Memory Task

To examine neuronal mechanisms of working memory processes, single-neuron activities recorded from DLPFC have been analyzed while monkeys performed a variety of working memory tasks. The working memory task often selected for neurophysiological experiments is the delayed-response task, because the delayed-response task requires working memory of spatial position (Goldman-Rakic, 1987; Funahashi and Kubota, 1994; Fuster, 1997) and because delayed-response deficits have been observed by the selective lesion of the cortex within and surrounding the principal sulcus (area 46) (Goldman-Rakic, 1987; Funahashi et al., 1993a; Petrides, 1994; Fuster, 1997). Fuster and Alexander (1971) and Kubota and Niki (1971) were the investigators who first examined single-neuron activity in DLPFC while monkeys performed a manual delayed-response or delayed alternation task. Since then, many studies have been performed to examine the characteristics of memory-related activity and other types of activity in the prefrontal cortex using the delayed-response task.

Funahashi et al. (1989) introduced an oculomotor version of the delayed-response (ODR) task (Fig. 1). In this task, the monkey was sat in a monkey chair and faced a TV monitor in a dark room. After about a 5 sec inter-trial interval, a central fixation target was presented at the center of the TV monitor. While the monkey maintained fixation at the fixation target, a visual cue was briefly (0.5 sec) presented randomly at one of the 4 or 8 predetermined peripheral positions. Then, the delay period (usually 3 sec) was introduced. During the visual cue presentation and the delay period, the monkey was required to maintain fixation at the fixation target. At the end of

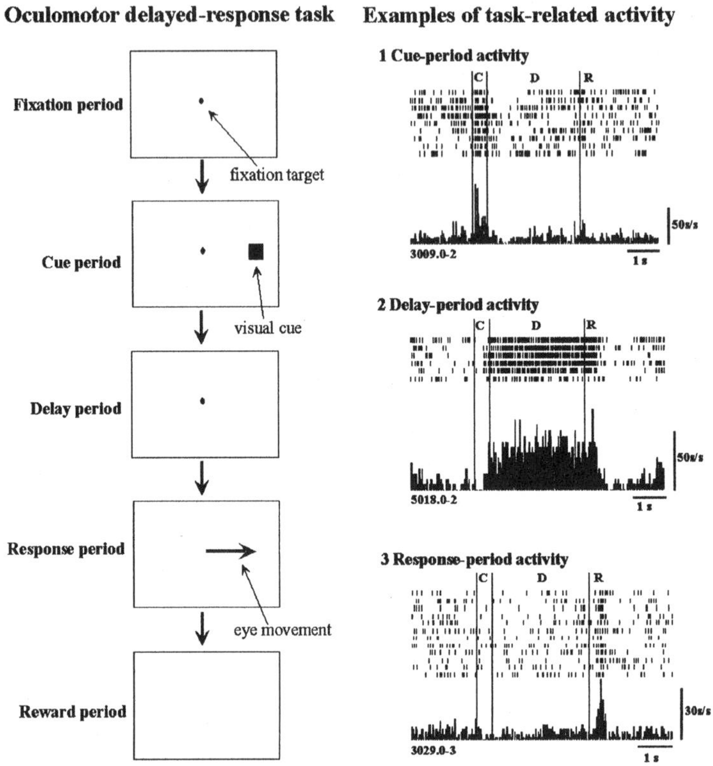

Figure 1: Schematic drawings of an oculomotor version of the delayed-response task (ODR task) and three examples of task-related activity recorded from the prefrontal cortex while monkeys performed the ODR task. C, D, R indicate cue period, delay period, and response period, respectively. Delay length was 3 sec.

the delay period, the fixation target was extinguished. This was the go signal for the monkey to make a saccade to the position where the visual cue had been presented. If the monkey made a correct saccade within a limited time (usually 0.4 to 0.5 sec), a liquid reward was given.

As is shown in Figure 1, we could observe three types of task-related activity (cue-period activity, delay-period activity, and response- period activity) while monkeys perfprmed the ODR task (Funahashi et al., 1989, 1990; Takeda and Funahashi, 2002).

2.2 Cue-Period Activity

A large number of neurons in DLPFC showed transient excitation or inhibition when the visual cues were presented during performances of working memory tasks. This transient activation responded to the visual cue presentation is called cue-period activity. Funahashi et al. (1990) found cue-period activity in 28% of task-related neurons while monkeys performed the ODR task. Most (93%) of cue-period activity showed transient excitation, but the remaining (7%) showed transient inhibition. The response latencies of excitatory cue-period activity were distributed from 37 to 309 ms with a median of 116 ms. In addition, most (96%) of cue-period activity were directional, such that cue-period activity was observed only when the visual cues were presented at certain areas in the visual field. Majority (71%) of cue-period activity had the best directions toward the visual field contralateral to the recording hemisphere.

The prefrontal cortex receives strong inputs from the posterior parietal cortex and the inferior temporal cortex (Goldman-Rakic, 1987; Fuster, 1997). Therefore, visual responses were also observed during the tasks in which visual stimuli had no behavioral significance to the monkey (e.g. a visual probe task). Funahashi et al. (1990) found no difference in the response magnitude and the tuning function of visual responses between the ODR task and the visual probe task. It has been shown that prefrontal visual neurons had visual receptive fields in the visual field (Mikami et al., 1982; Suzuki and Azuma, 1983). The centers of the visual receptive fields were located mainly in the visual field contralateral to the recorded hemisphere and the mean width of the receptive field was about 1/4 of the visual field. These characteristics agree with the characteristics of cue-period activity observed during delayed-response performances (Funahashi et al., 1990). Therefore, cue-period activity appears to be a visual response to the visual cue, and the characteristics of cue-period activity would correspond to the characteristics of visual receptive fields of prefrontal neurons.

2.3 Response-Period Activity

Response-period activity includes movement-related activity and post-trial activity. Since Kubota and Niki (1971) first reported movement- related activity in the prefrontal cortex, movement-related activity has been observed in tasks using manual responses (Niki and Watanabe, 1976; Kubota and Funahashi, 1982; Sawaguchi, 1987) as well as oculomotor responses (Joseph and Barone, 1987; Boch and Goldberg, 1989). Movement-related activity usually begins before the initiation of the behavioral response, and often persists during the behavioral execution. Moreover, this activity is often differential (Niki and Watanabe, 1976) or directional (Kubota and Funahashi, 1982), i.e. neuronal activation occurs when the movement directs one or a few particular directions. Therefore, it has been concluded that movement- related activity in the prefrontal cortex is related to the initiation or execution of the response behavior (Fuster, 1997).

However, the neurophysiological study using saccadic eye movements revealed that, although many prefrontal neurons showed saccade- related activity, the great majority of this activity was post-saccadic (Joseph and Barone, 1987; Funahashi et al., 1991). Post-saccadic activity observed in the prefrontal cortex had some features (Funahashi et al., 1991). First, post-saccadic activity was observed only during saccades for the task and not observed during spontaneous saccades outside the task. Second, a great majority of post-saccadic activity exhibited directional selectivity. The evidence that the distributions of preferred directions and tuning widths were similar between post-saccadic activity and pre-saccadic activity in the prefrontal cortex suggests that post-saccadic activity could be a feed-back signal from the oculomotor centers. Third, Goldman-Rakic et al. (1990) showed that the termination of excitatory delay-period activity coincided with the initiation of post-saccadic activity by population analyses of prefrontal activities. As is seen in Figure 1, excitatory delay-period activity usually terminated rapidly once response behavior was initiated. Therefore, post-saccadic activity has been considered to act as a reset signal to terminate delay-period activity, which becomes unnecessary information once the monkey performed a response behavior. Since hand and arm movement disorders and oculomotor disorders are not observed in prefrontal patients (Fuster, 1997; Stuss and Knight, 2002), the prefrontal cortex does not seem to directly participate in the initiation and the execution of motor behavior. Instead, since the prefrontal cortex has been considered to play an important role for executive control (Smith and Jonides, 1999; Funahashi, 2001), the prefrontal cortex might send regulatory signals to other cortical areas and receive feedback information from these cortical areas to perform multiple operations coordinately. Therefore, although some neurons having pre-saccadic activity actually participate in motor controls, we had better consider that a majority of neurons having movement-related response-period activity may participate in executive control processes.

3. DELAY-PERIOD ACTIVITY AS A NEURONAL CORRELATE OF TEMPORARY INFORMATION STORAGE

Fuster (1973) first showed memory-related single-neuron activity in DLPFC while monkeys performed a manual delayed-response task. He found that most prefrontal neurons showed sustained excitation during the delay period (delay-period activity). In addition, delay-period activity was not observed in trials without baits and in error trials. Therefore, Fuster (1973) suggested that delay-period activity is attributable to a role of the prefrontal cortex in mnemonic processes. Since then, delay-period activity has been observed frequently during delayed-response performances and its characteristics have been analyzed in detail (Funahashi et al., 1989, 1993a,b, 1997; Wilson et al., 1993; Rainer et al., 1998).

A majority of neurons exhibited tonic sustained excitation during the delay period, although some prefrontal neurons exhibited a gradual increase or decrease of their discharge rates during the delay period. The duration of delay-period activity depended on the length of the delay period. The duration of delay-period activity was prolonged or shortened, when the length of the delay period increased or decreased, respectively (Fuster, 1973; Funahashi et al., 1989). In addition, delay-period activity was not observed or was truncated when the subject made an error (Fuster, 1973; Funahashi et al., 1989). These results support the notion that delay-period activity is a neuronal correlate of a temporary storage mechanism of working memory.

In addition, a great majority (80%) of delay-period activity showed directional selectivity, such that delay-period activity was observed only when the visual cues were presented at a certain area in the visual field (Funahashi et al., 1989). Figure 2 is an example of the directional delay-period activity, in which maximum delay-period activity was observed at the 270° trial condition. To describe directional selectivity of delay-period activity quantitatively, Funahashi et al. (1989) constructed a tuning curve by fitting mean discharge rates across different cue directions on the Gaussian function and determined the preferred cue direction that the maximum delay-period activity was observed. The distribution of preferred cue directions revealed that all possible directions around the fixation target were represented by a population of delay-period activity. However, the distribution of preferred cue directions had contralateral bias, such that most of prefrontal neurons had preferred cue directions in the visual field contralateral to the recorded hemisphere. These results suggest that prefrontal neurons exhibiting directional delay-period activity have mnemonic receptive fields (memory fields) within the visual field. Prefrontal neurons' memory fields seem to have similar characteristics to the visual receptive fields of prefrontal visual neurons (Funahashi et al., 1989, 1990). The analysis of the tuning width suggests that the width of the memory field corresponds to a quarter of the visual field. The characteristics of memory fields have been confirmed by Rainer et al. (1998).

Delay-period activity has also been observed during other delay tasks such as visual discrimination tasks (Romo et al., 1999; Constantinidis et al., 2001a), a delayed matching-to-sample task (Wilson et al., 1993; Miller et al., 1996; Rao et al., 1997; Rainer et al., 1999), go/no-go tasks (Watanabe, 1986), and conditional tasks (Asaad et al., 2000; Hoshi et al., 2000). Romo et al. (1999) showed that the discharge rates of prefrontal neurons during the

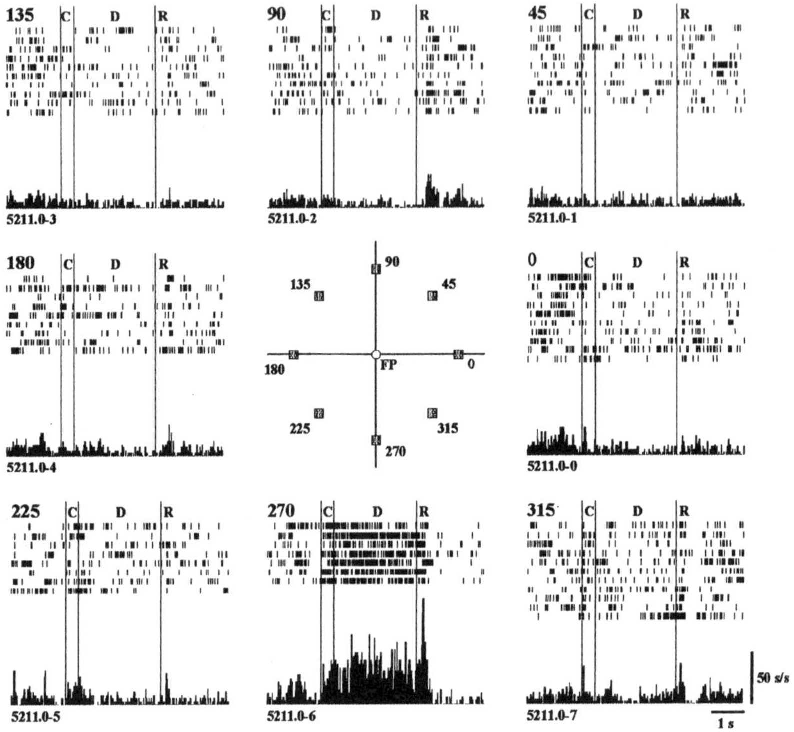

Figure 2: An example of directional delay-period activity. Delay length was 3 sec. Adopted from Funahashi et al. (1989).

delay period varied monotonically depending upon the stimulus frequency when monkeys performed a tactile frequency discrimination task. They found that the discharge rates during the delay period either increased or decreased monotonically when the vibrating frequency at the sample condition changed systematically. Based on these results, they suggested that this monotonic stimulus encoding could be the basic representation of one-dimensional sensory stimulus quantities in working memory system.

These results indicate that delay-period activity is a common feature for many prefrontal neurons. Several experiments indicate that prefrontal neurons having delay-period activity are highly specialized and are distinct components of working memory processes. Funahashi et al. (1990) found that delay-period activity was only task-related activity in 44% of the neurons that exhibited the delay-period activity. Carlson et al. (1997) also found that 73% of task-related neurons exhibited delay-period activity and that most (70%) of neurons having delay-period activity did not respond to any sensory stimuli examined.

In summary, delay-period activity has been observed in many prefrontal neurons while monkeys performed spatial as well as non-spatial working memory tasks. These prefrontal neurons showed differential activation depending upon the nature of the sensory cues or behavioral conditions. In spatial tasks, the magnitude of delay-period activity was tuned spatially, indicating that delay-period activity represents spatial information regarding the sensory cue or the behavioral response. In non-spatial tasks using one-dimensional sensory stimulus such as a frequency discrimination task, the magnitude of delay-period activity represents information regarding stimulus characteristics monotonically. These results indicate that delay-period activity represents distinct information necessary to perform the task correctly, and support that delay-period activity is a neuronal correlate of the temporary storage mechanism of information. Since many prefrontal neurons exhibit only delay-period activity as task-related activity, prefrontal neurons having delay-period activity seem to be highly specialized and distinct components of working memory processes.

4. HOW IS INFORMATION PROCESSED?

4.1 What Information Does Task-Related Activity Represent?

Although many prefrontal neurons exhibited directional delay-period activity during delayed-response performances, it has not been obvious whether directional delay-period activity represents the information regarding where the visual cue has been presented (retrospective information) or the information regarding where the response behavior will be directed (prospective information). Niki and Watanabe (1976) used two types of delayed-response task (the left-right and the up-down delayed-response tasks) and a conditional position task with delay, in which monkeys needed to press the right (or left) key when the visual cue was presented at the upward (or downward) position, and found that 70% of differential delay-period activity represented the location of the visual cue, whereas the remaining 30% represented the location to respond. Funahashi et al. (1993b) used delayed pro- and anti-saccade tasks, and found that 68% of directional delay-period activities were stimulus-direction-dependent, whereas 25% of them were response-direction-dependent.

Recently, Takeda and Funahashi (2002) examined whether directional delay-period activity was cue position-dependent or response direction-dependent, using the ODR task and a rotatory ODR (R-ODR) task. In the

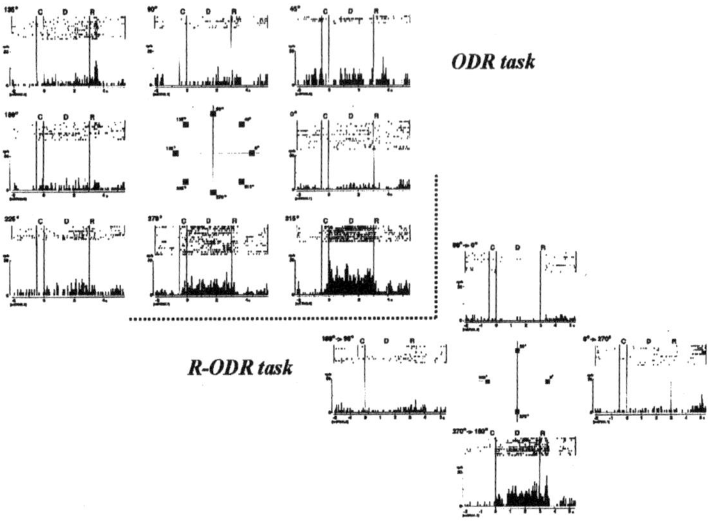

Figure 3: Directional delay-period activity encoding the location of the visual cue. Adopted from Takeda and Funahashi (2002).

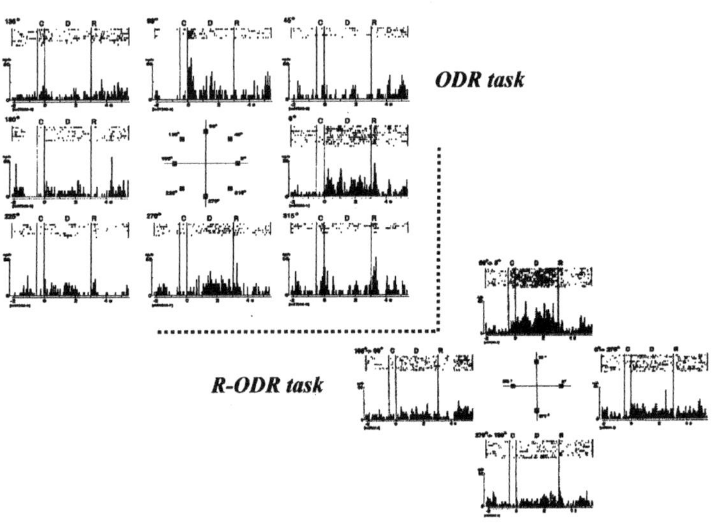

Figure 4: Directional delay-period activity encoding the direction of the saccade. Adopted from Takeda and Funahashi (2002).

latter task, monkeys were required to make a saccade 90° clockwise to the direction where the visual cue had been presented. They defined whether task-related activity encodes the location of the visual cue or the direction of the saccade by comparing the best directions of task-related activity obtained under these two tasks. If the best directions are the same, this activity represents the location of the visual cue. However, if the best direction under the ODR performances has 90° difference from the best direction under the R-ODR performances, this activity represents the direction of the saccade.

Figure 3 shows an example of prefrontal delay-period activity during the performances of these two tasks. This neuron exhibited excitatory delay-period activity only in the 270° and 315° trials of the ODR task and only in the 270° trial of the R-ODR task. In the 270° trial of the R-ODR task, the visual cue was presented at the 270° position, however saccades were directed to the 180° direction. Therefore, it is concluded that delay-period activity of this prefrontal neuron represents the location of the visual cue.

Figure 4 shows another example of prefrontal delay-period activity. Excitatory delay-period activity was observed only in the 0° trial of the ODR task and only in the 90° trial of the R-ODR task. In the 90° trial of the R-ODR task, the visual cue was presented at the 90° position, but the saccade was directed toward the 0° direction. Therefore, it is concluded that this neuron's directional delay-period activity represents the direction of the saccade.

Figure 5 (next page) indicates the distribution of differences of the peak directions between these two tasks. In this figure, the values closed to 0° mean that task-related activity represents the location of the visual cue, whereas the values closed to 90° mean that task-related activity represents the direction of the saccade. This figure reveals (1) that all cue-period activities encode the position of the visual cue, (2) that the majority (86%) of delay-period activity encodes the position of the visual cue whereas the minority (13%) encodes the direction of the saccade, and (3) that most (70%) of response-period activity encode the direction of saccade. These results suggest that the information processing to obtain saccade directions from visual cues is performed in the prefrontal cortex.

4.2 Visualization of Information Processing by Population Analysis of Prefrontal Activity

Information processing to obtain response directions from sensory cues is necessary to perform the R-ODR task correctly. A neuronal mechanism related to this information processing could be an ideal model to examine how information represented by prefrontal activities is processed during working memory processes. There are evidences that single particular

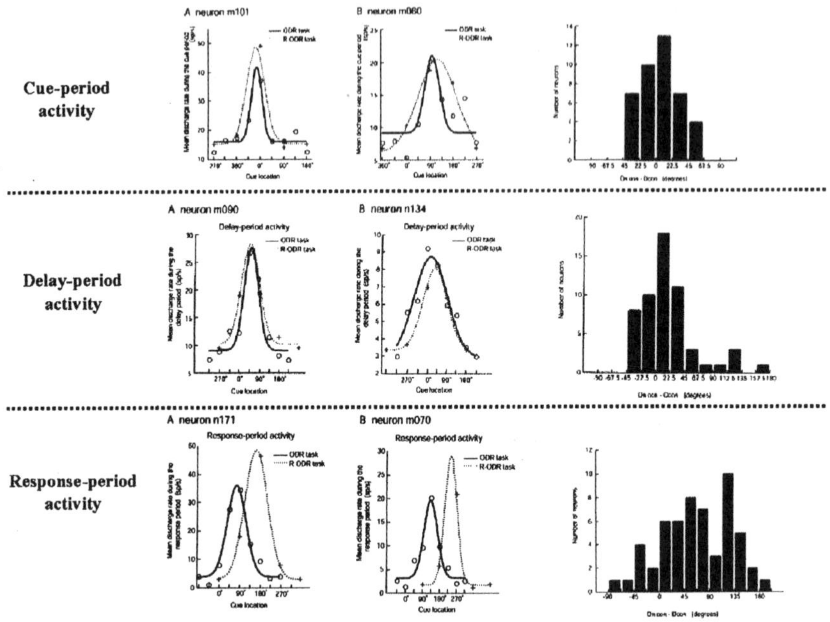

Figure 5: Examples of tuning curves of task-related activity under two task conditions and the distribution of differences of the peak directions. Adopted from Takeda and Funahashi (2002).

information is encoded by a population of neurons. Therefore, to understand how information is processed by prefrontal neurons, we need to consider the information represented by a population of prefrontal activities and its temporal change along the task performance.

We used a population-vector analysis to visualize information represented by a population of prefrontal neurons (Funahashi and Takeda, 2002b). The population-vector analysis was originally introduced by Georgopoulos et al. (1986) to analyze motor cortical activity. They calculated a "cell vector" for every neuron, whose length represented the magnitude of the neuron's activity during arm reaching movements toward a particular direction and whose direction corresponded to the neuron's preferred direction. Using cell vectors, they calculated a population vector, which is a weighted sum of all cell vectors calculated under the same behavioral condition (e.g. one particular direction of arm movements). All task-related activities recorded from the prefrontal cortex were examined under 8 radial directions. Most of task-related activity showed directional selectivity during ODR performances. The preferred direction of each task-related activity could be determined by the tuning curve constructed by fitting neuronal activity during the ODR task on the cosine function.

Cell vectors and population vectors of cue-period activity were shown in the top left figure of Figure 6. Population vectors under all cue conditions directed mostly toward the position of the visual cue, indicating that population vectors calculated from cue-period activity could predict the direction where the visual cue was presented. Similarly, cell vectors and population vectors of response-period activity were shown in the bottom left figure of Figure 6. Population vectors under all cue conditions also directed mostly toward the direction of saccadic eye movement, indicating that population vectors calculated from response-period activity could also predict the direction where the monkey made saccades during the response period.

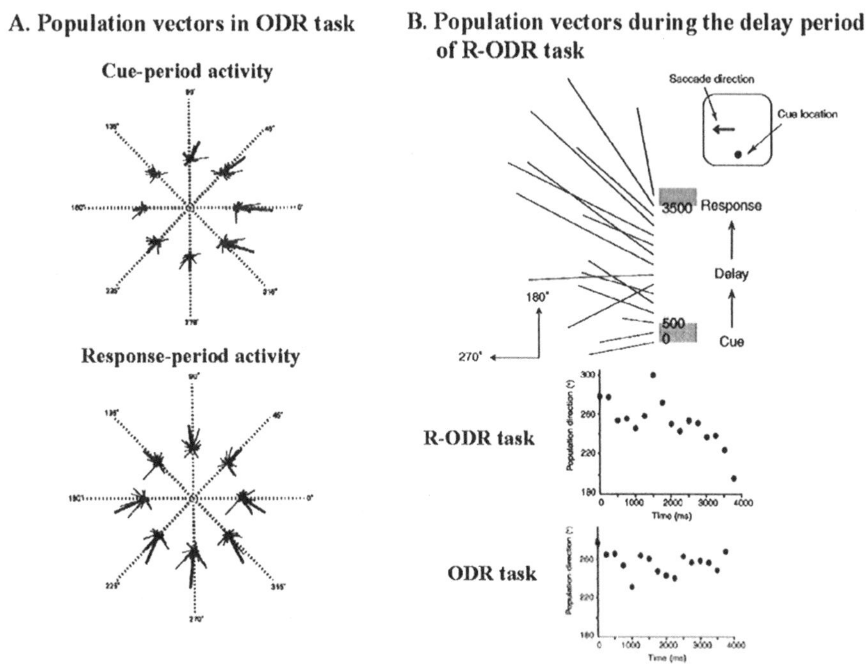

Figure 6: Population vectors calculated by cue- and delay-period activities (A) and population vectors calculated by delay-period activity (B). Adopted from Funahashi and Takeda (2002b).

Based on these results, we then tried to visualize information represented by a population of prefrontal activities during the delay period of the ODR and R-ODR tasks using population vectors. We divided one trial into sixteen 250 ms-periods, and calculated population vectors using mean discharge rates during every 250 ms-period for 110 prefrontal neurons. The top-right figure of Figure 6 shows population vectors during the delay period in the 270° trial of the R-ODR task. Directions of population vectors were directed toward the 270° direction during the cue period and at the beginning of the delay period. Then, directions of population vectors began to rotate at the middle of the delay period, continued to rotate slowly from the 270° direction to the 180° direction during the late half of the delay period, and finally directed mostly to the 180° direction at the response. However, as is shown in the bottom-right figure of Figure 6, in the 270° trial of the ODR task, directions of all population vectors were mainly directed toward the 270° direction. These results indicate that information processing to obtain saccade directions from visual cues takes place during the delay period of the R-ODR task and that information represented by a population of prefrontal activities during this process can be visualized by population vectors calculated from a population of prefrontal activities. The population-vector analysis not only visualizes information represented on internal neuronal processes but also visualizes the temporal change of represented information. Similar analysis has previously been performed by Georgopoulos et al. (1989) using motor cortical neurons. The neuronal population vector is a useful method to demonstrate temporal changes of information represented by a population of neuronal activities during a cognitive process.

5. NEURONAL MECHANISMS FOR WORKING MEMORY PROCESSES

By examining prefrontal task-related activity while monkeys performed various working memory tasks, evidences for neuronal mechanisms related to working memory have been accumulated. For example, delay-period activity has been considered to be a neuronal correlate of the temporary active storage mechanism of information. Delay-period activity has been shown to represent a variety of information including the spatial position, the physical feature of the stimulus, the forthcoming behavioral response, the quality of reward that the subject would receive, the difference of the task, and the rule of the task. Although delay-period activity could represent a variety of information, the information represented by delay-period activity is only the information relevant for the task performance. In addition, using complex conditional tasks, delay-period activity has been shown to represent several kinds of information simultaneously.

Although the evidences for the neuronal mechanisms of short-term storage of information have been accumulated, neuronal mechanisms for processing information is poorly understood. To understand neuronal mechanisms for processing information, we need to know what information is represented by prefrontal activity and how the information represented by a population of prefrontal activities changes along a trial. Among various kinds of information, the spatial information is easy to describe quantitatively and easy to manipulate by investigators. Therefore, we used two kinds of oculomotor delayed-response tasks (ODR and R-ODR tasks) and identified what information the task-related activity in each task represents. In addition, using a population vector analysis, we were able to visualize not only the information represented by a population of prefrontal activities but also the temporal change of the information represented by a population of prefrontal activities along the ODR and R-ODR trials. The exact neuronal mechanism for changing the information represented by a population of prefrontal activities remains to be solved. However, recent experiments have revealed extensive functional interactions among neighboring prefrontal neurons (Funahashi and Inoue, 2000; Constantinidis et al., 2001b; Funahashi, 2001). Therefore, further neurophysiological analyses for dynamic and flexible interaction and its temporal modulation are needed to understand neuronal mechanisms of processing information for working memory.

REFERENCES

Asaad WF, Rainer G, Miller EK (2000) Task-specific neural activity in the primate prefrontal cortex. J Neurophysiol 84:451-459.

Baddeley A (1986) Working Memory. Oxford: Oxford University Press.

Boch RA, Goldberg ME (1989) Participation of prefrontal neurons in the preparation of visually guided eye movements in the rhesus monkey. J Neurophysiol 61:1064-1084.

Carlson S, Rama P, Tanila H, Linnankoski I, Mansikka H (1997) Dissociation of mnemonic coding and other functional neuronal processing in the monkey prefrontal cortex. J Neurophysiol 77:761-774.

Collette F, Van der Linden M (2002) Brain imaging of the central executive component of working memory. Neurosci Biobehav Rev 26:105-125.

Constantinidis C, Franowicz MN, Goldman-Rakic PS (2001a) The sensory nature of mnemonic representation in the primate prefrontal cortex. Nature Neurosci 4:311-316.

Constantinidis C, Franowicz MN, Goldman-Rakic PS (2001b) Coding specificity in cortical microcircuits: a multiple-electrode analysis of primate prefrontal cortex. J Neurosci 21:3646-3655.

Duncan J, Owen AM (2000) Common regions of the human frontal lobe recruited by diverse cognitive demands. Trends Neurosci 23:475-482.

Funahashi S (2001) Neuronal mechanisms of executive control by the prefrontal cortex. Neurosci Res 39:147-165.

Funahashi S, Inoue M (2000) Neuronal interactions related to working memory processes in the primate prefrontal cortex revealed by cross-correlation analysis. Cereb Cortex 10:535-551.

Funahashi S, Kubota K (1994) Working memory and prefrontal cortex. Neurosci Res 21:1-11.

Funahashi S, Takeda K (2002a) Information processes in the primate prefrontal cortex in relation to working memory processes. Rev Neurosci 13:313-346.

Funahashi S, Takeda K (2002b) Population vector analysis by primate prefrontal neuron activities. J Biol Physics 28:527-537

Funahashi S, Bruce CJ, Goldman-Rakic PS (1989) Mnemonic coding of visual space in the monkey's dorsolateral prefrontal cortex. J Neurophysiol 61:331-349.

Funahashi S, Bruce CJ, Goldman-Rakic PS (1990) Visuospatial coding in primate prefrontal neurons revealed by oculomotor paradigms. J Neurophysiol 63:814-831.

Funahashi S, Bruce CJ, Goldman-Rakic PS (1993a) Dorsolateral prefrontal lesions and oculomotor delayed-response performance: evidence for mnemonic "scotomas." J Neurosci 13:1479-1497.

Funahashi S, Chafee MV, Goldman-Rakic PS (1993b) Prefrontal neuronal activity in rhesus monkeys performing a delayed anti-saccade task. Nature 365:753-756.

Funahashi S, Inoue M, Kubota K (1997) Delay-period activity in the primate prefrontal cortex encoding multiple spatial positions and their order of presentation. Behav Brain Res 84:203-223.

Fuster JM (1973) Unit activity in prefrontal cortex during delayed-response performance: neuronal correlates of transient memory. J Neurophysiol 36:61-78.

Fuster JM. (1997) The Prefrontal Cortex: Anatomy, Physiology, and Neuropsychology of the Frontal Lobe, (3rd Ed.), Philadelphia: Lippincott-Raven.

Fuster JM, Alexander GE (1971) Neuron activity related to short-term memory. Science 173:652-654.

Georgopoulos AP, Schwartz AB, Kettner RE (1986) Neuronal population coding of movement direction. Science 233:1416-1419.

Georgopoulos AP, Lurito JT, Petrides M, Schwartz AB, Massey JT (1989) Mental rotation of the neuronal population vector. Science 243:234-236.

Goldman-Rakic PS (1987) Circuitry of primate prefrontal cortex and regulation of behavior by representational memory. In: Handbook of Physiology: The Nervous System: Higher Functions of the Brain, Sect 1, vol V (Plum F, ed), pp 373-417. American Physiological Society.

Goldman-Rakic PS, Funahashi S, Bruce CJ (1990) Neocortical memory circuits. Cold Spring Harbor Symp Quant Biol 55:1025-1038.

Hoshi E, Shima K, Tanji J (2000) Neuronal activity in the primate prefrontal cortex in the process of motor selection based on two behavioral rules. J Neurophysiol 83:2355-2373.

Joseph JP, Barone P (1987) Prefrontal unit activity during a delayed oculomotor task in the monkey. Exp Brain Res 67:460-468.

Kubota K, Funahashi S (1982) Direction-specific activities of dorsolateral prefrontal and motor cortex pyramidal tract neurons during visual tracking. J Neurophysiol 47:362-376.

Kubota K, Niki H (1971) Prefrontal cortical unit activity and delayed alternation performance in monkeys. J Neurophysiol 34:337-347.

Mikami A, Ito S, Kubota K (1982) Visual response properties of dorsolateral prefrontal neurons during visual fixation task. J Neurophysiol 47:593-605.

Miller EK (2000) The prefrontal cortex and cognitive control. Nature Rev Neurosci 1:59-65.

Miller EK, Cohen JD (2001) An integrative theory of prefrontal cortex function. Annu Rev Neurosci 24:167-202.

Miller EK, Erickson CA, Desimone R (1996) Neural mechanisms of visual working memory in prefrontal cortex of the macaque. J Neurosci 16:5154-5167.

Miyake A, Shah P (1999) Models of Working Memory: Mechanisms of Active Maintenance and Executive Control. Cambridge University Press, Cambridge.

Niki H, Watanabe M (1976) Prefrontal unit activity and delayed response: relation to cue location versus direction of response Brain Res 105:79-88.

Petrides M (1994) Frontal lobes and working memory: evidence from investigations of the effects of cortical excisions in non-human primates. In: Handbook of Neuropsychology, vol 9 (Boller F, Spinnler H, and Hendler JA, eds), pp 59-82. Elsevier, Amsterdam.

Rainer G, Asaad WF, Miller EK (1998) Memory fields of neurons in the primate prefrontal cortex. Proc Natl Acad Sci USA 95:15008-15013.

Rainer G, Rao SC, Miller EK (1999) Prospective coding for objects in primate prefrontal cortex. J Neurosci 19:5493-5505.

Rao SC, Rainer G, Miller EK (1997) Integration of what and where in the primate prefrontal cortex. Science 276:821-824.

Romo R, Brody CD, Hernandez A, Lemus L (1999) Neuronal correlates of parametric working memory in the prefrontal cortex. Nature 399:470-473.

Sawaguchi T (1987) Properties of neuronal activity related to a visual reaction time task in the monkey prefrontal cortex. J Neurophysiol 58:1080-1099.

Smith EE, Jonides J (1999) Storage and executive processes in the frontal lobe. Science 283:1657-1661.

Stuss DT, Knight RT (2002) Principles of Frontal Lobe Function. Oxford University Press, New York.

Suzuki H, Azuma M (1983) Topographic studies on visual neurons in the dorsolateral prefrontal cortex of the monkey. Exp Brain Res 53:47-58.

Takeda K, Funahashi S (2002) Prefrontal task-related activity representing visual cue location or saccade direction in spatial working memory tasks. J Neurophysiol 87:567-588.

Watanabe M (1986) Prefrontal unit activity during delayed conditional go/no-go discrimination in the monkey. I. Relation to the stimulus. Brain Res 382:1-14.

Wilson FAW, O'Scalaidhe SP, Goldman-Rakic PS (1993) Dissociation of object and spatial processing domains in primate prefrontal cortex. Science 260:1955-1958.

Chapter 10

THE ROLE OF DOPAMINE IN COGNITION: INSIGHTS FROM NEUROPSYCHOLOGICAL STUDIES IN HUMANS AND NON-HUMAN PRIMATES

Roshan Cools[1] and Angela C. Roberts[2]
Departments of [1]Experimemtal Psychology and [2]Anatomy, University of Cambridge, Downing Street, Cambridge, UK

Keywords: Dopamine, cognition, attentional flexibility, spatial working memory, functional imaging, prefrontal cortex, striatum, Parkinson's disease, marmosets, 6-OHDA.

Abstract: The behavioral evidence for dopaminergic modulation of prefrontal cognitive functioning is reviewed. The experimental studies in rats, monkeys, and humans that highlight the inverted U-shaped relationship between dopamine and cognition are described and discussed in relation to theories of motivation and arousal. It is suggested that the disruptive effects of L-DOPA on reversal learning and decision making and the facilitatory effects of L-DOPA on task-switching and spatial working memory in patients with Parkinson's disease support a role for dopamine in modulating both ventral and dorsolateral-striatal circuits. The fact that certain prefrontal tasks may be more sensitive to dopaminergic modulation than others is considered in the light of findings from 6-OHDA lesion studies in monkeys. A specific role for dopamine in protecting prefrontal processing from interference is shown by the marked disruption of marmosets with 6-OHDA lesions of the prefrontal cortex (PFC) in performing a visual discrimination in the presence of distracting stimuli. The finding that 6-OHDA lesions of the caudate nucleus protect the performance of marmosets from such distracting stimuli is discussed in relation to the possible competition and co-ordination of the PFC and caudate nucleus and emphasizes the need for any theory of dopamine function in the PFC to take into account the role of dopamine at the level of the striatum.

1. INTRODUCTION

The prefrontal cortex (PFC) provides executive control over behavior, acting to optimize behavior in complex situations. Examination of behavior arising from damage to the PFC suggests that the PFC performs this executive role by defining goals and planning the necessary sequences of behaviors to achieve those goals. In so doing, the PFC engages and disengages cognitive components that are themselves mediated by dedicated processing modules in other regions of the brain. Central to its operation are the proposed working memory capabilities of the PFC, which facilitate the processing of task-relevant information (including operations that hold relevant information 'on-line' and 'update' that information when necessary) and inhibit the processing of task-irrelevant information, thereby protecting the task-relevant information from interference. Abundant empirical evidence from psychopharmacological and neuropsychological studies in both experimental animals and humans indicates that dopamine (DA) is critical for many of these prefrontal operations. Along with a number of other neuromodulatory transmitter systems including noradrenaline, serotonin, and acetylcholine, DA modulates prefrontal activity and is, in turn, regulated by feedback projections from the prefrontal cortex (PFC). Here we will consider the nature of the relationship between DA and prefrontal processing, focusing in particular on our own findings in patients with Parkinson's disease and marmosets with 6-OHDA lesions of the PFC.

2. DOPAMINE, COGNITION, AND AROUSAL: INSIGHTS FROM EXPERIMENTAL STUDIES FROM RATS AND MONKEYS

The first study to implicate DA in prefrontal processing was reported by Brozoski, Rosvold, and Goldman in 1979 (Brozoski et al., 1979). They showed that 6-OHDA infusions into the dorsolateral PFC, which destroyed catecholamine terminals, impaired performance on a spatial delayed alternation task. This impairment was nearly as severe as that induced by surgical ablation of the dorsolateral PFC itself, but unlike surgical ablation, the 6-OHDA lesion-induced impairment could be reversed by treatment with the mixed D1/D2 DA receptor agonist, apomorphine. This study has subsequently been replicated in marmosets (Roberts et al., 1994) and rats (Simon, 1981) with 6-OHDA lesions of the PFC. Consistent with an involvement of DA in spatial working memory, DA levels have been shown to increase in the PFC during working memory performance (Watanabe et al., 1997), and the neuronal 'memory field' activity of PFC neurons is strongly modulated by an application of DA (e.g. Sawaguchi et al., 1990; Williams and Goldman-Rakic, 1995; Yang and Seamans, 1996).

It appears that both excessive as well as insufficient DA receptor stimulation can be detrimental for working memory performance (Williams and Goldman-Rakic, 1995; Zahrt et al., 1997; Arnsten, 1998), suggesting that an 'inverted U' relationship exists between DA levels and cognitive performance. For example, Zahrt et al. (1997) examined the effects of infusing a D1 DA receptor agonist, SKF 81297, into the rat PFC during performance of a spatial delayed alternation task. The agonist produced a dose-related impairment on the task when infused into the PFC, which was reversed by pre-treatment with a D1 DA receptor antagonist, SCH 23390. Murphy et al. (1996) showed that infusion of SCH 23390 by itself impaired performance. Thus, cognitive processing, in particular spatial working memory, appears dependent upon an optimal level of DA, with too much or too little DA resulting in disruption of performance.

The principle of curvilinear behavioral dose response curves has long been established within theories of arousal and motivation (e.g. Yerkes and Dodson, 1908; Hebb, 1955; Eysenck, 1982). Optimal levels of arousal have been linked to maximally efficient cognitive performance and lesser or greater levels of arousal to impaired performance (Robbins and Everitt, 1987), and in keeping with these findings, 'PFC' cognitive functions are impaired when subjects are exposed to high levels of arousal in the form of 'stress' (Arnsten, 1998). Since arousing and motivating stimuli, including both stressors and rewards, are known to increase the release of DA and NA in the PFC (Thierry et al., 1976; Robbins and Everitt, 1987), it has been proposed that the catecholamines may well mediate some of the arousal and motivational influences on cognition. For example, administration of DA receptor antagonists have been shown to prevent stress-induced cognitive dysfunction in monkeys (Arnsten et al., 2000), suggesting that stress-induced impairments in working memory performance may be caused by an over-stimulation of catecholamine receptors (Arnsten, 1998). Interestingly, some of the deleterious effects of aging on cognitive performance may also be mediated by the catecholamine systems. Aging is associated with naturally occurring catecholamine depletion (Wenk et al., 1989; Arnsten et al., 1994), and the impairments in spatial working memory performance that accompany aging can be partially reversed by injections of the DA agonist, SKF38393. In contrast, SKF38393 has no effect on the performance of young control monkeys. The complete opposite pattern of results was seen following administration of the D1 DA antagonist, SCH23390, which significantly impaired the memory performance of young control monkeys but did not impair aged monkeys. All of these effects are entirely consistent with an inverted U-shaped behavioral response curve, which predicts that the effects of DA-ergic agonism or antagonism on performance will be dependent upon current baseline levels of DA. Thus, DA-ergic *stimulation* of a depleted baseline (as in aging) or DA-ergic *antagonism* of an already high baseline (as during stress) would be expected to bring levels towards 'optimum', thereby improving performance.

Under certain circumstances, DA-ergic stimulation has even been observed to improve performance in normal, control rats. Granon and colleagues examined the efficacy of a partial D1 DA receptor agonist to improve the accuracy of rats on an attentional task. They demonstrated that an infusion of SKF-38393 into the PFC improved the accuracy of rats with a relatively low baseline level of performance but had no effect in those rats performing at a superior level of performance (Granon et al., 2000). Similar findings have been reported in humans and, as will be discussed below, may be due, in part, to intrinsic differences in the baseline DA levels of individual subjects.

3. COGNITIVE PERFORMANCE AND ITS RELATIONSHIP TO DOPAMINE LEVELS IN HUMANS

Administration of the D2 DA receptor agonist bromocriptine to young healthy human volunteers was observed to *improve* performance on a test of executive function in those individuals with an apparently low working memory capacity while *impairing* performance in those individuals with a high working memory capacity (Kimberg et al., 1997). Similar results were described by Mattay et al. (2000), examining the effects of dextro-amphetamine on performance of an n-back working memory task. To test the hypothesis that such inter-subject variability in cognitive performance and the differential effects of DA-ergic stimulation may be a consequence of underlying differences in baseline DA levels of individual subjects, Egan et al. (2001) exploited the fact that a common functional polymorphism (Val $^{108/158}$ Met) in the catechol-O-methyltransferase (COMT) gene, can account for a 4-fold variation in the catabolism of DA in the human population. Thus, they examined the relationship between this polymorphism and performance on a test of prefrontal function, the Wisconsin Card Sorting Test (WCST; Grant and Berg, 1948), in a number of subject groups including healthy volunteers. A significant relationship was found between these two measures such that the low enzyme activity Met allele (predicted to produce high DA levels) was associated with enhanced performance and the high enzyme activity Val allele (predicted to produce low DA levels) was associated with impaired performance (replicated by Malhotra et al., 2002). Overall, the COMT genotype explained 4% of variance in frequency of perseverative errors on the WCST thus supporting the hypothesis that naturally occurring differences in baseline DA may contribute to variations in cognitive performance. A follow-up study (Mattay et al., 2002b) comparing the effects of dextroamphetamine in subjects with either the Met-Met polymorphism or Val-Val polymorphism suggested that the differential effects of DA-ergic manipulation on cognitive performance may also, in part, be dependent upon the same underlying differences in baseline DA. Dextroamphetamine improved performance on an n-back task in the lower functioning Val-Val subjects (presumed low DA levels), but impaired performance in the higher functioning Met-Met subjects (presumed high DA levels).

Functional magnetic resonance imaging (fMRI) analysis of subjects performing the n-back task has revealed that there is a relationship between this common functional polymorphism and overall PFC activity. Thus, subjects with the low enzyme activity Met allele have the greatest task-related signal increases while subjects with the high enzyme activity Val allele have the smallest task-related increases. Treatment with dextroamphetamine, which improved performance in Val-Val subjects, induced task-related *decreases* in their PFC activity compared with placebo. In contrast, the reduction in performance following dextroamphetamine in the Met-Met subjects was accompanied by task-related *increases* in the PFC compared with placebo (Mattay et al., 2002b). The authors' interpretation of these results was that administration of DA decreased PFC activity in subjects with low baseline DA levels (in Val-Val subjects), by increasing task-relevant neuronal activity, and suppressing task-irrelevant spontaneous activity, i.e. by increasing the signal-to-noise ratio, or the 'efficiency' in the PFC. Conversely, the authors suggested that dextroamphetamine decreased 'efficiency' in subjects with high baseline DA levels (in Met-Met subjects). Increased PFC 'efficiency' following administration of DA agents to subjects with low baseline levels of DA is consistent with previous neuroimaging studies in healthy volunteers (Mattay et al., 2000; Mehta et al., 2000) and also in patients with Parkinson's disease (Cools et al., 2002b, see next section; Mattay et al., 2002a).

4. THE 'OVERDOSE' HYPOTHESIS IN PARKINSON'S DISEASE: ITS RELEVANCE TO COGNITION AND THE INVERTED U-SHAPED DOPAMINE FUNCTION

An alternative approach to assessing the role of DA in human cognition is to study those clinical disorders in which the DA system is highly compromised. Parkinson's disease (PD) is a progressive neurodegenerative disorder, mainly characterized by motor symptoms. The primary pathology is cell loss in the substantia nigra leading to severe DA loss in the dorsal parts of the striatum as well as additional DA loss in the PFC. Although the disease is mainly a movement disorder, patients also exhibit significant cognitive deficits, even in the earliest stages of the disease. Given strong connections between the dorsal parts of striatum and the dorsal parts of the PFC (Alexander et al., 1986), it is not surprising that the pattern of cognitive deficits resembles that seen in patients with dorsal frontal lobe damage (Owen et al., 1992). Thus, like frontal lobe patients, mild PD patients exhibit significant impairment on tests of attentional set shifting (Downes et al., 1989; Owen et al., 1992), task-set switching (Hayes et al., 1998; Cools et al., 2001a), planning and spatial working memory (Owen et al., 1992, 1995). These deficits are sometimes remediated following administration of L-DOPA medication (Lange et al., 1992), a precursor affecting primarily levels of DA in the striatum (Hornykiewicz, 1974; Maruyama et al., 1996). Thus, for example, Lange et al. (1992) showed that withdrawal of L-DOPA exacerbated the deficits on the Tower of London planning task and tests of spatial working memory. In a recent functional neuroimaging study, we have examined the effects of L-DOPA treatment on the blood flow changes associated with these tasks (Cools et al., 2002b). Eleven patients with mild PD were scanned on two occasions, once 'on' L-DOPA and once 'off' L-DOPA, during the performance of the same Tower of London planning test and a related test of spatial working memory. Significant L-DOPA-induced task-related blood flow decreases in the dorsolateral PFC were observed. Thus, whilst patients 'off' L-DOPA exhibited increased blood flow levels during the memory and planning tasks compared with a visuomotor control task, blood flow levels were similar during all tasks in patients 'on' L-DOPA (see Fig. 1).

Comparison of blood flow data extracted from 6 age- and IQ-matched healthy volunteers, acquired during performance of the same task, revealed that L-DOPA normalized blood flow in this PFC area. Although no significant performance differences were observed, the drug-induced blood flow changes in the PFC correlated negatively with the drug-induced performance changes. Thus, the greater the performance improvement following L-DOPA, the greater the corresponding decrease in PFC blood flow, possibly reflecting increased cortical 'efficiency' (see also Mattay et al., 2000, 2002a,b; Mehta et al., 2000).

Although L-DOPA medication generally ameliorates motor symptoms, the effects on cognitive functions are more complex: both beneficial as well as detrimental effects have been observed (Gotham et al., 1988; Kulisevsky et al., 1996, 2000; Swainson et al., 2000). In this context, it is noteworthy that DA depletion in PD progresses over time from the dorsal parts to the ventral parts of the striatum and the mesocorticolimbic VTA (ventral tegmental

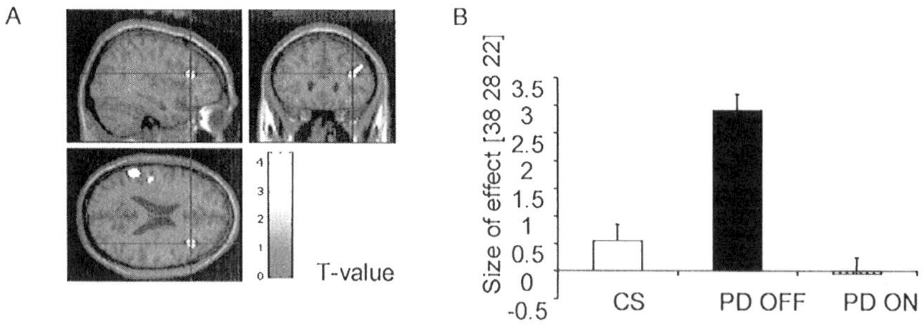

Figure 1: A. The average PET image (statistical parametric map) is shown as sagittal, coronal and transverse sections superimposed upon the MNI (Montreal Neurological Institute) template brain (individual brain considered most typical of the 305 brains used to define the MNI standard) for the drug by task interaction contrast (Spatial Working Memory task-related blood flow changes modulated by L-DOPA). A significant peak was observed in the right dorsolateral PFC (at coordinates x, y, z = 38 28 22; this image was rendered using an uncorrected threshold of $p < 0.005$). B. Blood flow was increased during the Spatial Working Memory task relative to the Visuomotor Control task in the 'off' condition, but blood flow did not differ between the tasks in the 'on' condition. Post hoc comparisons with baseline voxel values, extracted from healthy volunteers, revealed that the blood flow pattern in the 'on' state was similar to that in control subjects, whereas the pattern in the 'off' state was not.

area)-PFC system. Thus, compared with the putamen and the dorsal caudate nucleus, the ventral striatum and the PFC are relatively intact in the early stages of the disease (Kish et al., 1988). This spatio-temporal progression of DA depletion in PD, leading to differential baseline levels of DA within the forebrain of early PD patients, may underlie the opposing effects of L-DOPA on distinct cognitive tasks. Indeed, the 'DA overdose hypothesis' proposed by Gotham et al. (1988) states that L-DOPA doses necessary to ameliorate the lack of DA in severely depleted brain areas such as the dorsal striatum and its connections with the dorsolateral PFC, may 'overdose' any area where DA levels are relatively normal, such as the ventral striatum and its connections to the orbitofrontal cortex. To test this hypothesis, Swainson et al. (2000) compared the performance of never-medicated and medicated mild PD patients on a spatial memory test, associated with the dorsolateral PFC (Owen et al., 1996), and tasks of reversal learning, associated with ventral striatal-orbitofrontal brain circuitry in both monkeys (Divac et al., 1967; Dias et al., 1996) and humans (Rolls, 1999). They showed that never-medicated PD patients, although impaired on the spatial memory test, performed significantly better than medicated PD patients on tasks of reversal learning. Hence, L-DOPA appeared to ameliorate spatial memory performance, but impair reversal learning performance. However, the medicated patients in this study were more severely impaired clinically than the never-medicated patients. Thus, differences in cognitive performance between the two patient groups may have been due to differences in disease severity rather than reflecting the effects of L-DOPA medication.

Recently, we obtained more direct support for the 'overdose hypothesis' by testing a group of patients 'on' medication and a group of patients 'off' medication, which were well-matched for disease severity as well as age, verbal IQ, and usual L-DOPA doses (Cools et al., 2001b). The groups were tested on three tasks of cognitive flexibility, that have been associated with distinct fronto-striatal circuitry. First, a task-switching paradigm was used, in which subjects had to rapidly and continuously switch between two tasks A and B (letter-naming and number-naming), that had been well-learned beforehand (Rogers et al., 1998). The sequence of trials employed (AABBAA and so on) enabled the measurement of switching (i.e. A to B or B to A) against a baseline of non-switching (i.e. A to A or B to B), as captured by the computation of switch costs. Switch costs were obtained by subtracting performance (reaction times and errors) on switch trials from performance on non-switch trials. Second, subjects were tested on the Intra-dimensional/Extra-dimensional (ID/ED) set shifting task, which involves shifting between two dimensions of bi-dimensional compound stimuli in a visual discrimination learning context (see also Downes et al., 1989). This task was designed to decompose the WCST (Grant and Berg, 1948) into its constituent elements (Roberts et al., 1988) and, unlike the task-switching paradigm, the attentional sets are not learned beforehand. It started with a simple discrimination stage, after which it proceeded to eight further stages. In short, after a number of set formation and set maintenance stages in which subjects learned to attend to one of two dimensions, for example 'shape', they had to shift their responding to the other, newly relevant dimension, 'line', at the critical extra-dimensional set shifting (EDS) stage (see Downes et al., 1989 for further details). These first two tasks of high-level attentional shifting have been associated with the *dorsal* PFC and the dorsal caudate nucleus in various neuroimaging and primate studies (Dias et al., 1996; Meyer et al., 1998; Rogers et al., 2000; Sohn et al., 2000). Third, subjects were assessed on the probabilistic reversal learning task, also used by Swainson et al. (2000) (see also Lawrence et al., 1999). In this visual discrimination task, subjects had to discover by trial and error which one of two colors was correct by touching one color on each trial. After forty trials of this initial acquisition stage, the contingencies were reversed so that subsequently the other color was correct, and subjects had to shift responding to the other color. To prevent ceiling effects, a difficult probabilistic contingency was employed, in which correct responses were positively reinforced on 80% of trials but negatively reinforced on 20% of trials. A recent event-related fMRI study confirmed that successful

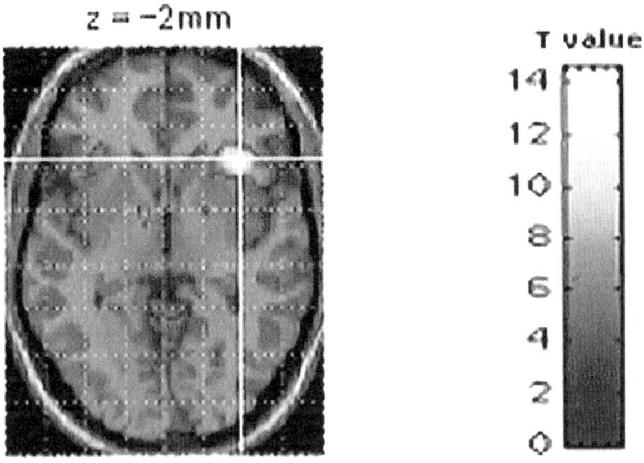

Figure 2: An event-related fMRI study revealed significant signal changes in ventral fronto-striatal circuitry when subjects reversed their responding. Signal changes in the bilateral ventrolateral PFC (peak at coordinates x, y, z = 38, 24, -2) and ventral striatum are superimposed on the MNI (Montreal Neurological Institute) template brain (individual brain considered most typical of the 305 brains used to define the MNI standard).

performance on this task involves *ventral* fronto-striatal circuitry in healthy volunteers (see Fig. 2; Cools et al., 2002a), consistent with studies in non-human primates (Divac et al., 1967; Jones and Mishkin, 1972; Dias et al., 1996). L-DOPA was predicted to induce beneficial effects on the high-level task-switching and ID/ED shifting tasks, associated with dorsal fronto-striatal circuitry, but detrimental 'overdose' effects on the reversal learning task, associated with ventral fronto-striatal circuitry.

Consistent with this 'overdose hypothesis', withdrawal of L-DOPA had a detrimental effect on task-switching, but a beneficial effect on reversal learning. Thus, whilst patients 'off' medication exhibited significantly increased switch costs on the task-switching paradigm compared with patients 'on' medication and controls (see Fig. 3A), significantly more patients 'on' medication failed to complete the reversal learning task than patients 'off' medication and controls (see Fig. 3B). The latter finding of a detrimental effect of L-DOPA on probabilistic reversal learning concurs with findings from Mehta et al. (2001) that administration of the D2 DA receptor agonist bromocriptine to healthy volunteers also impairs performance on the same probabilistic reversal learning task. The finding that L-DOPA did not affect performance at the extra-dimensional shift stage of the ID/ED shifting task may indicate that shifting in the context of new rule learning is less sensitive to L-DOPA withdrawal than shifting between already well-established task-sets. The finding is consistent with a recent study in marmosets (see also below), in which DA depletion from the caudate nucleus impaired a shift to an already well-established, but not a novel, attentional set (Collins et al., 2000). In contrast, the lack of effect of DA on the ID/ED task appears, at first sight, not to be consistent with two previous studies in which administration of the DA D2 antagonist sulpiride and the DA enhancer methylphenidate, respectively, did significantly modulate performance on the task in healthy volunteers (Mehta et al., 1999; Rogers et al., 1999b). However, a more sensitive three-dimensional version of the ID/ED task was employed in the latter studies.

Preliminary data from a pharmacological fMRI study in which we administered the D2 DA receptor antagonist sulpiride and the monoamine enhancer methylphenidate to young healthy volunteers, using a cross-over, placebo-controlled design, are consistent with the findings in PD. In this study, 12 subjects were scanned 3 times during the performance of the same probabilistic reversal learning task and a visual checkerboard presentation, acting as a visual control task. Significant task by drug interactions were observed in ventrolateral, but not dorsolateral, PFC. The results confirmed that the locus of DA-ergic modulation of reversal-related brain areas was the ventrolateral PFC and not dorsolateral PFC (Clark L, Cools R, Bullmore E, Robbins T W, unpublished findings). The lack of modulation by DA of visual areas during checkerboard presentation indicates that the DA agents did not modulate just any brain area that is significantly activated by a task. This finding argues against an interpretation in terms of non-specific vascular effects of DA. Rather, we hypothesize that the finding of DA modulation of ventral, but not dorsal, fronto-striatal brain circuitry during reversal learning was due to modulation at the neuronal level.

A follow-up neuropsychological study on effects of L-DOPA withdrawal in mild PD patients extended our findings in PD patients to other tasks associated with ventral fronto-striatal brain circuitry (Cools et al., 2003). In this cross-over, within-subjects study, PD patients were tested, once 'on' and once 'off' medication, on a decision making task (see Rogers et al. 1999c), also associated with ventral PFC (Rogers et al., 1999a), and on a revised version of the task-switching paradigm. In the decision making task, 10 red or blue boxes were presented on a computer screen, with the red:blue ratio varying from trial to trial. Subjects were told that a yellow token was hidden under one of these colored boxes and that they had to decide whether it was hidden under a red or blue box. They were required, first, to choose the most likely outcome (that is, red or blue) and, second, to bet a certain amount of points which, subsequently, depending on the actual outcome, was added to, or subtracted from, a total points score. The task consisted of two conditions that affected the way subjects placed their bets and acted as a control for impulsivity effects. In the 'ascending' condition, a 'bet' box, initially containing 5% of the total points score gradually filled up with points until it contained 95% of the total points score. Conversely, in the 'descending' condition, the 'bet' box, initially containing 95% of the total score gradually decreased in points until the box contained 5% of the total score. Subjects touched this 'bet' box as soon as it contained the desirable amount of points. Hence, if subjects wished to make a large bet, they had to wait until the box contained the desired amount of points in the 'ascending' condition but make a quick response in the 'descending' condition.

The results replicated previous findings that L-DOPA medication remediated an impairment on task-switching (see Fig. 3C; Hayes et al., 1998; Cools et al., 2001b). In addition, the data revealed that this amelioration of task-switching by L-DOPA was accompanied by significantly increased impulsivity on the decision making task: patients 'on' medication placed their bets more quickly than both patients 'off' medication and controls (see Fig. 3D; Cools et al., 2003). This abnormal betting strategy was hypothesized to reflect either a form of motor impulsivity or delay aversion, i.e. an intolerance for waiting, that can manifest as a tendency to select an immediate reward over a delayed reward (see e.g. Castellanos and Tannock 2002). Sonuga-Barke (2002) has argued that delay aversion is based on more fundamental abnormalities in reward mechanisms, which in turn have been associated with limbic-striatal circuitry, including the nucleus accumbens and the ventral PFC (Mogenson, 1987; Robbins et al., 1989; Schultz et al., 1992; Delgado et al., 2000; Knutson et al., 2001). Consistent with this hypothesis, Cardinal et al. (2001) showed that selective lesions of the rat nucleus accumbens core induced persistent impulsive choice on a delayed reinforcement task. The impulsive, or 'affective' nature of the impairment is critical for the interpretation of the results in terms of the 'overdose hypothesis', as clearly not any decision making impairment is indicative of ventral frontal dysfunctioning (Ernst et al., 2002; Manes et al., 2002). The detrimental effect of L-DOPA on impulse control parallels the deficit, induced by L-DOPA, on reversal learning, which was interpreted to reflect an impairment in the inhibitory control of affective information (Dias et al., 1996). The dual cognitive effects of DA-ergic medication in PD patients on task-switching on the one hand and reversal learning and decision making on the other are consistent with recent models of segregated PFC function, in which distinct dorsal and ventral areas are proposed to underlie attentional and emotional processing respectively (Yamasaki et al., 2002).

In summary, our data reveal both beneficial and detrimental effects of DA on cognitive function *within* the same group of PD patients and provide

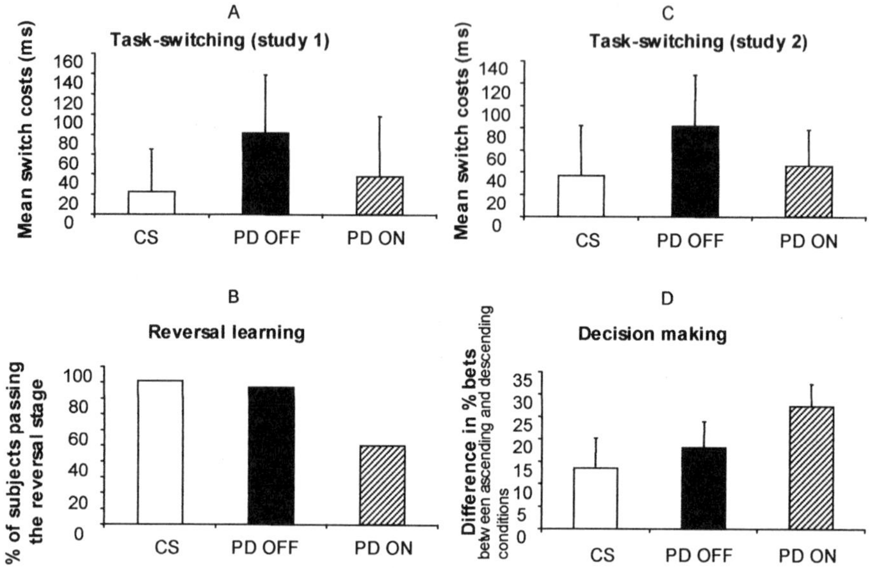

Figure 3: A. **Task-switching (study 1).** Mean switch costs (ms) and standard errors of the difference. Switch costs were calculated by subtracting mean reaction times on non-switch trials from mean reaction times on switch trials. PD patients 'off' medication exhibited significantly increased switch costs compared with patients 'on' medication (F (1,27)=13.2, p=0.001) and controls (F (1,54)=10.7, p=0.002). B. **Reversal learning.** The same group of patients 'on' medication performed significantly more poorly on a task of probabilistic reversal learning than the group of patients 'off' medication and controls. Thus, significantly more patients 'on' L-DOPA failed to reach the learning criterion of 8 consecutive correct trials at the reversal stage than patients 'off' medication (χ^2 (1) 3.8, p=0.05). No differences were observed at the initial acquisition stage. C. **Task-switching (study 2).** A follow-up within-subjects study replicated the finding that patients exhibit increased switch costs when 'off' medication compared with 'on' medication (F (1,11)=5.2, p=0.04). D. **Decision making.** Difference scores in terms of proportions of bets placed between the 'ascending' and 'descending' conditions. Patients were more impulsive when they were 'on' L-DOPA compared with when they were 'off' L-DOPA. Thus, patients placed larger bets in the 'descending' condition, but smaller bets in the 'ascending' condition when they were 'on' medication compared with when they were 'off' medication (F (1,11)=5.9, p=0.03) and compared with control subjects (F (1,22)=5.4, p=0.03). Error bars represent standard errors of the difference.

additional evidence that DA-ergic effects on cognition depend on baseline DA levels in underlying neural substrates. Moreover, it highlights the wide ranging modulatory influence that DA has on prefrontal processing, affecting both dorsolateral and orbital PFC functioning. Some tests of prefrontal function, for example task switching, do appear to be more sensitive to DA modulation than others, for example the ID/ED paradigm; a theme that will be discussed further in the following section. Of course, the effects of L-DOPA treatment on cognition in PD probably reflect L-DOPA's action both at the level of the striatum and at the level of the PFC, changes in PFC activity following L-DOPA treatment being a consequence of either the direct effects of L-DOPA on PFC function or indirect effects via the striatum.

5. FURTHER INSIGHT INTO THE NATURE OF DOPAMINERGIC MODULATION OF PREFRONTAL PROCESSING: EFFECTS OF 6-OHDA LESIONS OF THE PFC IN MARMOSETS ON ATTENTIONAL SET-SHIFTING AND SPATIAL SEARCH

A striking demonstration that certain prefrontal tasks are more sensitive to DA depletion than others comes from a comparison of the effects of 6-OHDA lesions of the PFC of the marmoset on performance of a spatial delayed response task and a self-ordered spatial search task. While 6-OHDA lesions of the PFC in marmosets were shown to disrupt spatial delayed response performance, they had no effect on performance of the spatial search task (Collins et al., 1998; see Fig. 4A). In the latter task, monkeys had to make a series of self-ordered responses to an array of 2, 3, 4, or 5 boxes presented on a computer screen, touching each box only once in order to gain access to reward. Excitotoxic lesions of the PFC produced a robust perseverative impairment at all levels of difficulty, a deficit that disappeared, however, if animals were prevented from repeating their immediately preceding response (see Fig. 4B). The intact performance of 6-OHDA lesioned marmosets on this spatial search task, in contrast to their impaired performance on a classic test of spatial working memory, could be interpreted to suggest that certain prefrontal processes are more sensitive to DA-ergic modulation than others. A related explanation may be that distinct prefrontal processes may depend upon different optimal levels of DA.

An important question that remains to be answered is what the precise nature and function of the DA-ergic modulation of prefrontal processing are. One common theme to emerge from experimental and computational modeling studies of prefrontal DA is that DA may act to stabilize task-

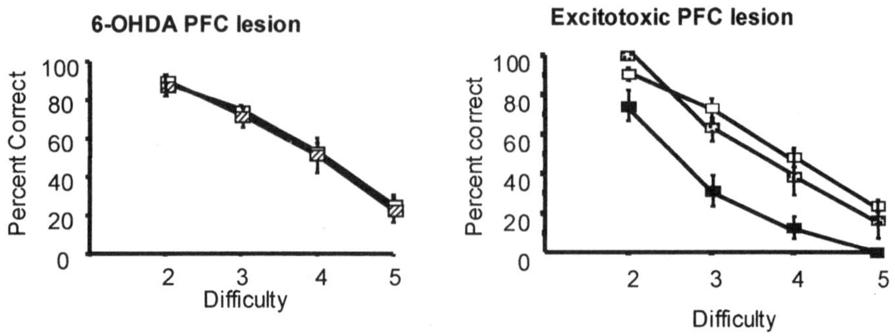

Figure 4: Effects of 6-OHDA and excitotoxic lesions of the PFC on performance of a spatial search task. Percentages of correct responses on the 2, 3, 4 and 5 box problems are presented for the 6-OHDA lesioned group before surgery (open squares) and after surgery (hatched squares) in A and for the excitotoxic lesioned group before surgery (open squares) and after surgery (black squares) in B. Only the excitotoxic lesioned group is significantly impaired, $P < 0.01$. However, by preventing the repetition of the immediately preceding response (see text for details), the performance of the excitotoxic lesioned group (black and white squares) is significantly improved such that it no longer differs from pre-operative levels.

relevant representations within the PFC and protect them from task-irrelevant information (Braver and Cohen, 2000; Durstewitz et al., 2000; Crofts et al., 2001; Dreher et al., 2002). Insufficient DA may therefore lead to temporally and spatially unfocused signal transfer, whilst excessive DA levels may lead to over-focused and thereby blocked signal transfer (Yang and Seamans, 1996). Consistent with this hypothesis, Zahrt et al. (1997) showed that local injection of D1 agonists in the rat PFC induced increased perseveration on the delayed alternation task possibly reflecting blocked signal transfer and thus continuation of previously relevant responses. Conversely, injection of D1 antagonists led to a chance level of performance, possibly reflecting too many response options.

Direct evidence that a reduction in DA leads to increased distractibility has come from our own studies examining the effects of 6-OHDA lesions of the PFC of marmosets on the acquisition and shifting of an attentional set. Marmosets were trained on a series of bi-dimensional compound visual discriminations, similar to that described in the previous section (Roberts et al., 1994). Prior to, and immediately following surgery, the same stimulus dimension, e.g. 'shape', was relevant across each of the discriminations and all animals, regardless of whether they had received a 6-OHDA lesion or sham procedure, displayed a similar level of performance. In contrast, when required to learn a discrimination in which the previously irrelevant dimension became relevant, i.e. to perform a shift of attentional set, the lesioned animals made fewer errors than controls. A follow-up study (Crofts et al., 2001) revealed that this improved performance in attentional set shifting was in fact accompanied by a *disruption* in developing an attentional set. Usually, the development of an attentional set over a series of discriminations is reflected by an accompanying improvement in performance as animals learn to attend to exemplars from the relevant dimension and ignore exemplars from the irrelevant dimension. However, comparison of the first and last discrimination of the series showed no such improvement in 6-OHDA lesioned monkeys (see Fig. 5A). Moreover, when animals had reached criterion on the final discrimination of the series, replacing exemplars from the irrelevant dimension with two novel exemplars disrupted performance in the lesioned monkeys more so than in the controls, confirming that the lesioned animals' performance was subject to increased susceptibility to distraction from task-irrelevant information (see Fig. 5B). Thus, a failure to acquire and subsequently maintain an attentional set in the face of distracting stimuli induced an apparent increased flexibility when shifting to another, newly relevant dimension.

The finding that excitotoxic lesions of the PFC induced the opposite pattern of impairment on the attentional set-shifting task, that is, robust perseveration on a discrimination requiring a shift of attentional set (Dias et al., 1996, 1997) also concurs with the computational model by Dreher et al. (2002), as that model assumes that excitatory inputs on PFC pyramidal cells trigger response switches. The disruption of pyramidal cells by excitatory lesions would prevent these excitatory inputs from inducing any response switches hence resulting in perseveration.

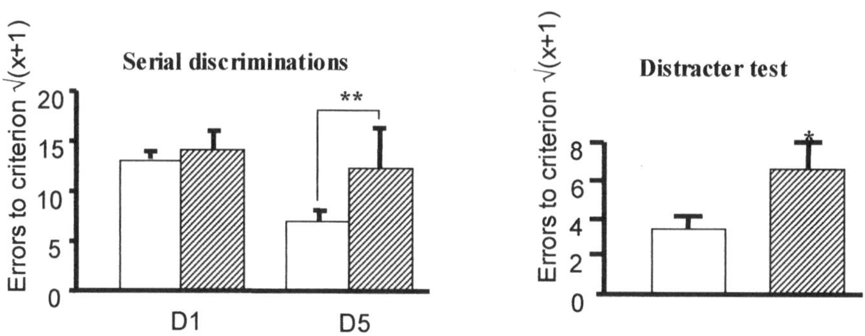

Figure 5: Effects of 6-OHDA lesions of the PFC on the acquisition of a series of compound visual discriminations in which the same perceptual dimension remains relevant across the discriminations A. The significant improvement in performance of the last discrimination (D5) compared to the first discrimination (D1) in the control group (open bars) is not seen in the lesioned group (hatched bars). B. In the subsequent distracter test in which exemplars from the irrelevant dimension are replaced with novel exemplars the lesioned group makes many more errors before regaining criterion relative to the control group. ** $P<0.01$; * $P<0.05$.

The deficit in acquiring and maintaining an attentional set towards a perceptual dimension in the face of distracting information induced by 6-OHDA lesions of the PFC contrasted with the effects that have been observed following 6-OHDA lesions of the caudate nucleus. The latter subcortical DA lesions appeared to induce a greater focusing on the relevant perceptual dimension during the development and maintenance of an attentional set as these animals were significantly less distractible than control monkeys (see Fig. 5B). Thus, reduced DA levels in the caudate nucleus resulted in an animals' responding being controlled more strongly by the currently rewarded stimulus. Although this study did not reveal significant inflexibility at the extra-dimensional shift stages of the ID/ED task, marmosets with 6-OHDA lesions from the caudate nucleus were previously shown to exhibit deficits when shifting back to or re-engaging an already well-established attentional set at a second extra-dimensional shift stage (Collins et al., 2000); akin to the impairments seen in switching between established attentional sets in PD patients. It is noteworthy that an 'inverse' relationship exists between DA metabolism in the two terminal fields of the striatum and the PFC (Pycock et al., 1980; Roberts et al., 1994), and thus the opposing effects of striatal and frontal 6-OHDA lesions underline the possible competition and co-ordination between the PFC and the basal ganglia.

The experimental literature on perseveration versus distractibility in PD is more controversial. Thus, whilst PD patients exhibit robust switching deficits (e.g. Hayes et al., 1998; Cools et al., 2001a), it is unclear whether this impairment is due to inflexibility or instability. Although some authors have suggested overly rigid, inflexible attention in PD (Gauntlett-Gilbert et al., 1999), accumulating evidence indicates that PD patients have difficulties with maintaining an attentional or response set in the face of distraction (Flowers and Robertson, 1985; Sharpe, 1990; Wright et al., 1990; Filoteo et al., 1994; Maddox et al., 1996; Partiot et al., 1996; Filoteo et al., 1997; Filoteo and Maddox, 1999). For example, patients have been shown to be impaired on the Stroop task, in which subjects have to name the color of 'color' words and ignore the actual word itself (Henik et al., 1993; Stam et al., 1993). Patients have also been found to exhibit abnormally rapid disengagement, as evidenced by a reduced cost of invalid cueing and a normal cost of valid cueing in a covert attentional orienting paradigm (Wright et al., 1990; Filoteo et al., 1994). Similar rapid disengagement has been found following administration of a DA receptor antagonist to normal healthy volunteers (Clark et al., 1989). Furthermore, although some studies have reported significantly increased perseveration on the WCST (Grant and Berg, 1948) in PD patients (Lees and Smith, 1983; Dimitrov et al., 1999), studies employing more sophisticated versions of the ID/ED shifting task, enabling the measurement of the contribution of perseveration and 'learned irrelevance' to the shifting deficit, revealed that the deficit at the EDS (extra-dimensional set shifting) stage cannot solely be explained by increased perseveration (Owen et al., 1993; Gauntlett-Gilbert et al., 1999). This evidence, together with results from a number of recent studies indicating that PD patients exhibit switching deficits but *only* when competition is present in the stimulus display (Cools et al., 2001a; Ravizza and Ciranni, 2002), suggests that PD and its associated DA loss are better characterized by instability than inflexibility.

6. SUMMARY AND CONCLUSIONS

In considering the behavioral evidence for DA-ergic modulation of cognition, it has been made clear that DA can influence a variety of prefrontal processes that are dependent upon distinct prefrontal circuitry. These processes appear to depend upon an optimum level of DA, and thus the administration of DA can disrupt or facilitate performance depending upon baseline levels of DA. The finding that different tests of prefrontal function exhibit differential sensitivity to DA manipulation supports further the hypothesis that different processes require different optimum levels of DA. DA levels can be altered by the presence of motivating and arousing stimuli including both reward-related stimuli and stressors. Thus, any theory of PFC DA function has to take into account not only the relationship between DA and reward, and its proposed role as a teaching signal for reinforcement learning (Schultz, 1997), but also the relationship between dopamine, stress, and the potential role that DA may have (along with noradrenaline) in switching the PFC 'on' or 'off-line' (Arnsten, 1998). For a complete understanding of the contribution of DA to prefrontal functioning, it is also necessary to take into account the role of DA at the level of the striatum.

REFERENCES

Alexander G, DeLong M, Stuck P (1986) Parallel organisation of functionally segregated circuits linking basal ganglia and cortex. Ann Rev Neurosci 9:357-381.

Arnsten AF, Murphy FC, Merchant K (2000) The selective dopamine D4 receptor antagonist, PNU-101387G, prevents stress-induced cognitive deficits in monkeys. Neuropsychopharmacology 23:405-410.

Arnsten AF, Cai JX, Murphy FC, Goldman-Rakic PS (1994) Dopamine D1 receptor mechanisms in the cognitive performance of young adult and aged monkeys. Psychopharmacology 116:143-151.

Arnsten AFT (1998) Catecholamine modulation of prefrontal cortical cognitive function. Trends Cogn Sci 2:436-446.

Braver TS, Cohen JD (2000) On the Control of Control: The Role of Dopamine in Regulating Prefrontal Function and Working Memory. In: Control of Cognitive Processes. Attention and Performance XVIII. (Monsell S and Driver J, eds), pp 713-737. The MIT Press, Cambridge, Massachusetts.

Brozoski TJ, Brown R, Rosvold HE, Goldman PS (1979) Cognitive deficit caused by regional depletion of dopamine in the prefrontal cortex of rhesus monkeys. Science 205:929-931.

Cardinal RN, Pennicott DR, Sugathapala CL, Robbins TW, Everitt BL (2001) Impulsive choice induced in rats by lesions of the nucleus accumbens core. Science 292:2499-2501.

Castellanos FX, Tannock R (2002) Neuroscience of attention-deficit/hyperactivity disorder: the search for endophenotypes. Nat Rev Neurosci 3:617-628.

Clark CR, Geffen GM, Geffen LB (1989) Catecholamines and the covert orientation of attention in humans. Neuropsychologia 27:131-139.

Collins P, Roberts AC, Dias R, Everitt BJ, Robbins TW (1998) Perseveration and strategy in a novel spatial self-ordered sequencing task for nonhuman primates: effects of excitotoxic lesions and dopamine depletions of the prefrontal cortex. J Cogn Neurosci 10:332-354.

Collins P, Wilkinson LS, Everitt BJ, Robbins TW, Roberts AC (2000) The effect of dopamine depletion from the caudate nucleus of the common marmoset (callithrix jacchus) on tests of prefrontal cognitive function. Behav Neurosci 114:3-17.

Cools R, Barker RA, Sahakian BJ, Robbins TW (2001a) Mechanisms of cognitive set flexibility in Parkinson's disease. Brain 124:2503-2512.

Cools R, Barker RA, Sahakian BJ, Robbins TW (2001b) Enhanced or impaired cognitive function in Parkinson's disease as a function of dopaminergic medication and task demands. Cereb Cortex 11:1136-1143.

Cools R, Clark L, Owen AM, Robbins TW (2002a) Defining the neural mechanisms of probabilistic reversal learning using event-related functional Magnetic Resonance Imaging. J Neurosci 22:4563-4567.

Cools R, Stefanova E, Barker RA, Robbins TW, Owen AM (2002b) Dopaminergic modulation of high-level cognition in Parkinson's disease: the role of the prefrontal cortex revealed by PET. Brain 125:584-594.

Cools R, Barker RA, Sahakian BJ, Robbins TW (2003) L-DOPA medication remediates cognitive inflexibility, but increases impulsivity in patients with Parkinson's disease. Neuropsychologia (in press).

Crofts HS, Dalley JW, Van Denderen JCM, Everitt BJ, Robbins TW, Roberts AC (2001) Differential effects of 6-OHDA lesions of the frontal cortex and caudate nucleus on the ability to acquire an attentional set. Cereb Cortex 11:1015-1026.

Delgado MR, Nystrom LE, Fissell C, Noll DC, Fiez JA (2000) Tracking the hemodynamic responses to reward and punishment in the striatum. J Neurophysiol 84:3072-3077.

Dias R, Robbins TW, Roberts AC (1996) Dissociation in prefrontal cortex of affective and attentional shifts. Nature 380:69-72.

Dias R, Robbins TW, Roberts AC (1997) Primate analogue of the Wisconsin Card Sorting Test: effects of excitotoxic lesions of the prefrontal cortex in the marmoset. Behav Neurosci 110:872-886.

Dimitrov M, Grafman J, Soares AHR, Clark K (1999) Concept formation and concept shifting in frontal lesion and Parkinson's disease patients assessed with the california card sorting test. Neuropsychology 13:135-143.

Divac I, Rosvold HE, Szwarcbart MK (1967) Behavioral effects of selective ablation of the caudate nucleus. J Comp Physiol Psychol 63:184-190.

Downes JJ, Roberts AC, Sahakian BJ, Evenden JL, Morris RG, Robbins TW (1989) Impaired extra-dimensional shift performance in medicated and unmedicated Parkinson's disease: evidence for a specific attentional dysfunction. Neuropsychologia 27:1329-1343.

Dreher J-C, Guigon E, Burnod Y (2002) A model of prefrontal cortex dopaminergic modulation during the delayed alternation task. J Cogn Neurosci 14:853-865.

Durstewitz D, Seamans J, Sejnowski T (2000) Dopamine-mediated stabilization of delay-period activity in a network model of prefrontal cortex. J Neurophysiol 83:1733-1750.

Egan MF, Goldberg TE, Kolachana BS, Callicott JH, Mazzanti CM, Straub RE, Goldman D, Weinberger DR (2001) Effect of COMT Val108/158 Met genotype on frontal lobe function and risk for schizophrenia. Proc Natl Acad Sci USA 98:6917-6922.

Ernst M, Bolla K, Mouratidis M, Contoreggi C, Matochik JA, Kurian V, Cadet J-L, Kimes AS, London ED (2002) Decision-making in a risk-taking task: A PET study. Neuropsychopharmacology 26:682-691.

Eysenck MW (1982) Attention and Arousal. Cognition and Performance. Springer-Verlag, Berlin Heidelberg New York.

Filoteo JV, Maddox WT (1999) Quantitative modeling of visual attention processes in patients with Parkinson's disease: effects of stimulus integrality on selective attention and dimensional integration. Neuropsychology 13:106-222.

Filoteo JV, Delis DC, Demadura TL, Salmon DP, Roman MJ, Shults CW (1994) Abnormally rapid disengagement of covert attention to global and local stimulus levels may underlie the visuoperceptual impairment in Parkinson's patients. Neuropsychology 8:210-217.

Filoteo JV, Delis DC, Salmon DP, Demadura T, Roman MJ, Shults CW (1997) An examination of the nature of attentional deicits in patients with Parkinson's disease: Evidence from a spatial orienting task. J Int Neuropsychol Soc 3:337-347.

Flowers KA, Robertson C (1985) The effect of Parkinson's disease on the ability to maintain a mental set. J Neurol Neurosurg Psychiatr 48:517-529.

Gauntlett-Gilbert J, Roberts RC, Brown VJ (1999) Mechanisms underlying attentional set-shifting in Parkinson's disease. Neuropsychologia 37:605-616.

Gotham AM, Brown RG, Marsden CD (1988) 'Frontal' cognitive function in patients with Parkinson's disease 'on' and 'off' levodopa. Brain 111:299-321.

Granon S, Passetti F, Thomas KL, Dalley JW, Everitt BJ, Robbins T (2000) Enhanced and impaired attentional performance after infusion of D1 dopaminergic receptor agents into rat prefrontal cortex. J Neurosci 20:1208-1215.

Grant DA, Berg EA (1948) A behavioral analysis of degree of reinforcement and ease of shifting to new responses in a Weigl-type card sorting problem. J Exp Psychol 38:404-411.

Hayes AE, Davidson MC, Keele SW (1998) Towards a functional analysis of the basal ganglia. J Cogn Neurosci 10:178-198.

Hebb D (1955) Drives and CNS (conceptual nervous system). Psychol Rev 62:243-254.

Henik A, Singh J, Beckley DJ, Rafal RD (1993) Disinhibition of automatic word reading in Parkinson's disease. Neurology 17:427-424.

Hornykiewicz O (1974) The mechanisms of action of L-DOPA in Parkinson's disease. Life Sci 15:1249-1259.

Jones B, Mishkin M (1972) Limbic lesions and the problem of stimulus-reinforcement associations. Exp Neurol 36:362-377.

Kimberg DY, D'Esposito M, Farah MJ (1997) Effects of bromocriptine on human subjects depend on working memory capacity. Neuroreport 8:3581-3585.

Kish SJ, Shannak K, Hornykiewicz O (1988) Uneven patterns of dopamine loss in the striatum of patients with idiopathic Parkinson's disease. New Eng J Med 318:876-880.

Knutson B, Adams CM, Fong G, Hommer D (2001) Anticipation of increasing monetary reward selectively recruits nucleus accumbens. J Neurosci 21:1-5.

Kulisevsky J (2000) Role of dopamine in learning and memory: implications for the treatment of cognitive dysfunction in patients with Parkinson's disease. Drugs Aging 16:365-379.

Kulisevsky J, Avila A, Barbano M, Antonijoan R, Berthier ML, Gironell A (1996) Acute effects of levodopa on neuropsychological performance in stable and fluctuating Parkinson's disease patients at different levodopa plasma levels. Brain 119:2121-2132.

Lange KW, Robbins TW, Marsden CD, James M, Owen AM, Paul GM (1992) L-DOPA withdrawal in Parkinson's disease selectively impairs cognitive performance in tests sensitive to frontal lobe dysfunction. Psychopharmacology 107:394-404.

Lawrence AD, Sahakian BJ, Rogers RD, Hodges JR, Robbins TW (1999) Discrimination, reversal, and shift learning in Huntington's disease: mechanisms of impaired response selection. Neuropsychologia 37:1359-1374.

Lees AJ, Smith E (1983) Cognitive deficits in the early stages of Parkinson's disease. Brain 106:257-270.

Maddox WT, Filoteo JV, Delis DC, Salmon DP (1996) Visual selective attention deficits in patients with Parkinson's disease: A quantative model-based approach. Neuropsychology 10:197-218.

Malhotra AK, Kestler LJ, Mazzanti CM, Bates JA, Goldberg TE, Goldman D (2002) A functional polymorphism in the COMT gene and performance on a test of prefrontal cognition. Am J Psychiatry 159:652-654.

Manes F, Sahakian BJ, Clark L, Rogers RD, Antoun N, Aitkin M, Robbins TW (2002) Decision-making processes following damage to the prefrontal cortex. Brain 125:624-639.

Maruyama W, Naoi M, Narabayashi H (1996) The metabolism of L-DOPA and L-threo-3,4-dihydroxyphenylserine and their effects on monoamines in the human brain: analysis of the intraventricular fluid from parkinsonian patients. J Neurol Sci 139:141-148.

Mattay VS, Callicot JH, Bertolino A, Heaton I, Frank JA, Coppola R, Berman KF, Goldberg TE, Weinberger DR (2000) Effects of dextroamphetamine on cognitive performance and cortical activation. Neuroimage 12:268-275.

Mattay VS, Tessitore A, Callicott JH, Bertonlino A, Goldberg TE, Chase TN, Hyde TM, Weinberger DR (2002a) Dopaminergic modulation of cortical function in patients with Parkinson's disease. Ann Neurol 58:630-635.

Mattay VS, Fera F, Hariri A, Tessitore A, Lee S, Goldberg TE, Kolachana BS, Callicott JH, Egan MF, Weinberger DR (2002b) Effect of the COMT VAL158MET polymorphism on the neuromodulatory effects of amphetamine. In: Society for Neuroscience 32nd Annual Meeting. Orlando, Florida, US.

Mehta M, Owen AM, Sahakian BJ, Mavaddat N, Pickard JD, Robbins TW (2000) Methylphenidate enhances working memory by modulating discrete frontal and parietal lobe regions in the human brain. J Neurosci 20: RC65.

Mehta MA, Sahakian BJ, McKenna PJ, Robbins TW (1999) Systematic sulpiride in young adult volunteers simulates the profile of cognitive deficits in Parkinson's disease. Psychopharmacology 146:162-174.

Mehta MA, Swainson R, Ogilvie AD, Sahakian BJ, Robbins TW (2001) Improved short-term spatial memory but impaired reversal learning following the dopamine D2 agonist bromocriptine in human volunteers. Psychopharmacology 159:10-20.

Meyer DE, Evans JE, Lauber EJ, Gmeindl L, Rubinstein J, Junck L, Koeppe RA (1998) The role of dorsolateral prefrontal cortex for executive cognitive processes in task switching. In: Annual Meeting of the Cognitive Neuroscience Society. San Francisco, CA.

Mogenson G (1987) Limbic-motor integration. Prog Psychobiol Physiol Psychol 12:117-169.

Murphy BL, Arnsten AFT, Jentsch JD, Roth RH (1996) Dopamine and spatial working memory in rats and monkeys: pharmacological reversal of stress-induced impairment. J Neurosci 16:7768-7775.

Owen AM, James M, Leigh JM, Summers BA, Marsden CD, Quinn NP, Lange KW, Robbins TW (1992) Fronto-striatal cognitive deficits at different stages of Parkinson's disease. Brain 115:1727-1751.

Owen AM, Roberts AC, Hodges JR, Summers BA, Polkey CE, Robbins TW (1993) Contrasting mechanisms of impaired attentional set-shifting in patients with frontal lobe damage or Parkinson's disease. Brain 116:1159-1179.

Owen AM, Sahakian BJ, Hodges JR, Summers BA, Polkey CE, Robbins TW (1995) Dopamine-dependent frontostriatal planning deficits in early Parkinson's disease. Neuropsychology 9:126-140.

Owen AM, Evans AC, Petrides M (1996) Evidence for a two-stage model of spatial working memory processing within the lateral frontal cortex: a positron emission tomography study. Cereb Cortex 6:31-38.

Partiot A, Verin M, Pillon B, Teixeira-Ferreira C, Agid Y, Dubois B (1996) Delayed response tasks in basal ganglia lesions in man: further evidence for a striato-frontal cooperation in behavioral adaptation. Neuropsychologia 34:709-721.

Pycock CJ, Kerwin RW, Carter CJ (1980) Effect of lesion of cortical dopamine terminals on subcortical dopamine receptors in rats. Nature 286:74-77.

Ravizza S, Ciranni M (2002) Contributions of the prefrontal cortex and basal ganglia to set shifting. J Cogn Neurosci 14:472-483.

Robbins TW, Everitt BJ (1987) Psychopharmacological studies of arousal and attention. In: Cognitive Neurochemistry (Stahl SM, Iversen SD, and Goodman EC, eds). Oxford University Press, Oxford.

Robbins TW, Cador M, Taylor JR, Everitt BJ (1989) Limbic-striatal interactions in reward-related processes. Neurosci Biobehav Rev 13:155-162.

Roberts AC, Robbins TW, Everitt BJ (1988) The effects of intradimensional and extradimensional shifts on visual discrimination learning in humans and non-human primates. Q J Exp Psychol 40b:321-341.

Roberts AC, De Salvia MA, Wilkinson LS, Collins P, Muir JL, Everitt BJ, Robbins TW (1994) 6-Hydroxydopamine lesions of the prefrontal cortex in monkeys enhance performance on an analog of the Wisconsin Card Sort Test: possible interactions with subcortical dopamine. J Neurosci 14:2531-2544.

Rogers RD, Sahakian BJ, Hodges JR, Polkey CE, Kennard C, Robbins TW (1998) Dissociating executive mechanisms of task control following frontal lobe damage and Parkinson's disease. Brain 121:815-842.

Rogers RD, Owen AM, Middleton HC, Williams EJ, Pickard JD, Sahakian BJ, Robbins TW (1999a) Choosing between small, likely rewards and large, unlikely rewards activates inferior and orbitofrontal cortex. J Neurosci 20:9029-9038.

Rogers RD, Blackshaw AJ, Middleton HC, Matthews K, Hawtin K, Crowley C, Hopwood A, Wallace C, Deakin JFW, Sahakian BJ, Robbins TW (1999b) Tryptophan depletion impairs stimulus-reward learning while methylphenidate disrupts attentional control in healthy young adults: implications for the monoaminergic basis of impulsive behavior. Psychopharmacology 146:482-491.

Rogers RD, Everitt BJ, Baldacchino A, Blackshaw AJ, Swainson R, Wynne K, Baker NB, Hunter J, Carthy T, Booker E, London M, Deakin JFW, Sahakian BJ, Robbins TW (1999c) Dissociable deficits in the decision-making cognition of chronic amphetamine abusers, opiate abusers, patients with focal damage to prefrontal cortex, and tryptophan-depleted normal volunteers: evidence for monoaminergic mechanisms. Neuropsychopharmacology 20:322-339.

Rogers RD, Andrews TC, Grasby PM, Brooks DJ, Robbins TW (2000) Contrasting cortical and subcortical activations produced by attentional-set shifting and reversal learning in humans. J Cogn Neurosci 12:142-162.

Rolls ET (1999) The functions of the orbitofrontal cortex. Neurocase 5:301-312.

Sawaguchi T, Matsumura M, Kubota K (1990) Effects of dopamine antagonists on neuronal activity related to a delayed response task in monkey prefrontal cortex. J Neurophysiol 63:1401-1412.

Schultz W, Apicella P, Scarnati E, Ljungberg T (1992) Neuronal activity in monkey ventral striatum related to the expectation of reward. J Neurosci 12:4595-4610.

Schultz W (1997) Dopamine neurons and their role in reward mechanisms. Curr Opin Neurobiol 7:191-197.

Sharpe HS (1990) Distractibility in early Parkinson's disease. Cortex 26:239-246.

Simon H (1981) Dopaminergic A10 neurons and the frontal system. J Physiol 77:81-95.

Sohn M, Ursu S, Anderson JR, Stenger VA, Carter CS (2000) The role of prefrontal cortex and posterior parietal cortex in task switching. Proc Natl Acad Sci USA 97:13448-13453.

Sonuga-Barke EJ (2002) Psychological heterogeneity in AD/HD - a dual pathway model of behavior and cognition. Behav Brain Res 130.

Stam CJ, Visser SL, Op de Coul AAW, De Sonneville LMJ, Schellens RILA, Brunia CHM, De Smet JS, Gielen G (1993) Disturbed frontal regulation of attention in Parkinson's disease. Brain 116:1139-1158.

Swainson R, Rogers RD, Sahakian BJ, Summers BA, Polkey CE, Robbins TW (2000) Probabilistic learning and reversal deficits in patients with Parkinson's disease or frontal or temporal lobe lesions: possible adverse effects of dopaminergic medication. Neuropsychologia 38:596-612.

Thierry AM, Tassin JP, Blanc G, Glowinski J (1976) Selective activation of mesocortical DA system by stress. Nature 263:242-244.

Watanabe M, Kodama T, Hikosaka K (1997) Increase of extracellular dopamine in primate prefrontal cortex during a working memory task. J Neurophysiol 78:2795-2798.

Wenk GL, Pierce DJ, Struble RG, Price DL, Cork LC (1989) Age-related changes in multiple neurotransmitter systems in the monkey brain. Neurobiol Aging 9:11-19.

Williams GV, Goldman-Rakic PS (1995) Modulation of memory fields by dopamine D1 receptors in prefrontal cortex. Nature 376:572-575.

Wright MJ, Burns RJ, Geffen GM, Geffen LB (1990) Covert orientation of visual attention in Parkinson's disease: an impairment in the maintenance of attention. Neuropsychologia 28:151-159.

Yamasaki H, LaBar KS, McCarthy G (2002) Dissociable prefrontal brain systems for attention and emotion. Proc Natl Acad Sci USA 99:11447-11451.

Yang CR, Seamans JK (1996) Dopamine D1 receptor actions in layers V-VI rat prefrontal cortex neurons in vitro: modulation of dendritic-somatic signal integration. J Neurosci 16:1922-1935.

Yerkes RM, Dodson JD (1908) The relation of the strength of stimulus to the rapidity of habit formation. J Comp Neurol Psychol 18:459-482.

Zahrt J, Taylor JR, Mathew RG, Arnsten AFT (1997) Supranormal Stimulation of D1 Dopamine Receptors in the Rodent Prefrontal Cortex Impairs Spatial Working Memory Performance. J Neurosci 17:8528-8535.

Acknowledgements

This work was supported by a programme grant from the Wellcome Trust to T.W. Robbins, B.J. Everitt, A.C. Roberts, and B.J. Sahakian. R. Cools was supported by the Parkinson's Disease Society of the U.K., is currently a Dorothy Hodgkin Royal Society Research Fellow, and holds a junior research fellowship from St John's College, Cambridge. We also acknowledge the support from the MRC Centre, Behavioral and Clinical Neuroscience.

Chapter 11

THE ROLE OF HUMAN PREFRONTAL CORTEX IN MOTIVATED PERCEPTION AND BEHAVIOR: A MACROSCOPIC PERSPECTIVE

Andreas Keil
Department of Psychology, University of Konstanz, Konstanz, Germany.

Keywords: PFC networks, neuroplasticity, EEG, MEG, emotion, motivation, visual perception, human.

Abstract: This chapter discusses a framework for the understanding of perception and action regulation in the presence of behaviorally relevant information. The role of large-scale prefrontal cortex (PFC) networks in dynamic and variable neuronal architecture underlying this particular function is highlighted. Evidence is presented supporting the view that macroscopic (neural mass) oscillations are crucial for signal transmission and plastic changes in these networks. Thus, important aspects of PFC functioning can be studied using large-scale electrocortical measures with respect to time course and topography, employing frequency-domain analyses. These issues are illustrated with experimental data from studies of selective attention, operant conditioning, gestalt perception, and emotional perception. The chapter concludes with a discussion of elements for a model of plastic perception-action regulation as mediated by dynamic cortical networks.

1. INTRODUCTION

In electrophysiological research on the functioning of the brain in action, studies on multiple levels of observation have contributed significantly to our understanding of prefrontal cortex (PFC) functioning. Microscopic-level approaches have explored the activity of single animal neurons in response to external stimuli as well as in anticipation of internal and external events. Paired with behavioral paradigms, important findings of this well-known research have led to models of stimulus processing, memory, and overt action that focus on the role of specific cells during certain animal behaviors (Fuster, 1990). On the mesoscopic level, authors have examined neural activity as measured by means of local field potentials or dense intracranial electrode arrays in human and non-human subjects (Freeman, 1994). These approaches have stimulated models of inter-laminar interaction and neuronal communication between nearby regions. By contrast, macroscopic parameters of neural activity have been used to assess the characteristics of large neuronal assemblies that fire in synchrony and thus generate detectable voltage or magnetic gradients on the surface of the scalp. These are usually recorded by means of electroencephalography (EEG) or magneto-encephalography (MEG). It is this latter type of measures that will be focused on in this chapter. As a consequence, the theoretical perspective presented here will emphasize the role of PFC as a key structure in a plastic, variable large-scale network that receives highly processed sensory information and organizes motivated perception and behavior. To this end, I first turn to some theoretical and methodological issues related to using large-scale electro-cortical PFC activity as a dependent variable. The following sections discuss findings of macroscopic brain oscillations in different behavioral and cognitive domains. Finally, elements for a dynamic model of motivated perception and behavior are presented. This model aims at accounting for the large amount of behavioral and physiological data that have been related to different aspects of affective/motivational perception and action.

1.1 Evoked and Induced Brain Responses in Human EEG/MEG Data

1.1.1 The Neurophysiological Basis of Large-Scale Brain Oscillations

One requirement for the extra-cranial measurement of voltage currents generated by macroscopic brain processes is the synchronous activity of a large number of cortical neurons. These must have common orientation in space to generate an open electrical field that is strong enough to be measured from outside the brain. Apical dendrites of cortical pyramidal cells, the cell type that represents the majority of cortical neurons, are often aligned in parallel and thus meet this requirement. Furthermore, histological and electrophysiological works have suggested that excitatory synapses at apical dendrites of cortical pyramidal cells mediate a large proportion of intra-cortical signal exchange (Braitenberg and Schüz, 1998). Importantly, these synapses comprise both local adjunctions as well as connections with distant cortical neurons via longer-range cortico-cortical fibers. Thus, the voltage gradients as reflected in EEG/MEG recordings are commonly believed to reflect meaningful information on neuronal mass activity

(Freeman, 1994). In terms of EEG recordings, the number of cortical pyramidal cells necessary for generating extracranially detectable electric fields depends on recording technology and anatomical features such as skull thickness. Hence, estimates vary from 10 million to 1 billion neurons (Nunez, 2000). This number may be substantially smaller for MEG recordings (Elbert, 1998). Given that communication within and between coupled neuronal populations of neurons involves excitatory as well as inhibitory connections, the overall activity of the network in time will likely be oscillatory in nature. This property represents a general feature of other biological large-scale systems (Haken, 1983). Given its salience even with simple recording setups, the oscillatory character of EEG and the relationship between different types of oscillations and mental processes was the focus of pioneering work in electroencephalography (Berger, 1938).

1.1.2 Functional Relevance of Large-Scale Oscillatory Activity in the Human Brain

Taken together, these considerations suggest the sensitivity of EEG/MEG measures to coherently oscillating, spatially extended cortical networks of neurons. On a psychological level, mechanisms based on oscillations account for behavioral states that require stability of information processing across time as is the case, for instance, in working memory tasks, but also allow rapid switching between states (Vaadia et al., 1995). Oscillatory coupling serves the latter purpose because small changes of inhibition/excitation can easily and rapidly change the overall system dynamics. The view is adopted here that spatio-temporal properties of neuronal masses represent relevant aspects of external stimuli as well as internal dispositions toward these stimuli (Freeman, 1994). PFC neuronal networks have a special role in this process, contributing to the dynamic adaptation of behavior to external requirements. A glance at PFC afferents suggests that this function is based on multi-modal sensory information at the highest level of processing. PFC receives highly processed sensory information from visual, auditory, and somatosensory regions. Therefore, approaches that focus on integrative processing appear most useful in the study of PFC networks. One major group of such theories, accounting for coherent oscillatory activity in the brain, build on Hebb's well-known idea of neuronal cell assemblies (Hebb, 1949). Hebb described these as being formed by temporally correlated action potentials. Drawing mainly from speculation, he assumed that activity 'reverberating' in assemblies and changes of synaptic connectivity may underlie learning, memory, and perception. Recent theoretical proposals extended Hebb's model by the dynamical dimension of synchronized neuronal firing rates (e.g. Singer et al., 1997). The majority of this theoretical work aimed at explaining perceptual feature binding and perceptual integration. The following sections of the present chapter will discuss in more detail the implications of these studies for other than perceptual functions.

Oscillatory large-scale neuronal activity can be examined using a variety of recording and data reduction techniques, each shedding light on different facets of this activity. A useful distinction between these different aspects has been made by Galambos (1992). He proposed to distinguish between (1) spontaneous rhythms, which are not related to external stimuli, (2) evoked responses, which are elicited and precisely time-locked to the onset of an external stimulus, (3) emitted oscillations, which are time-locked to a stimulus that has been omitted, and (4) induced oscillations that are initiated by, but not time- and phase-locked to, the onset of a stimulus. One of the most widely used electro-cortical parameters in the cognitive neurosciences, the event-related potential (ERP), is based on time-domain averaging across stimulus events. In Galambos' terms, the ERP thus measures time- and phase-locked processes, i.e. evoked activity. In contrast, measures involving frequency-domain averaging across events are sensitive to induced, in addition to evoked, neuronal activity. They reflect amplitude and phase changes in response to stimuli, while they do not depend on phase-locking. Further advantages of frequency-domain approaches to the study of oscillatory activity, compared to ERP parameters, lie in their more immediate relationship with physiological mechanisms as well as with theories stated on a neurophysiological level. Among other methods, time-frequency representations of EEG/MEG signals have allowed a description of temporal dynamics in different frequency ranges. Time-frequency data can be obtained by means of wavelet-transform, a method that allows the researcher to use variable temporal resolution at different frequencies. This helps identifying brief epochs of high-frequency oscillations (above 20 Hz), which are assumed to occur during formation of percepts or memories, during activation of learned associations, as well as during preparation and execution of actions. These high-frequency phenomena have often been referred to as gamma-band activity (GBA). Given their relationship with the functions listed above, GBA measures may be a useful tool to examine the network aspects of PFC functioning. Parameters of low-frequency electro-cortical dynamics are useful to complete this picture. For instance, modern views of alpha oscillations (8-12 Hz) relate these rhythms not only to idling states of the brain but also to diverse functions comprising sensory, motor, and memory processes. In section 2 of this chapter, I will discuss evidence showing that a variety of specific functions involve large-scale modulations of PFC networks both in terms of amplitude and phase. Examples are presented from the studies of operant conditioning, selective attention, and emotional perception. As a consequence of the empirical evidence, a dynamic network model of motivated perception and action regulation is outlined.

1.2 Behavioral Studies Relating Perception and Action to PFC Functioning

Behavioral studies have a crucial role when it comes to generation of hypotheses on brain functioning. In this context, behavioral work with brain-lesioned patients has provided important insights in terms of structures being necessary conditions for the execution of experimental tasks. Specifically, experimental designs are important that demonstrate overlaps between perception and action coding, because they suggest representation in a common network. Indeed, a number of recent studies in healthy participants have provided evidence for overlapping coding of perceptual and motor information in experimental situations (Prinz, 1997). Using a variety of tasks, this work has suggested that motor actions may alter the way percepts are organized and vice versa. In a typical experiment (Hecht et al., 2001), participants were either trained to perform a motor task involving execution of two cyclic arm movements at a given speed (motor group), or alternatively were trained to give estimates regarding the timing of a visually displayed moving stimulus (perception group). Testing the effects of training on the domain not involved, i.e. testing accuracy of the motor task in the perceptual group and vice versa, Hecht and colleagues observed improved performance, compared to a non-trained control group. Thus, benefit by training was not specific to the respective domain, supporting the view that changes coded in one domain (motor or perceptual skills) were available for the other one. In conclusion, these data point towards integration of information coming from different modalities and domains into a common network. They are also in line with views based on animal electrophysiology (Fuster, 1990), emphasizing the relevance of temporal organization for sensorimotor integration.

As stated above, PFC is a candidate structure for the regulation and organization of these higher-order representations as well as their behavioral manifestations. One possible source of experimental evidence for a network perspective of PFC plasticity in perception and behavior relies on behavioral and cognitive dysfunction in brain-lesioned patients. Many authors in this field have demonstrated a strong decline in perceptual and behavioral organizations along temporal and spatial dimensions, when PFC was impaired by traumatic injury or other pathological processes. This body of evidence emphasizes the critical role of PFC in a variety of tasks, which have been related to numerous functional domains. The wealth of findings cannot be re-iterated here. Recent comprehensive reviews on this literature are given e.g. by Wood and Grafman (2003). It is important for this chapter however to note that most neuropsychological results point to a general deficit in perception-action organization in time and space, when PFC lesions are present. Many authors have described deficits following damage to PFC in terms of affective and motivational processes. For the classic case of Phineas Gage, this has nicely been illustrated by Damasio and his group (Damasio et al., 1994), who related Gage's PFC lesion to impairments in decision making and emotional processing. Similarly, Rolls and colleagues (Rolls et al., 1994) reported that patients with damage to their ventral PFC displayed reduced ability in adapting their behavior to changing reward contingencies. Ever since the report on Phineas Gage's lesion-related deficits, the role of PFC in inhibition of impulsive behavior has also been discussed. Thus, many studies have focused on the consequences of impaired action-perception regulation for social behavior. There is abundant evidence on PFC patients showing marked problems in social functioning such as anti-social behavior, while cognitive abilities are spared (Brazzelli et al., 1994; Grafman et al., 1996). But prefrontal patients also show a variety of specific problems in typical laboratory tasks related to integration of action and perception representations. Reviewing this literature, Knight and Grabowecky (2000) suggested that PFC deficits manifest themselves in the paradigms such as inhibitory control of perception, excitatory control of perception, sustained attention, stimulus-elicited behavior, planning tasks, and temporal coding tasks.

2. LARGE-SCALE ELECTROCORTICAL CORRELATES OF PFC FUNCTIONING

2.1 Acquisition of New Behaviors and Memory Formation

Having illustrated that i) behavioral and perceptual representations interact bi-directionally, suggesting integrative processing, and that ii) PFC is a necessary structure to temporally and spatially organize behavior with respect to a perceived context, I will now address the question of large-scale brain oscillations during integrative processing. It was suggested earlier in this chapter that macroscopic PFC networks act to link highly processed sensory information to action representations and vice versa. Consequently, PFC should play a central role in learning contingencies between behaviors and reinforcing events. One approach to test this hypothesis relies on operant conditioning paradigms, as they involve the acquisition of associations between a specific behavior and certain stimuli having reinforcing value. In addition, the temporal relationship between stimuli and behavior can be varied in such a task, thus allowing to assess the effects of time representation on macroscopic PFC activity. Especially during shaping, which involves continuous refinement of the learned response, ongoing changes in PFC activity can be expected. On the level of neural systems, several studies have demonstrated a critical role of dorsolateral, medial, and orbital prefrontal cortices for the acquisition of reinforcement contingencies. Reviewing human electrophysiological studies of orbitofrontal function, Zald and Kim (1996) concluded that orbitofrontal networks are essential in assembling memories and behavioral patterns that enable the individual to react appropriately on the basis of past experience. Likewise, reports from neuroimaging studies demonstrated prefrontal activation during verbal memory encoding and retrieval (Wagner et al., 1998), as well as during associative learning (Molchan et al., 1994). In line with these neuroimaging results, Tallon-Baudry and colleagues (Tallon-Baudry et al., 1998, 1999) have observed modulations of high-frequency EEG activity during a delayed matching to sample task. This task involved the activation of a learned representation during a retention interval as well as initiation of a correct behavioral response, depending on stimulus parameters. Taken together, these studies point to a role of high-frequency modulations of neuronal masses in memory formation and learning. Paralleling animal experiments and in vitro cellular works, an increase in synchronization of neuronal activity may contribute to the formation/reorganization of cell assemblies (Miltner et al., 1999) and possibly to the retrieval of information stored in such assemblies.

In a recent study, Keil et al. (2001b) examined oscillatory PFC activity in human dense array EEG during operant conditioning, using a fixed-interval reinforcement schedule with a variable limited hold period. The task involved a motor action (key press) that earned maximum reward when performed within a given time interval. This interval was narrowed in the course of the experiment, depending on performance, which resulted in shaping (see Fig. 1).

Thus, participants' temporal accuracy was continuously shaped throughout the experimental session. Each response elicited a screen display of numbers indicating the money value of that response, serving as reinforcers. Random reinforcement and self-paced button pressing without reward were added as control conditions. These were designed to reflect processing of reward stimuli having no perceived relationship with behavior (random reinforcement) and to account for the effects of action execution without necessity to integrate action and perception (self-paced pressing). EEG was recorded from 128 electrodes, and time-frequency analyses of EEG were conducted for the period following the presentation of reward stimuli. Figure 1 shows topographical maps of EEG high frequency power in response to the presentation of the feedback stimulus. Frontal EEG gamma activity (20 –

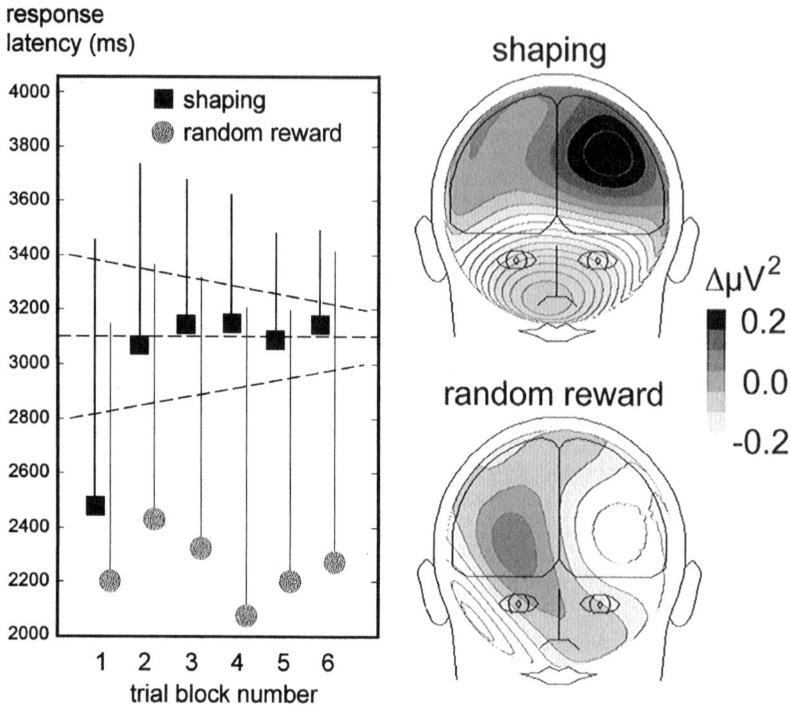

Figure1: Prefrontal high-frequency oscillations as a correlate of shaping. Left panel: Grand mean (n=11) latencies of key presses for the shaping condition (black squares) and the random reinforcement condition (gray circles), shown as a function of time. Each bin represents an average across 10 trials. Bars indicate standard deviation. Note that the shaping regime was effective in enhancing likelihood of response latencies within the limited hold interval, which was narrowed with improving performance (dashed lines). Right panel: Topography of high-frequency EEG spectral power associated with processing of the reward stimulus (400-650 ms after stimulus onset) during shaping (top) and random reinforcement (bottom).

30 Hz) showed a differentiation between operant learning and the two control conditions both in terms of amplitude and topography. In particular, shaping was associated with a pronounced left frontal GBA increase in response to feedback stimuli, whereas this pattern was observed neither in the random reinforcement nor in the self-paced condition. Thus, adaptation of behavior to changing external contingencies was specifically related to differential activation at prefrontal recording sites. These findings are concordant with the notion that macroscopic high-frequency dynamics of neuronal cell assemblies may be regarded as a mechanism involved in learning and memory. More precisely, they suggest that continuous changes

of PFC neural mass activity are involved in tasks requiring integrative processing of sensory information and behavioral strategies.

2.2 Selective Attention and Top-Down Feature Processing

Integrated neuronal representations of temporal and spatial order are not only essential in the acquisition of new behaviors, as is the case in conditioning paradigms. Many aspects of selective attention are related to the ability to identify relevant stimuli in a temporal stream of information, or alternatively to adapt a motor response to the task requirements present at a specific location in the field of view. Sometimes, it is also necessary for the organism to specifically react to combinations of features. In experimental psychology, these phenomena have been examined using target detection tasks. For example, in feature-based attention paradigms, participants are asked to respond when a task-relevant set of stimulus features appears in a series of sensory events. It is not surprising that electrophysiological, neuropsychological, and neuroimaging data have consistently pointed to an important role of PFC in this domain of tasks. In particular, involvement of large-scale fronto-parietal networks has been established for attentional processes such as spatial selection with and without eye movements (Corbetta, 1998), feature selection (Anllo-Vento and Hillyard, 1996), sustained attention (Harmony et al., 1999), mismatch detection (Alho et al., 1994), as well as orienting in time (Nobre, 2001).

Measures of large-scale oscillatory brain activity have also been used to describe PFC functioning during target detection tasks. In a series of EEG experiments, Tomberg and Desmedt (1998; also Tomberg, 1999) reported enhancement of frontal gamma phase locking during somatosensory target detection. The timing of this response suggested parieto-frontal interaction, organizing perceptual input and action readiness. Using a similar paradigm in the visual modality, Hermann and collaborators (Hermann et al., 1999) reported evoked (phase-locked) gamma responses at frontal electrode sites, which were greatest when stimulus features were task-relevant. These authors concluded that frontal gamma activity is associated with top-down processes, comparing incoming information with templates in working memory. In line with these results, auditory targets were effective in enhancing phase-locked GBA over frontal leads, when paired with a motor task (Yordanova et al., 2001). In this study, sensory-motor integration was proposed as a possible function of PFC neuronal assemblies. This function can be mediated by coherently networks oscillating not only at high (gamma-range) but also at lower frequencies such as alpha (Kolev et al., 2001) or theta (Klimesch et al., 1999). Most interestingly, the time course of frontal oscillatory brain responses during spatial selective attention suggests a reciprocal activation of parieto-occiptal and prefrontal cortical regions (Gruber et al., 1999). Accordingly, oscillatory coupling between those structures might reflect memory processes that act to facilitate future executions of the respective task by altering synaptic connectivity. This speculation is consistent with Desimone's view (Desimone, 1996) that in learning and memory, top-down modulatory input from PFC may alter sensory stimulus processing in both a short-term and long-term manner, changing the network architecture underlying a representation. According to Desimone, the neuronal mechanisms related to these modulations may include a variety of processes, such as refinement for learned representations (repetition suppression), increase of amplitude for behaviorally relevant information (enhancement), and continuation of sensory neuronal response after stimulus termination (delay activity). From animal work, it is well known that oscillatory activity is capable of mediating these processes, allowing for immediate changes as well as long-term plasticity. Hence, a complete picture of perceptual plasticity in attentional tasks should take this type of activity into account (Keil et al., 2001a).

2.3 Gestalt Perception

In the activities of daily life, gestalt perception is closely tied to the differential initiation of action. As an example, most social situations require differential responding to faces and objects. This is also true for most experimental paradigms of gestalt closure, which require differential key press or similar operationalizations. Previous work has shown that oscillatory brain activity was enhanced in specific frequency ranges and time windows following the presentation of coherent, but not incoherent visual stimuli (Tallon et al., 1995; Tallon-Baudry et al., 1997). Likewise, periods of EEG synchronization in the gamma range across electrodes were reported in response to identifiable vs. inverted face figures (Rodriguez et al., 1999). In an attempt to study acquisition of meaningful gestalt representations, Gruber and co-workers (Gruber et al., 2002) used a rapid perceptual learning design. They presented fragmented pictures, which were selected in such a way that subjects were unable to identify them. When confronted with a non-fragmented version of these pictures, participants learned to identify the fragmented versions in subsequent trials. Results showed an increase of spectral gamma power at parietal electrode sites after rapid perceptual learning. In addition, neural activity in the gamma band was highly synchronized between posterior electrode sites.

While these studies focused on visual cortex, involvement of PFC networks in gestalt perception has also been reported. Herrmann and collaborators (Hermann et al., 1999), who varied both targetness and gestalt

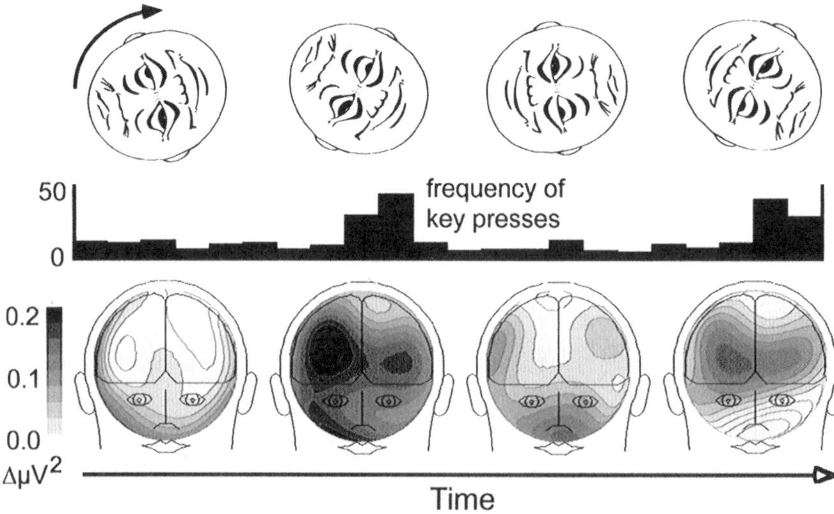

Figure 2: Grand mean (N = 11) topographical distribution of high-frequency EEG spectral power during observation of a rotating ambiguous stimulus. Top: four orientations of rotating sad/happy face figure corresponding to the time segments represented by topographical maps. Middle: frequency of partcipants' response (key press) indicating perceptual switching as a function of time. Probability of perceptual switch was highest at orientations 2 and 4, as shown on top. Bottom: front view of the topographical distribution of spectral power in the 29-45 Hz band during the four stimulus orientations shown above.

properties of Kanizsa triangles, observed frontal high-frequency amplitude enhancement in human EEG as an additive function of targetness and gestalt properties. No differences between gestalt and non-gestalt visual objects were found however, when none of the stimuli was relevant for the experimental task (Herrmann and Mecklinger, 2000), a finding that lends support to the important role of PFC for perception-action regulation. The question arises as to how differential percepts in a changing visual environment are related to specific motor actions. This aspect of gestalt perception was examined in an experiment using bistable figures (Keil et al., 1999).

A rotating black and white face drawing was perceived either as a sad or happy face when oriented vertically. Unambiguous perception was not possible when the drawing was oriented horizontally. The timing of perceptual switches was task-relevant in that participants indicated perceived changes from sad to happy and vice versa by a key press in certain trials. As a main result, a significant increase of induced gamma power was found at electrode sites over visual cortical areas, when a face could be perceived, but not when the stimulus was rotated horizontally. Regarding PFC activity, a more complex pattern was observed. A schematic representation of the experimental setup and topographical distributions of induced GBA are given in Figure 2. Frontal enhancement in induced gamma power occurred when a changing percept was identified and perceptual switching was indicated by means of key press. Analysis of timing of these responses showed that anterior and posterior electrodes did not display temporally parallel GBA modulation. This suggests that object recognition and perceptual switching as well as their integration with motor actions differentially involve oscillatory activity in visual and prefrontal cortex. Taken together, these results support the view endorsed here, namely that processing of meaningful visual objects together with their behavioral implications activates large-scale PFC networks.

2.4 Emotional Perception

In addition to the instruction by experimenters in the laboratory, the motivational relevance of a stimulus may be related to intrinsic affective/motivational properties, causing action readiness in the observer. Theoretical accounts and empirical work in the field of affective perception have suggested that emotions can be viewed as action dispositions (Frijda, 1988; Lang et al., 1998a). Thus, perception of emotionally arousing stimuli would enable an individual to react efficiently and successfully in a situation indexed as motivationally significant by properties of the visual scene. This concept is therefore relevant to a view of PFC networks as mediators of perception-action regulation. In particular, the theory proposed by Lang and co-workers (Lang, 1994) suggests that emotional perception be organized in neuronal networks connecting stimulus representations, response representations, memory, and motivational circuitry. Accordingly, presentation of affective external stimuli represents one paradigm to examine the processes integrating affective networks. Perception of affectively arousing stimuli, compared to calm control stimuli, has been reported to be associated with activation (enhancement) in a variety of structures. Using neuroimaging techniques, signal increases varying with affective arousal have been observed in the bilateral amygdaloid complex (Schneider et al., 1997), prefrontal cortex (Dolan et al., 1996; Royet et al., 2000), anterior cingulate (Lane et al., 1997), and visual cortex (Lang et al., 1998b). Human electrophysiological research has reliably observed a modulation of late deflections of the ERP as a function of motivational significance (Schupp et al., 2000; Keil et al., 2002). Specifically, greater magnitude of the P300 deflection as well as a sustained later positivity characterizes the electro-cortical response to emotionally salient (i.e.

pleasant or unpleasant), compared to neutral, pictures. This effect has been theoretically related to the concept of motivated attention, in which motivationally relevant stimuli naturally and perhaps automatically arouse and direct attentional resources (Lang et al., 1997; Öhman et al., 2001).

In a series of experiments, we have demonstrated that visual processing may also be altered by affective stimulus properties at several processing stages, starting at the level of the P1/N1 component of the ERP and at mid GBA responses around 80-100 ms post-stimulus (Keil et al., 2001c; Keil et al., 2002). For instance, using a hemifield design with colored affective pictures, Keil and collaborators (Keil et al., 2001c) reported differences of the N1 amplitude for arousing, compared to neutral, stimuli. In analogy to findings in the field of selective attention (Hillyard and Anllo-Vento, 1998), a 'sensory gain' mechanism has been hypothesized to amplify sensory processing according to the importance of the stimulus for the organism. Similar ideas have been expressed on the grounds of reaction time (Hartikainen et al., 2000; Öhman et al., 2001) and memory performance data (Bradley et al., 1992), suggesting that perceptual tasks are associated with enhanced behavioral performance when emotional intensity of stimuli is high. Thus, several parameters of visual processing showed a modulation as function of emotional arousal, together with a topographical distribution that is consistent with generators in higher-order visual, as well as right-parietal cortices (Junghöfer et al., 2001; Keil et al., 2001a). These effects have been interpreted as manifestations of re-entrant modulation, which may be effected by deep cortical structures such as the amygdala, or by prefrontal cortical structures, among others (Lang et al., 1998a). Is there any evidence for large-scale PFC involvement in emotional perception? Especially for the modulation of late cortical potentials (>300 ms post-stimulus), there are multiple reports of frontal ERP effects. These have been theoretically related to inhibition of a motor response in the experimental environment, where avoiding unpleasant, or approaching pleasant stimuli is not possible (Diedrich et al., 1997). Even when the motor response was held constant by experimental requirements as is the case in a simple reaction task, there was differential prefrontal activation, depending on aversiveness/pleasantness of visual stimuli (Northoff et al., 2000).

Keil and Ihssen (under revision) used the attentional blink (AB) paradigm, an experimental design allowing for manipulation of emotional arousal as well as the time between perception and availability of the percept to response processes, to further examine the question of interactions between higher-order processing and behavioral relevance. The attentional blink is a period of reduced awareness occurring, for instance, during "rapid serial visual presentation" (RSVP), when a first target (T1) is followed by a second target (T2) in a stream of distractors. It has been repeatedly shown that at high rates of visual presentation (e.g. at 6 Hz or higher), T2s being presented in an interval between 180 ms and 500 ms after a given T1 are reported less accurately (Raymond et al., 1992). Psychophysiological work suggests a post-perceptual, prefrontal locus of the AB (Vogel et al., 1998). We found that with short T1-T2 intervals, affectively arousing (pleasant and unpleasant verbs) T2s were better identified than affectively neutral T2s. Thus, motivationally/affectively relevant material was selected preferentially from a temporal stream of verbal information. This finding raises the question as to which mechanism would lead to enhanced post-perceptual processing for arousing, compared to neutral, T2s. Investigating into affective word recognition, Kitayama (1990) has proposed that facilitation observed for emotionally arousing verbal stimuli may reflect changes in the threshold of activation for a 'target code' of a given affective word. This code represents the specific lexical and semantic nodes that are activated when a given word is being identified. Thus, network theories of affective perception and action such as the model by Lang and collaborators (Lang et al., 1998a) may complement these accounts by adding predictions regarding stimulus variations along affective dimensions. If affectively arousing stimuli activate a network connecting stimulus representations, response representations, and related semantic memories, then sensory input could be weighted according to the motivational significance of the perceived stimulus for the organism. This in turn would cause amplified sensory processing of relevant stimulus features. Functional imaging work as well as electrophysiological studies suggest that this process can be mediated by re-entrant input to sensory cortex (Lang et al., 1998b; Keil et al., 2001c). As a consequence, stronger and faster activation in the affective network may propagate through the stages of processing and lead to increasing facilitation. Analyzing evoked oscillatory brain responses locked to the rapidly presented stimuli, Keil and Ihssen (in preparation) found enhanced coupling between visual and prefrontal regions. This points not only to re-entrant modulation of visual cortex by prefrontal areas, but also to bi-directional connectivity integrating perceptual and associative memory components in a common network.

3. CORTICAL PLASTICITY IN EMOTIONAL LEARNING: ELEMENTS FOR A LARGE-SCALE MODEL

As outlined above, a growing literature supports the notion that visual processing is modulated by the behavioral relevance of a given stimulus. In many cases, stimuli have high motivational relevance because they are associated with reward or punishment. Another example of stimuli having very high behavioral relevance is fear-related stimuli (see also Herry and Garcia, Chapter 6). Investigations comparing healthy controls and patients with anxiety disorders have repeatedly shown that differences between these groups can be observed on a variety of perceptual, behavioral, and electrophysiological parameters. These results have suggested that viewing fear-related stimuli enhances early perception. However, fear may not be the only affective disposition leading to an enhanced perceptual performance. As shown above, several studies investigating affective perception in healthy human participants have also converged, showing reliable enhancement of psychophysiological and behavioral parameters in early ranges around 100-200 ms post-stimulus as a function of emotional stimulus intensity. These results have been replicated many times, using different experimental designs, stimulus categories, and recording techniques. While re-entrant modulation can account for late sensory amplification of arousing stimulus features, the nature of early perceptual enhancement remains unclear. Here I suggest that cortical plasticity may mediate changes in neuro-architecture that are underlying such early perceptual effects (Fig. 3).

Models based on auditory fear conditioning work in rodents have suggested that a fast route exist for rough stimulus evaluation, being based on thalamo-amygdaloid circuitry, thus complementing fine-grained but slow visual analysis (LeDoux, 2000). The use of this model to account for fast visual processing in anxiety has been challenged however, both on the grounds of rodent research and theoretical considerations. Shi and Davis (2001) have observed in the rodent model that this 'low route' may only be in effect with highly salient stimuli, or if no other route is available. This is concordant with neuropsychological data (de Gelder et al., 1999). Alternative perspectives predicting fast identification and processing of fear-relevant visual stimuli have been based on network theories of emotional processing (Öhman et al., 2001). One basic tenet of the model proposed in the present chapter is that affective perception/action information is integrated in dynamic and plastic neuronal networks (Keil et al., 2001a). Thus, visual processing may be adaptable to the motivational significance of a stimulus category or certain features signalling significance, in order to optimize processing of behaviorally relevant information. Rather than using a 'low', amygdala-based route for the processing of *each* stimulus in a given context, the information about features being associated with threatening stimuli may be embodied in a cortical network, including visual cortex. As a short-term mechanism, oscillatory re-entrant modulation of visual cortex may change network thresholds to selectively enhance the relevant response. This has been repeatedly demonstrated by EEG studies of affective perception as summarized above. In addition, persisting top-down modulation by coherently oscillating networks can have long-term effects on

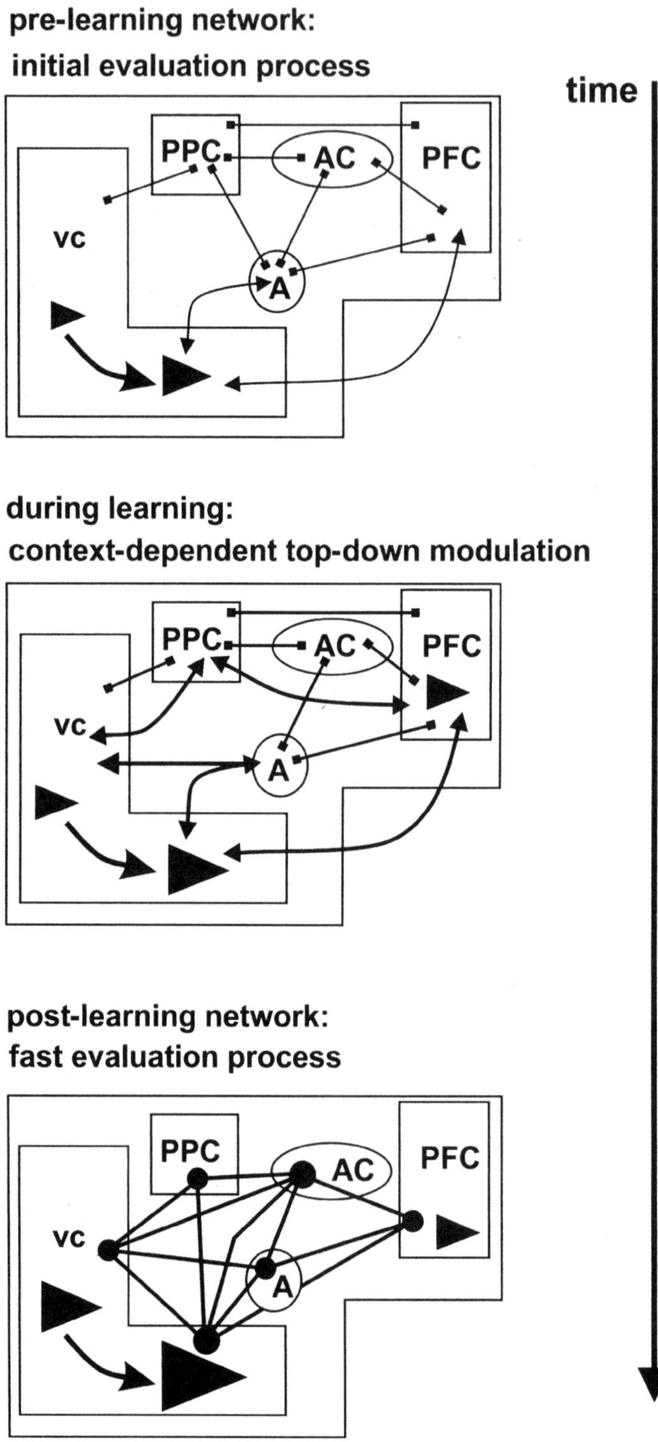
pre-learning network:
initial evaluation process
time
PPC
AC
PFC
VC
A
during learning:
context-dependent top-down modulation
PPC
AC
PFC
VC
A
post-learning network:
fast evaluation process
PPC
AC
PFC
VC
A

network architecture. If motivational relevance remains high across time, oscillatory top-down regulation may also induce more permanent plastic changes of network architecture or synaptic connectivity (Gilbert et al., 2001). Hence, a cell assembly emerges, which is specifically sensitive to arousing features in a given context, even at very early levels of perception.

This notion is consistent with the findings that showed plastic changes in sensory processing as a consequence of classical conditioning procedures. Such changes have been demonstrated both in the auditory and visual modalities (Recanzone, 1998; Knight et al., 1999; Bao et al., 2003). Macroscopic EEG data presented in this chapter suggest that PFC oscillations at high frequency play an important role both in mediating these plastic changes and in providing the short-term re-entrant input to sensory cortices preceding such changes. As a consequence of the dynamics described above, early sensory amplification of stimuli may occur as a function of learning the relationship between elementary stimulus features and their behavioral relevance. Pilot results from our laboratory indeed support this prediction, showing that when a conditioned visual stimulus is paired with an unconditioned unpleasant event, the conditioned stimulus will gradually enhance very early electro-cortical responses (60-90 ms post-stimulus) of visual cortex. Paralleling this process in time, PFC networks showed increased amplitude in high-frequency oscillatory activity, suggesting that they contribute to the effects observed.

Figure 3: Dynamics of emotional learning in the visual domain. The figure schematically displays some of the large-scale brain structures involved in affective perception and action regulation as well as relevant connections between them: anterior cingulate (AC), posterior parietal cortex (PPC), amygdaloid complex (A), prefrontal cortex (PFC). Top: A novel stimulus (represented by the black triangle) having behavioral significance is perceived. Following visual analysis, the network is activated, and based on memories stored in other areas, re-entrant oscillatory activity enhances visual cortex (VC) processing of the relevant features. At PFC level, behavioral dispositions, memories, and stimulus representations are integrated to regulate action. Middle: Repeated presentations of the relevant stimulus in a specific context are associated with coherent oscillatory activity within and between networks representing contextual, stimulus, and action information. Memories representing and integrating these aspects are stored in PFC-based networks. Coherent neuronal activity (i) lowers thresholds in the relevant networks, (ii) enhances effectivity of sensory processing, e.g. by increased sensitivity in visual cortex, and (iii) induces long-term changes. Bottom: As a long-term consequence, connectivity between structures is enhanced for a specific stimulus, context, and action, thus facilitating adaptive behavior. Visual cortex is part of these networks, responding more rapidly and effectively to this stimulus.

4. CONCLUSIONS

Animal and human data converge, suggesting a central role of PFC in establishing widespread neuronal networks for the integration and the organization of perception and motivated action. These widespread networks may use oscillatory activity to provide short-term top-down modulation to sensory systems as well as long-term changes of neuronal architecture. Macroscopic oscillatory brain activity as can be measured using EEG/MEG represents a useful correlate of the processes underlying this function. We have reported evidence that oscillatory mass activity of PFC neurons can be observed in a variety of experimental tasks. In particular, tasks requiring the acquisition of new behaviors and the activation of action dispositions in a specific situational context are well suited to elicit enhancement of PFC high-frequency oscillations. Likewise, perception of stimuli having behavioral relevance is accompanied by marked modulations of frontal high-frequency activity. PFC macroscopic electrocortical changes are also abundant during selective attention and feature integration, highlighting the role of PFC in multi-sensory and multi-modal memory representations. Elements for a view of dynamic cortical networks in motivated perception and action must therefore include accounts for short-term and long-term adaptation of neuronal systems to changing external requirements. Macroscopic oscillations in wide-spread neuronal assemblies are capable of inducing dynamic changes and plastic processes on different time scales and thus should be considered as one candidate mechanism for integrative functioning of the central nervous system.

REFERENCES

Alho K, Woods DL, Algazi A, Knight RT, Naatanen R (1994) Lesions of frontal cortex diminish the auditory mismatch negativity. Electroencephalogr Clin Neurophysiol 91:353-362.

Anllo-Vento L, Hillyard SA (1996) Selective attention to the color and direction of moving stimuli: electrophysiological correlates of hierarchical feature selection. Percept Psychophys 58:191-206.

Bao S, Chan VT, Zhang LI, Merzenich MM (2003) Suppression of cortical representation through backward conditioning. Proc Natl Acad Sci USA.

Berger H (1938) Über das Elektrenkephalogramm des Menschen (1.-9. Mitteilg. 1929-1938). Nova Acta Leopoldina NF 6:173-309.

Bradley MM, Greenwald MK, Petry MC, Lang PJ (1992) Remembering pictures: pleasure and arousal in memory. J Exp Psychol Learn Mem Cogn 18:379-390.

Braitenberg V, Schüz A (1998) Cortex: Statistics and Geometry of Neuronal Connectivity. Springer, Berlin.

Brazzelli M, Colombo N, Della Sala S, Spinnler H (1994) Spared and impaired cognitive abilities after bilateral frontal damage. Cortex 30:27-51.

Corbetta M (1998) Frontoparietal cortical networks for directing attention and the eye to visual locations: identical, independent, or overlapping neural systems? Proc Natl Acad Sci USA 95:831-838.

Damasio H, Grabowski T, Frank R, Galaburda AM, Damasio AR (1994) The return of Phineas Gage: clues about the brain from the skull of a famous patient. Science 264:1102-1105.

de Gelder B, Vroomen J, Pourtois G, Weiskrantz L (1999) Non-conscious recognition of affect in the absence of striate cortex. NeuroReport 10:3759-3763.

Desimone R (1996) Neural mechanisms for visual memory and their role in attention. Proc Natl Acad Sci USA 93:13494-13499.

Diedrich O, Naumann E, Maier S, Becker G (1997) A frontal positive slow wave in the ERP associated with emotional slides. J Psychophysiol 11:71-84.

Dolan RJ, Fletcher P, Morris J, Kapur N, Deakin JF, Frith CD (1996) Neural activation during covert processing of positive emotional facial expressions. Neuroimage 4:194-200.

Elbert T (1998) Neuromagnetism. In: Magnetism in medicine (Andrä H, Nowak W, eds), pp 190-262. New York: Wiley & Sons.

Freeman WJ (1994) Role of chaotic dynamics in neural plasticity. Prog Brain Res 102:319-333.

Frijda NH (1988) The laws of emotion. Am Psychol 43:349-358.

Fuster JM (1990) Behavioral electrophysiology of the prefrontal cortex of the primate. Prog Brain Res 85:313-323, discussion 323-324.

Galambos R (1992) A comparison of certain gamma-band (40 Hz) brain rhythms in cat and man. In: Induced Rhythms in the Brain (Basar E and Bullock T, eds), pp 103-122. Springer, Berlin.

Gilbert CD, Sigman M, Crist RE (2001) The neural basis of perceptual learning. Neuron 31:681-697.

Grafman J, Schwab K, Warden D, Pridgen A, Brown HR, Salazar AM (1996) Frontal lobe injuries, violence, and aggression: a report of the Vietnam Head Injury Study. Neurology 46:1231-1238.

Gruber T, Müller MM, Keil A, Elbert T (1999) Selective visual-spatial attention alters induced gamma band responses in the human EEG. Clin Neurophysiol 110:2074-2085.

Gruber T, Muller MM, Keil A (2002) Modulation of induced gamma band responses in a perceptual learning task in the human EEG. J Cogn Neurosci 14:732-744.

Haken H (1983) Synergetics: An Introducton. Springer, Berlin.

Harmony T, Fernandez T, Silva J, Bosch J, Valdes P, Fernandez-Bouzas A, Galan L, Aubert E, Rodriguez D (1999) Do specific EEG frequencies indicate different processes during mental calculation? Neurosci Lett 266:25-28.

Hartikainen KM, Ogawa KH, Knight RT (2000) Transient interference of right hemispheric function due to automatic emotional processing. Neuropsychologia 38:1576-1580.

Hebb D (1949) The organization of behavior; a neuropsychological theory. Wiley, New York.

Hecht H, Vogt S, Prinz W (2001) Motor learning enhances perceptual judgment: a case for action-perception transfer. Psychol Res 65:3-14.

Herrmann CS, Mecklinger A (2000) Magnetoencephalographic responses to illusory figures: early evoked gamma is affected by processing of stimulus features. Int J Psychophysiol 38:265-281.

Herrmann CS, Mecklinger A, Pfeifer E (1999) Gamma responses and ERPs in a visual classification task. Clin Neurophysiol 110:636-642.

Hillyard SA, Anllo-Vento L (1998) Event-related brain potentials in the study of visual selective attention. Proc Natl Acad Sci USA 95:781-787.

Junghöfer M, Bradley MM, Elbert TR, Lang PJ (2001) Fleeting images: a new look at early emotion discrimination. Psychophysiology 38:175-178.

Keil A, Ihssen N (under revision) Identification facilitation for emotionally arousing verbs during the early period of the attentional blink.

Keil A, Muller MM, Ray WJ, Gruber T, Elbert T (1999) Human gamma band activity and perception of a gestalt. J Neurosci 19:7152-7161.

Keil A, Gruber T, Müller MM (2001a) Functional correlates of macroscopic high-frequency brain activity in the human visual system. Neurosci Biobehav Rev 25:527-534.

Keil A, Müller MM, Gruber T, Wienbruch C, Elbert T (2001b) Human large-scale oscillatory brain activity during an operant shaping procedure. Cogn Brain Res 12:397-407.

Keil A, Müller MM, Gruber T, Stolarova M, Wienbruch C, Elbert T (2001c) Effects of emotional arousal in the cerebral hemispheres: a study of oscillatory brain activity and event-related potentials. Clin Neurophysiol 112:2057-2068.

Keil A, Bradley MM, Hauk O, Rockstroh B, Elbert T, Lang PJ (2002) Large-scale neural correlates of affective picture processing. Psychophysiology 39:641-649.

Kitayama S (1990) Interaction between affect and cognition in word perception. J Pers Soc Psychol 58:209-217.

Klimesch W, Doppelmayr M, Schwaiger J, Auinger P, Winkler T (1999) 'Paradoxical' alpha synchronization in a memory task. Cogn Brain Res 7:493-501.

Knight DC, Smith CN, Stein EA, Helmstetter FJ (1999) Functional MRI of human Pavlovian fear conditioning: patterns of activation as a function of learning. Neuroreport 10:3665-3670.

Knight RT, Grabowecky M (2000) Prefrontal cortex, time, and consciousness. In: The New Cognitive Neurosciences (Gazzaniga MS, ed). MIT Press, Cambridge MA

Kolev V, Yordanova J, Schurmann M, Basar E (2001) Increased frontal phase-locking of event-related alpha oscillations during task processing. Int J Psychophysiol 39:159-165.

Lane RD, Reiman EM, Bradley MM, Lang PJ, Ahern GL, Davidson RJ, Schwartz GE (1997) Neuroanatomical correlates of pleasant and unpleasant emotion. Neuropsychologia 35:1437-1444.

Lang PJ (1979) A bioinformational theory of emotional imagery. Psychophysiology 16:495-512.

Lang PJ (1994) The motivational organization of emotion: Affect-reflex connections. In: Emotions: Essays on emotion theory. (Van Goozen SHM, Van de Poll NE, Sergeant JE, eds), pp 61-93. Hillsdale, NJ: Lawrence Erlbaum Associates.

Lang PJ, Bradley MM, Cuthbert BN (1997) Motivated attention: affect, activation, and action. In: Attention and Orienting: Sensory and Motivational Processes (Lang PJ, Simons RF, Balaban MT, eds), pp 97-135. Hillsdale, N.J.: Lawrence Erlbaum Associates.

Lang PJ, Bradley MM, Cuthbert BN (1998a) Emotion, motivation, and anxiety: brain mechanisms and psychophysiology. Biol Psychiatry 44:1248-1263.

Lang PJ, Bradley MM, Fitzsimmons JR, Cuthbert BN, Scott JD, Moulder B, Nangia V (1998b) Emotional arousal and activation of the visual cortex: an fMRI analysis. Psychophysiology 35:199-210.

LeDoux JE (2000) Emotion circuits in the brain. Annu Rev Neurosci 23:155-184.

Miltner WH, Braun C, Arnold M, Witte H, Taub E (1999) Coherence of gamma-band EEG activity as a basis for associative learning. Nature 397:434-436.

Molchan SE, Sunderland T, McIntosh AR, Herscovitch P, Schreurs BG (1994) A functional anatomical study of associative learning in humans. Proc Natl Acad Sci USA 91:8122-8126.

Nobre AC (2001) Orienting attention to instants in time. Neuropsychologia 39:1317-1328.

Northoff G, Richter A, Gessner M, Schlagenhauf F, Fell J, Baumgart F, Kaulisch T, Kotter R, Stephan KE, Leschinger A, Hagner T, Bargel B, Witzel T, Hinrichs H, Bogerts B, Scheich H, Heinze HJ (2000) Functional dissociation between medial and lateral prefrontal cortical spatiotemporal activation in negative and positive emotions: a combined fMRI/MEG study. Cereb Cortex 10:93-107.

Nunez PL (2000) Toward a quantitative description of large-scale neocortical dynamic function and EEG. Behav Brain Sci 23:371-398; discussion 399-437.

Öhman A, Flykt A, Esteves F (2001) Emotion drives attention: detecting the snake in the grass. J Exp Psychol Gen 130:466-478.

Prinz W (1997) Perception and action planning. Eur J Cogn Psychol 9:129-154.

Raymond JE, Shapiro KL, Arnell KM (1992) Temporary suppression of visual processing in an RSVP task: an attentional blink? J Exp Psychol Hum Percept Perform 18:849-860.

Recanzone GH (1998) Rapidly induced auditory plasticity: the ventriloquism aftereffect. Proc Natl Acad Sci USA 95:869-875.

Rodriguez E, George N, Lachaux JP, Martinerie J, Renault B, Varela FJ (1999) Perception's shadow: long-distance synchronization of human brain activity. Nature 397:430-433.

Rolls ET, Hornak J, Wade D, McGrath J (1994) Emotion-related learning in patients with social and emotional changes associated with frontal lobe damage. J Neurol Neurosurg Psychiatry 57:1518-1524.

Royet JP, Zald D, Versace R, Costes N, Lavenne F, Koenig O, Gervais R (2000) Emotional responses to pleasant and unpleasant olfactory, visual, and auditory stimuli: a positron emission tomography study. J Neurosci 20:7752-7759.

Schneider F, Grodd W, Weiss U, Klose U, Mayer KR, Nagele T, Gur RC (1997) Functional MRI reveals left amygdala activation during emotion. Psychiatry Res 76:75-82.

Schupp HT, Cuthbert BN, Bradley MM, Cacioppo JT, Ito T, Lang PJ (2000) Affective picture processing: the late positive potential is modulated by motivational relevance. Psychophysiology 37:257-261.

Shi C, Davis M (2001) Visual pathways involved in fear conditioning measured with fear-potentiated startle: behavioral and anatomic studies. J Neurosci 21:9844-9855.

Singer W, Engel AK, Kreiter AK, Munk MHJ, Neuenschwander S, Roelfsema PR (1997) Neuronal assemblies: necessity, signature and dectability. Trends Cogn Sci 1:252-261.

Tallon C, Bertrand O, Bouchet P, Pernier J (1995) Gamma-range activity evoked by coherent visual stimuli in humans. Eur J Neurosci 7:1285-1291.

Tallon-Baudry C, Bertrand O, Delpuech C, Pernier J (1997) Oscillatory gamma-band (30-70 Hz) activity induced by a visual search task in humans. J Neurosci 17:722-734.

Tallon-Baudry C, Bertrand O, Peronnet F, Pernier J (1998) Induced gamma-band activity during the delay of a visual short-term memory task in humans. J Neurosci 18:4244-4254.

Tallon-Baudry C, Kreiter A, Bertrand O (1999) Sustained and transient oscillatory responses in the gamma and beta bands in a visual short-term memory task in humans. Vis Neurosci 16:449-459.

Tomberg C (1999) Cognitive N140 electrogenesis and concomitant 40 Hz synchronization in mid-dorsolateral prefrontal cortex (area 46) identified in non-averaged human brain potentials. Neurosci Lett 266:141-144.

Tomberg C, Desmedt JE (1998) Human perceptual processing: inhibition of transient prefrontal-parietal 40 Hz binding at P300 onset documented in non-averaged cognitive brain potentials. Neurosci Lett 255:163-166.

Vaadia E, Haalman I, Abeles M, Bergman H, Prut Y, Slovin H, Aertsen A (1995) Dynamics of neuronal interactions in monkey cortex in relation to behavioural events. Nature 373:515-518.

Vogel EK, Luck SJ, Shapiro KL (1998) Electrophysiological evidence for a postperceptual locus of suppression during the attentional blink. J Exp Psychol Hum Percept Perform 24:1656-1674.

Wagner AD, Poldrack RA, Eldridge LL, Desmond JE, Glover GH, Gabrieli JD (1998) Material-specific lateralization of prefrontal activation during episodic encoding and retrieval. NeuroReport 9:3711-3717.

Wood JN, Grafman J (2003) Human prefrontal cortex: processing and representational perspectives. Nat Rev Neurosci 4:139-147.

Yordanova J, Banaschewski T, Kolev V, Woerner W, Rothenberger A (2001) Abnormal early stages of task stimulus processing in children with attention-deficit hyperactivity disorder--evidence from event-related gamma oscillations. Clin Neurophysiol 112:1096-1108.

Zald DH, Kim SW (1996) Anatomy and function of the orbital frontal cortex, II: Function and relevance to obsessive-compulsive disorder. J Neuropsychiatry Clin Neurosci 8:249-261.

Chapter 12

TRANSCRANIAL MAGNETIC STIMULATION OF THE PREFRONTAL CORTEX: A COMPLEMENTARY APPROACH TO INVESTIGATE HUMAN LONG-TERM MEMORY

Simone Rossi[1], Carlo Miniussi[2], Paolo Maria Rossini[2,3,4], Claudio Babiloni[2,5], and Stefano Cappa[6]

[1]*Dipartimento di Neuroscienze, Sezione Neurologia, Università di Siena, Policlinico le Scotte, Viale Bracci, I-53100, Siena, Italy*

[2]*IRCCS S. Giovanni di Dio, Fatebenefratelli, Brescia, Italy*

[3]*Neurologia, Università Campus Biomedico, Roma, Italy*

[4]*AFaR-Dipartimento Neuroscienze, Ospedale Fatebenefratelli Isola Tiberina, Roma, Italy*

[5]*Dipartimento di Fisiologia Umana e Farmacologia, Università La Sapienza, Roma, Italy*

[6]*Centro di Neuroscienze Cognitive, Università Salute-Vita S. Raffaele, Milano, Italy*

Keywords: Long-term memory, transcranial magnetic stimulation, prefrontal cortex, humans, encoding, retrieval.

Abstract: Repetitive transcranial magnetic stimulation (rTMS) can noninvasively and focally stimulate the cerebral cortex, inducing a transient and safe interruption of brain function. Although its detailed mechanisms of action still need to be fully elucidated, it has been successfully applied to investigate encoding and retrieval phases during episodic long-term memory tasks, both in the visuospatial and verbal domains. The effects of rTMS are behaviorally measurable; therefore, it seems to be a good complementary approach to more traditional neuroimaging and electroencephalographic techniques for the investigation of the working brain.

1. INTRODUCTION

Why "knock-out" the brain to study memory? Is it not sufficient to "watch" what is happening in the working brain using conventional neuroimaging techniques, such as positron emission tomography (PET) and functional magnetic resonance (fMR)? To answer these questions, we will first briefly explain what transcranial magnetic stimulation (TMS) is and how it is supposed to operate on the brain. We will then review the available evidence concerning the use of rTMS as a complementary approach for probing brain functions in cognitive neuroscience, including long-term episodic memory processes.

1.1 Basic Principles of TMS and Repetitive TMS (rTMS)

The story dates back to the middle 80's, when Anthony Barker and colleagues from the University of Sheffield, U.K. (Barker et al., 1985) built the first magnetic stimulator able to non-invasively excite cortical neurons from the scalp surface. That revolutionary technique delivered a brief and strong magnetic field, which was generated by the current discharged within a coil of copper wires by a bank of capacitors. Such a magnetic field induced an electric current circulating up to a few centimeters away from the coil's external edge. Its direction was opposite to that of the current flowing into the coil, while its intensity was proportional to the magnetic field flux. The induced electric current was blind to the influence of extra-cerebral layers (scalp, skull and meninges), with no activation of their pain receptors, thus resulting in a fully tolerable procedure for the non-invasive stimulation of cortical neurons. Differently from the painful transcranial electrical stimulation of the scalp, which directly excites pyramidal axons at their hillock or even more deeply in the subcortical white matter, TMS is thought to activate superficial cortical neurons trans-synaptically (Caramia et al., 1989; Amassian et al., 1990). This is suggested by the observation that TMS of the scalp overlying the motor cortex produces indirect corticospinal volleys (the so called *indirect waves*), rather than *direct waves* (Di Lazzaro et al., 1998). However, there is not yet sound evidence that cortical neurons located in areas different from the motor cortex react to TMS in a similar way.

Soon after the introduction of the technique, it became clear that the applications of TMS could embrace several fields of clinical neurology, neurophysiology, and neuroscience. A first application derived from the observation that, applying the coil over the scalp overlying the motor cortex, subjects had involuntary contralateral muscle twitches that appeared, after each stimulus, with a latency compatible to the transit time along the fastest fibres of the corticospinal tract (Barker et al., 1985). Nowadays, this procedure is routinely used to probe the "conducibility" of central motor tracts and the local level of motor cortical excitability, both in research and clinical settings. Moreover, depending on stimulation parameters, TMS can also inhibit some brain areas either directly reached by the stimulus or at distance via indirect interplay of cortico-cortical inhibitory fibres (see Rossini and Rossi (1998) and Hallett (2000) for exhaustive reviews on conventional clinical TMS applications).

Among the different measures of cortical excitability, what is relevant to the use of TMS in neurocognitive studies is the concept of "Excitability Threshold" (ET) of individual brain areas, since the intensity of stimulation in any kind of investigation is generally expressed as a percentage of the ET value. As established by International Guidelines (Rossini et al., 1994), the resting ET of the motor cortex is the minimal intensity of the stimulator output required to elicit contralateral electromyographic responses of about 50 μV in a given relaxed muscle with a probability of 50%. Conventionally, hand muscles are chosen as a reference for the determination of the individual ET value.

Later in the 90's, the development of technology allowed to build magnetic stimulators able to deliver rhythmic trains of several stimuli during a defined period of time. This kind of TMS, therefore, is called repetitive TMS (rTMS). There is general consensus to consider as *slow-frequency* those rTMS trains below 1 Hz, and as *high-frequency* those rTMS trains ≥ 1 Hz. (up to 20-30 Hz) (Wassermann, 1998). Treating slow and high-frequency rTMS as separate items is essential, since these two kinds of stimulations produce extremely distinct effects on brain activity, which are partly measurable by means of neurophysiological or functional imaging techniques. As exhaustively reviewed in a recent article, converging evidences indicate that rTMS ≤1 Hz reduces cortical excitability both locally as well as in functionally related cortical regions, with effects that may outlast hours beyond the time of stimulation (Hoffmann and Cavus, 2002). Experimental studies in animals and pilot studies in humans point also to possible therapeutic use of such long-lasting effects, especially in patients with higher than normal levels of cortical excitability, such as focal dystonia, epileptic seizures, and auditory hallucinations in schizophrenia (Hoffmann and Cavus, 2002). Conversely, high-frequency rTMS, particularly when delivered in trains over 5 Hz, has opposite, mainly facilitatory effects on cortical excitability (Chen, 2000). A growing interest is emerging also for this kind of rTMS as an add-on therapeutic tool in a number of psychiatric conditions (reviews on this particular topic, Wassermann and Lisanby, 2001; George et al., 2002). However, users of rTMS should be aware of the potential risk to induce seizures even in normal individuals, in particular when high frequency stimulation is applied. Safety guidelines must always be followed in any TMS experiment to prevent such problems (Wassermann, 1998).

2. RATIONALE FOR THE USE OF TMS IN COGNITIVE NEUROSCIENCE

The relevance that rTMS has gained in the last years in the field of cognitive neuroscience depends mainly on its ability to transiently interfere with the functions of the stimulated cortical network, rather than on its effects on modulation of cortical excitability. As reported elsewhere in this volume, many complementary methods, such as PET, fMR, magneto-encephalography and high-resolution EEG, can be used to study the contribution of specific cortical networks to cognitive challenges. Each approach has obviously some "pros" and some "cons". Basically, the neuroimaging techniques based on *in vivo* measurements of local changes of blood flow such as PET and fMR, despite showing the best available spatial resolution, lack sufficient temporal resolution (from seconds to minutes) to covary with cognitive processes, such as memory, occurring in tens or hundreds of milliseconds. Because of these limitations, they do not allow to disentangle the functional hierarchy of a given area when several cortical regions are active simultaneously. More importantly, neither brain rhythms or event-related potentials, nor PET and fMR can unequivocally determine whether an active area is necessary for a particular function or behavior (Price and Friston, 1999). As pointed out in recent reviews (Pascual-Leone et al., 2000; Walsh and Cowey, 2000), what makes the rTMS approach unique in cognitive studies is its potential to induce a transient (and safe, provided that the guidelines are adhered) functional disruption of the stimulated cortical area.

A direct demonstration of detailed mechanisms of rTMS interference is still lacking, although it is reasonable to believe that they most likely rely on the induction of a *random neural noise* into the process under study. In other words, the rTMS (usually an high-frequency train) induces an external activity in the brain which is random with respect to the goal-state of the area stimulated, thereby disrupting task performance (Walsh and Cowey, 2000). This explanation, despite being the only available at the moment, is however subject to a lively discussion. Indeed, it seems not to be entirely satisfactory from a mechanistic point of view, since it is difficult to translate the concept of "increased noise" into precise terms that make sense from a computational perspective (Fitzpatrick and Rothman, 2000).

The possibility to transiently interfere with neural information processing led some researchers to propose the concept of the so called *virtual patient*, that is of a "virtual lesion" induced by rTMS in an otherwise healthy subject (Pascual-Leone et al., 2000). Although intriguing, it is somewhat misleading to translate *tout court* this concept to rTMS effects on the brain: indeed, emerging evidences indicate that if appropriately timed to the demands of the cognitive task, rTMS may produce facilitatory effects rather than disrupt functions (e.g. Boroojerdi et al., 2001; Cappa et al., 2002).

The spatial resolution of rTMS is not yet fully clarified. For cognitive studies as those dealing with memory, which aim at identifying eloquent areas for a given task, the use of focal eight-shaped coils is mostly required. For physical reasons, which go beyond the purposes of this chapter, such "butterfly" coils definitely guarantee a more focal stimulation than round coils, although the precise site and spatial spread of cortical neuronal depolarization is not completely understood. However, using focal coils and low intensities (i.e. near the individual ET) of stimulation of the motor cortex, the spatial resolution is high enough to obtain reliable cortical maps of individual hand muscles (e.g. Rossi et al., 1998) and to follow in time these maps to disclose cortical reorganizations following discrete lesions or motor learning (reviews, Hallett, 2000; Rossini and Pauri, 2000). There is also a bulk of evidence on the spatial congruence of centers of gravity of fMR activations relative to hand motor tasks and TMS mapping of hand muscles (Boroojerdi et al., 1999; Herwig et al., 2002). The possibility remains, however, that cortical neurons located outside of the motor cortex react in a different way to TMS, as in the case of motor and parietal cortices for recovery of excitability to paired stimuli (Oliveri et al., 2000). To date, the available technology does not allow the TMS to reach neural structures located deeply in the cortex or subcortical nuclei.

A common problem of most TMS studies is the localization of the stimulating coil with respect to the anatomy of the stimulated cortex. In this regard, the use of neuronavigational systems similar to those used for neurosurgical procedures seems to be promising (Herwig et al., 2001 for an extensive review). However, the procedure based on the digitalization of the stimulated scalp position coupled to the use of template brain MR models (available on the Web: for example, in SPM96 the model is obtained by averaging the magnetic resonance images of 152 subjects; Babiloni et al., 2001; Rossi et al., 2001, 2002; Sandrini et al., 2003) represents a good compromise among the localization accuracy, the high economical demands of neuronavigation systems, and the availability of these systems for those structures not exclusively devoted to research but also to daily clinical practice.

Another important aspect of rTMS application for inducing transient functional impairments of the brain is the use of controlled, or "sham", conditions. This is required for the fact that the stimulating procedure is associated with a number of sensory perceptions. For instance, the discharging coil produces a click sound that, in the case of rTMS around 20 Hz as for trains used in neurocognitive studies, may induce strong arousals. It disrupts the task performance *per se*, irrespective of the exact demands of the experimental design. Sham rTMS stimulation is generally carried out by keeping the coil perpendicular to the scalp position to be stimulated: the sound and the scalp contact are roughly similar to the active stimulation, but the magnetic field is not able to reach neither cortical neurons nor cutaneous receptors or superficial muscles (Rossi et al. 2001; Sandrini et al. 2003). Nowadays, specially-designed sham-coils are commercially available. They still produce the same sound during the stimulation, but no magnetic field is generated. They can rest tangential to the scalp surface exactly as they are during the active stimulation. This guarantees a more accurate experimental control than perpendicular coils, since it is more difficult for the subject in this preparation to realize the difference between sham and active rTMS, especially when low intensity stimulations are applied.

3. TMS STUDIES ON EPISODIC (LONG-TERM) MEMORY

Despite the increasing interest in TMS as one of the most attractive tools for neuroscientists (Chicurel, 2002), its application to the investigation of episodic long-term memory is still limited. A Medline search in the first week of the year 2003, using "TMS"/"rTMS", "prefrontal cortex", and "memory" as keywords, detected the articles most of which deal with verbal/visual working memory (n=5) and saccadic eye movements (n=3). Among remaining papers, only four articles (one of which is a review) fully matched our request and dealt with episodic memory. These relevant papers will be reviewed here. In addition, we will discuss some recent works from our research team, which are in press (Sandrini et al., 2003) or have been published only in abstract form (Rossi et al., 2003).

Episodic memory is a neurocognitive system that enables human beings to form a permanent record of everyday events, and is therefore essential for daily life (Tulving, 2002 for an exhaustive review). The ability to consciously remember an experience requires its initial encoding, its long-term storage, and the possibility of its subsequent retrieval. This is the form of memory affected in human amnesia, following bilateral damage to medial temporal regions or diencephalic structures (Squire et al., 2001). Neuroimaging investigations on the mechanisms underlying encoding and retrieval have confirmed the crucial role of these structures in episodic memory. In addition, they have disclosed the unexpected activation of many regions of the prefrontal cortex (Fletcher and Hanson, 2001), previously thought to be involved mainly in working memory and self-monitoring processes (Petrides, 1994).

A related, intriguing finding was the observation of a pattern of hemispheric asymmetry in memory encoding and retrieval. The left prefrontal cortex appears to be involved in the encoding of information about novel events into episodic memory and in the retrieval of information from semantic memory, whereas the right prefrontal cortex was found to play a crucial role in the retrieval of information from episodic memory. These findings formed the basis of the hemispheric encoding and retrieval asymmetry (HERA) model (Tulving et al., 1994). In its original formulation, the HERA model was limited to the interpretation of prefrontal activity associated with verbal memory tasks. Later, it was extended to nonverbal materials by Nyberg et al. (1996).

The idea of an absolute left hemispheric prefrontal specialization for encoding and right hemispheric for retrieval has come under considerable criticism. Many recent studies have indicated that multiple factors, such as the nature of the material to be remembered, interact with the pattern of hemispheric asymmetry, in particular during encoding (Wagner et al., 1998). Since the HERA model is essentially based on neuroimaging studies and is still subject to a lively discussion, it seems particularly suitable for interferential studies with rTMS (Walsh and Cowey, 2000). This method, which does not depend solely on the metabolic or hemodynamic response to the cognitive challenge, seems to be an excellent test for the predictions of the HERA theory.

3.1 Uncontrolled Studies (No Sham Stimulation)

The first mention of an induction of a free recall deficit (and of the direct demonstration of the prefrontal cortex involvement in long-term memory process) induced by rTMS dates back to 1994, when Grafman et al. (1994) investigated whether verbal recall could be selectively interfered by the site and timing of the TMS. Recall was tested in five healthy subjects immediately after the presentation of a list containing 12 words. Focal rTMS was applied in trains of 5 pulses at 20 Hz, but lasting 500 ms, delivered immediately after the display of each word or at delays of 250, 500 or 1000 ms, over several scalp positions including dorsolateral prefrontal, left mid-temporal, parietal, and occipital regions. These regions were identified according to the electrode locations of the 10-20 International EEG System. Subjects, who were generally able to correctly remember the words, significantly worsened their performance after rTMS to left mid-temporal and bilateral dorsofrontal regions at both 0 and 250 ms delays. However, partial flaws of this study were the lack of a sham-controlled condition and the stimulation of several scalp sites during the memory task. The latter factor makes it hard to disentangle whether the decremental effect on the memory performance was due to the stimulation of a specific region or to the summation of the effects of several rTMS trains sequentially applied to several scalp positions along the whole task.

These findings, together with other possible applications of TMS to study short-term or working memory and learning mechanisms, have been reviewed by Grafman and Wassermann (1999). In this review, they discussed the significant decline in story recall in normal subjects after rTMS. However, the referred study (Flitman et al., 1998) used such a high intensity and a long duration of rTMS trains that some adverse effects, rather than functional interference, might explain the transient memory decline. Moreover, in this study, a confounding factor due to possible additive effects of repeated trains of rTMS on several scalp sites along the course of the task could not be ruled out.

3.2 Controlled Studies (with Sham Stimulation)

The first controlled study that systematically addressed the "interference" rTMS approach to the functional asymmetries of the prefrontal cortex during encoding and retrieval phases is that of Rossi and co-workers (Rossi et al., 2001). This study included 13 healthy right-handed young volunteers (4 males and 9 females, mean age 30.1, age range 22-41) who were naïve to the material presented and to the experimental purposes. These subjects were asked to remember a set of complex colored pictures (six encoding blocks, each block containing 8 indoor images and 8 outdoor landscapes). One hour later, six paired retrieval blocks were again presented, each containing 16 pictures. Eight of the 16 pictures were novel indoor pictures (distracters), and the other 8 were the indoor pictures presented in the previous phase (tests). By pressing a mouse key, subjects had to recognize between "test" and "distracter" pictures. The six encoding/retrieval blocks were labeled according to the type (active or sham) and the side (left/right) of the rTMS applied to dorsolateral prefrontal areas (DLPFCs, as determined by the overly of the stimulated scalp position with a template brain model, see Fig. 1). These were: R-Enc (= right rTMS in encoding, no stimulation in retrieval); L-Enc (= left rTMS in encoding, no stimulation in retrieval); Sham (= left rTMS in encoding and right in retrieval); R-Ret (= no stimulation in encoding and right rTMS in retrieval); L-Ret (= no stimulation in encoding and left rTMS in retrieval); and Baseline, which served as reference condition and consisted of the absence of stimulation in encoding and retrieval. The experimental timing is shown in Figure 1. To prevent spread of the currents outside the PFC, trains of rTMS (20 Hz, 500 ms) were delivered at 10% below individual ET, immediately after each picture presentation.

Behavioral findings of this study directly demonstrated that the prefrontal

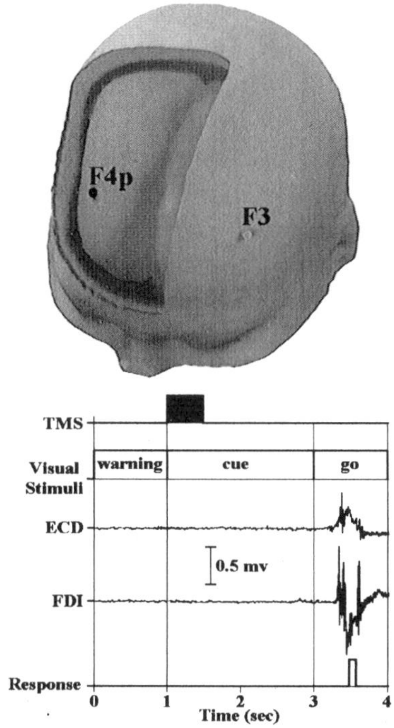

Figure 1: Top: template MR brain model indicating the site of rTMS application on the scalp and its projection on the underlying anatomy, corresponding to the DLPFC. Bottom: experimental timing. The rTMS lasts the first 500 ms from the picture presentation. Note that no motor responses are elicited by rTMS. The voluntary electromyographic activity (i.e. the subject's response) is derived simultaneously from the Extensor Communis Digitorum (ECD) and First Dorsal interosseous (FDI) muscles.

cortex was actively participating in both encoding and episodic retrieval, and basically reproduced what was predicted by the neuroimaging-derived HERA model. Indeed, the highest number of recognition errors was induced in the R-Ret block, where the right DLPFC was stimulated during retrieval. This suggested that the disrupting effect of rTMS was direct, since it took place immediately after the stimulation period when the retrieval effort was operating. The effect persisted for the duration required to complete motor response (about 1.5 sec; see Fig. 1). Such right-sided prevalence of the PFC during episodic retrieval is in line with most neuroimaging findings (Fletcher and Henson, 2001). Another finding, not as expected, was the left functional prevalence during the encoding phase. The probability to correctly remember the encoded information was significantly lower (versus sham and baseline blocks) when the left DLPFC had been stimulated during the encoding. Given the material used (essentially a visuospatial stimulus), a right side functional prevalence could be expected (e.g. Kirkhoff et al., 2000), although individual strategies of stimulus verbalization could not be excluded *a priori*. Whatever the case, this finding might suggest a less efficient encoding (or more "shallow" processing) and/or a faster decay of the information due to the concomitant interference by the rTMS.

It may be argued that, since PFC regions involved in working memory processes are largely overlapping with those involved in episodic memory (Fletcher and Henson, 2001), rTMS might have behaviorally influenced both. However, our experimental design minimized this possibility. In our experiment, the pictures were always available on the screen, so that if anything, working memory processes could have occurred only in self-monitoring of the responses emitted in the previous trials, or in conjunction with the influence on executive frontal functions such as the management of instructions, visuomotor transformation, and the response selection /execution. However, the lack of association between reaction times (which were significantly *shortened* by either left, right or sham rTMS) and the response correctness suggests that this possibility is unlikely.

Somewhat different findings were subsequently reported by Epstein et al. (2002), who studied 10 fluent-in-Kanji Japanese adults (mean age 34.4, age range 25-56) in an associative memory task involving pairs of Kanji (Chinese) pictographs and unfamiliar abstract patterns, so that only the abstract pattern represented novel material (Fig. 2).

In this study, focal TMS stimuli at intensities above the individual ET were applied during the encoding phase over the left and right DLPFC: these regions were identified by moving the coil 5 cm forward to the scalp position suited to elicit motor responses at individual ET level (i.e. the motor cortex). TMS on the cranial vertex was used as active control. Sham stimulation was performed in five additional subjects, by keeping the discharging coil near the head, but avoiding its scalp contact. During the blackout interval following the presentation of each Kanji character and its matching pattern, two TMS pulses were triggered at delays of 140 and 180 ms. The order of TMS sites and trials was counterbalanced. Subjects were instructed to remember all the three pairs of word-pattern association. Immediately after the end of each presentation set, the subject was handed three cards labeled with the three Kanji characters and a sheet containing six abstract patterns, half of which, randomly distributed on the sheet, had previously appeared during the test. Finally, the subject was asked to lay the Kanji cards directly on the patterns with which they had been paired.

The percentage of correct responses in subsequent recall of new

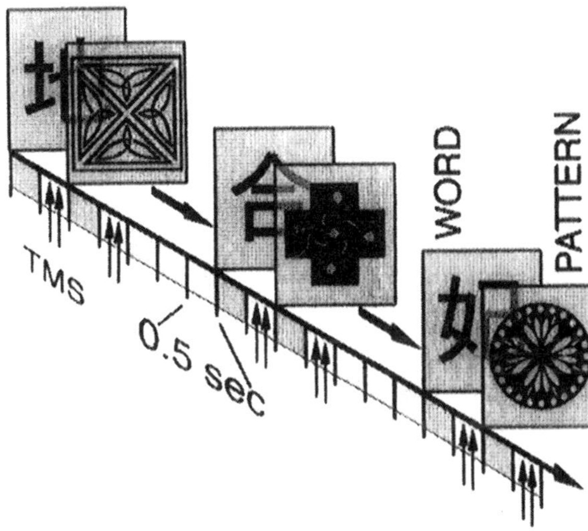

Figure 2: For each trial of associative memory encoding, sets of six paired stimuli were presented according to the order Kanji-abstract pattern, Kanji-abstract pattern, Kanji-abstract pattern. Each item was displayed for 500 ms; the items in a pair were separated by a blackout interval of 500 ms, and the different abstract patterns were separated by blackout intervals of 2000 ms. Instructions for the retrieval phase are in the text. Adopted from Epstein et al. (2002) with permission.

associations was significantly lower only after right DLPFC stimulation (about 40%), whilst active TMS delivered on left DLPFC or vertex and the sham TMS in the remaining subjects led to similarly good performances (about 60% of correct responses). On the basis of these findings, the authors (Epstein et al., 2002) conclude that the interference of TMS on behavioral performance is an evidence for the importance of the right DLPFC in encoding mechanisms of non-verbal material. An alternative hypothesis is that the encoding might have occurred elsewhere, and the functional relevance of the right DLPFC might be in relation to working memory processes required for the maintenance of paired associations until the recall phase. Some methodological factors could account for the discrepancies between these results and other findings on episodic memory encoding in healthy humans, both in the visuospatial and verbal domains (Rossi et al., 2001; Sandrini et al., 2003). Thus, the material displayed for the cognitive task, the type of TMS approach (double pulse stimulation versus 20 Hz rTMS) and the timing of TMS with respect to the picture presentation were all different. Also, the effects of age might have influenced results (Rossi et al., 2003), considering that the age range in the Epstein's study varied between 25 and 56 years (see next paragraph). The identification of the scalp region overlying the DLPFC might also have played a role, since reference systems were not used in their study.

Among factors that may influence the degree of functional PFC involvement in episodic memory, ageing is one of the most effective one, especially considering that the ability to learn and remember new information progressively declines along physiological ageing (review, Grady and Craick, 2000). Indeed, it has been suggested that one common finding in normal ageing is a reduction of the asymmetry predicted by the HERA model (so-called HAROLD model, the acronym for Hemispheric Asymmetry Reduction in OLDer adults) (Cabeza, 2002).

We therefore attempted to investigate this possibility with the interferential rTMS approach (Rossi et al., 2003). The whole experimental design, including the material displayed, rTMS sites of stimulations, timing and intensity, and blocks of pictures were identical to the previous study (Rossi et al., 2001). The behavioral performance of young subjects was compared with two other groups, one including 8 middle-aged subjects (between 50 and 60 years of age) and the other consisting of 8 subjects over 65 years of age. All subjects were right-handed and had a high educational level. A preliminary analysis of these data suggested that the functional relevance of the right DLPFC in retrieval mechanisms, which was evident in young individuals, progressively decreased across the life-span. In contrast, the reliance on the left DLPFC in encoding was preserved both in middle-aged and older individuals. Behavioral performances in the reference blocks (sham and baseline conditions) remained acceptable. From the "interferential" perspective, therefore, the rTMS of the right DLPFC was no longer effective in middle aged and older subjects in producing a significant number of errors during retrieval. Conversely, rTMS delivered on left DLPFC during encoding decreased the probability of being able to remember equally across the different age groups. These results are at first sight in agreement with the HAROLD model (Cabeza, 2002), especially regarding the age-dependency of the right PFC functional prevalence during retrieval operations. However, these findings cannot entirely explain why, at least within the current experimental setting, older subjects utilized the left PFC in retrieval as they did in encoding. Although putative mechanisms explaining the age-related modifications of frontal contributions to cognitive processing remain hypothetical, it cannot be excluded that additional brain regions, not reached by the rTMS interference, might be recruited in older subjects during episodic memory retrieval. Another explanation which is not mutually exclusive is that in older subjects, the degree of stimulus verbalization could be higher, thereby producing mainly left-lateralized activations during episodic encoding (Cabeza et al., 2000). Finally, the complexity levels of retrieval could be higher than in encoding, so that a bilateral contribution of DLPFC cortices was required in aged subjects: a somewhat similar phenomenon (i.e. activation of neural networks more widespread than in normal, including ipsilateral motor cortices) has been demonstrated in patients in the "early" stages of probable Alzheimer's disease performing a simple motor task (Babiloni et al., 2000).

In another study (Sandrini et al., 2003), we tested the functional PFC asymmetries in a verbal episodic memory task, in order to validate with the rTMS approach the HERA model in its original, verbal domain. Twelve healthy young Italian-speaking subjects, aged between 20 and 34 years old, were tested using the same experimental design as in our previous studies (Rossi et al., 2001, 2003). In the current setting, however, pairs of Italian nouns with high imagery content replaced the complex colored pictures. For each block of the six encoding phase blocks, 16 word pairs (8 semantically related and 8 unrelated) were randomly presented on the monitor for 2000 ms, with two inter-trial intervals (7000 or 8000 ms). During the encoding phase, the subjects were asked to classify word pairs as highly associated (e.g. bread – butter, garlic – onion), or non-associated (e.g. cow – table), according to norms collected for this experiment. One hour later, they were presented with the first word of each pair, and a choice was made between the second and a novel word (distracter). Events of the retrieval and encoding phases occurred with the same timing. Subjects were instructed to press the left or right key according to the position of the word which had been seen previously. The correct responses during the encoding and retrieval phases were evenly distributed across the left and right button presses (eight each), thus avoiding rules effects.

Both left and right DLPFC rTMS interfered with the encoding phase, by increasing the probability of error when subjects were asked to retrieve cue words. Moreover, the effect was specific for semantically unrelated word pairs, therefore suggesting a specific role of DLPFC(s) only when novel information had to be memorized. Such a bilateral reliance on DLPFCs for encoding process was somewhat unexpected, since most of previous neuroimaging literature pointed to a left-lateralization, at least for verbal materials (Fletcher and Henson, 2001). We therefore speculated that the current task might have required participants to perform a deep manipulation, in agreement with the predictions of the so-called "dual-coding theory". According to that, processing of abstract nouns would rely almost exclusively on left-sided verbal code representations, whereas concrete nouns additionally would gain access to a second image-based processing system in the right hemisphere (Paivio, 1986). The involvement of the right DLPFC during encoding (that is, increased errors induced by the right rTMS) of word pairs of high imagery content could parallel this strategy. Therefore, the bilateral DLPFC involvement might result from the combined engagement of verbal as well as non-verbal strategies in the context of episodic encoding. The results for retrieval were in agreement with the predictions of the classic HERA model, with a definite right-side prevalence of the rTMS interference on the behavioral performance.

3.3 Future Developments: Towards a Multi-Modal Functional Approach

The existing literature dealing with rTMS on episodic memory did not fully confirm the predictions of the neuroimaging-based HERA model. Therefore, one of our recent issues of interest was to investigate whether results compatible with the HERA model could be also obtained from an analysis of brain rhythmicity, which is implicated in the processes of

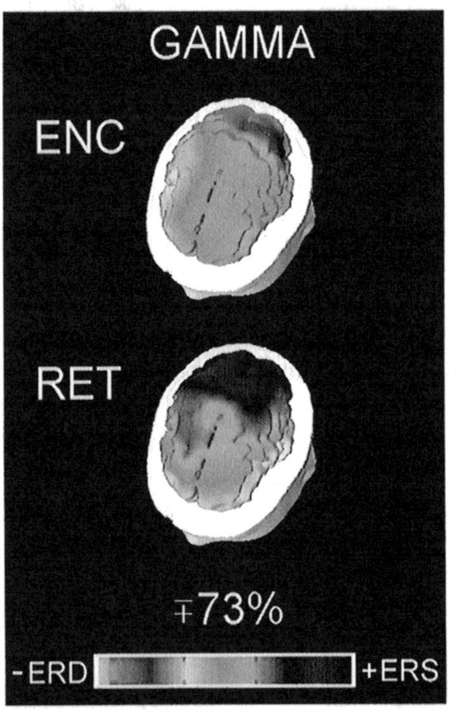

Figure 3: Three-dimensional color maps of cross-subjects gamma (about 40 Hz) Event Related Synchronization (ERS) peaks during encoding (ENC) and retrieval (RET) conditions in a visuospatial episodic long-term memory task.

facilitation and inhibition of information encoding and retrieval. Thus in separate recording sessions, healthy volunteers participated in EEG recordings using the same visuospatial stimuli (the indoor/outdoor paradigm) as in the previous rTMS studies (Rossi et al., 2001, 2003). Briefly, EEG data were recorded from 46 scalp electrodes (augmented 10-20 International EEG System) and were spatially enhanced by surface Laplacian estimation over a template MR head model. This estimation minimizes the effects of head volume conduction and annuls the influence of the reference electrode. Compared to rTMS evidence, EEG results showed no change of frontal rhythmicity consistent with HERA predictions. However, the HERA prediction (i.e. left prevalence in encoding and right prevalence in retrieval) was fitted by EEG gamma responses (about 40 Hz) in posterior parietal areas (Fig. 3), thus disclosing a possible role of "binding phenomena" in the dorsal stream subserving processes of visuospatial episodic memory (Del Percio et al., 2003).

Combining different neurophysiological techniques (i.e. exploring brain activity from different perspectives) during successful episodic long-term memory may therefore contribute to better address the parallel frontoparietal processes at the basis of cortical functional asymmetries. As a matter of fact, the idea of a unique pattern of "cortical activation/deactivation" explaining long-term memory processes seems too restrictive and moreover too dependent on the technique utilized to investigate brain functions. An alternative view might be that a comprehensive model of cognitive brainwork should take into account several indexes of parallel functional processing, including brain rhythmicity, hemodynamic and blood flow responses, and rTMS interference to investigate the cognitive challenge, according to a true multidimensional approach to the working brain.

4. CONCLUDING REMARKS

The number of TMS studies addressing episodic long-term memory in humans is still limited. However, all of them converge to indicating that the PFC plays an important role in long-term episodic memory. These results, therefore, extend the concept that the prefrontal regions are mainly devoted to short-term and working memory processes. An important qualification is that the functional role of PFC appears to be limited to the memory processing of novel information.

In addition, they consistently show that functional hemispheric asymmetries can be observed for encoding and retrieval processes. In line with the results of recent investigations using conventional neuroimaging procedures, however, some discrepancies concerning these functional asymmetries are emerging, which are not in full agreement with the predictions from the HERA model. To summarize, the important role of the right prefrontal areas in the retrieval of verbal and non-verbal material is confirmed by TMS. This functional asymmetry appears to be affected by healthy aging. In the case of encoding, the pattern of hemispheric asymmetry appears to be less consistent. We have observed a left sided, age-independent prevalence in the case of complex visual scenes, while a bilateral interference was present for word pairs. These results not only disagree with the HERA theory, but also do not fit with predictions based only on the verbal – non verbal nature of the memoranda. Other variables, such as task demands and individual strategies of memorization and retrieval may account for the inconsistent encoding effects. Other variables which should be taken into account are those related to the experimental design, to the timing and parameters of TMS, and to the accuracy of coil positioning with respect to the underlying cortical anatomy.

Fundamental questions that are still awaiting definite answers concern the detailed mechanisms of the action of rTMS interference with neural information processing, and the exact spatial accuracy of such effects. Nevertheless, among the armamentarium of methodologies that neuroscientists can use to investigate the "working brain", rTMS represents at the moment the unique tool that can non-invasively reproduce in the healthy brain the effects of cooling and microstimulation techniques used in animal experimental settings.

REFERENCES

Amassian VE, Qirck GJ, Stewart M (1990) A comparison of corticospinal activation by magnetic coil and electrical stimulation of monkey motor cortex. Electroencephalogr Clin Neurophysiol 77:390-401.

Babiloni C, Babiloni F, Carducci F, Cincotti F, Del Percio C, De Pino G, Maestrini S, Priori A, Tisei P, Zanetti O, Rossini PM (2000) Movement-related electroencephalographic reactivity in Alzheimer disease. Neuroimage 2:139-146.

Babiloni F, Carducci F, Cincotti F, Del Gratta C, Pizzella V, Romani GL, Rossini PM, Tecchio F, Babiloni C (2001) Linear inverse source estimate of combined EEG and MEG data related to voluntary movements. Hum Brain Mapp 14:197-209.

Barker AT, Jalinous R, Freeston H (1985) Non-invasive magnetic stimulation of human motor cortex. Lancet 1:1106-1107.

Boroojerdi B, Foltys H, Krings T, Spetzger U, Thron A, Topper R (1999) Localization of the motor hand area using transcranial magnetic stimulation and functional magnetic resonance imaging. Clin Neurophysiol 110:699-704.

Boroojerdi B, Phipps M, Kopylev L, Wharton CM, Cohen LG, Grafman J (2001) Enhancing analogic reasoning with rTMS over the left prefrontal cortex. Neurology 56:526-528.

Cappa SF, Sandrini M, Rossini PM, Sosta K, Miniussi C (2002) The role of the left frontal lobe in action naming: rTMS evidence. Neurology 59:720-723.

Caramia MD, Pardal AM, Zarola F, Rossini PM (1989) Electric vs magnetic trans-cranial stimulation of the brain in healthy humans: a comparative study of central motor tracts 'conductivity' and 'excitability'. Brain Res 6:98-104.

Cabeza R (2002) Hemispheric asymmetry reduction in older adults: the HAROLD model. Psychol Ageing 17:85-100.

Cabeza R, Anderson ND, Houle S, Mangels JA, Nyberg L (2000) Age-related differences in neural activity during item and temporal-order memory retrieval: a positron emission tomography study. J Cogn Neurosci 12:197-206.

Chen R (2000) Studies of human motor physiology with transcranial magnetic stimulation. Muscle Nerve (Suppl) 9:S26-32.

Chicurel M (2002) Magnetic mind games. Nature 417:114-116.

Del Percio C, Babiloni C, Babiloni F, Cappa S, Carducci F, Cincotti C, Miniussi C, Moretti DV, Pasqualetti P, Rossi S, Rossini PM (2003) Cortical functional asymmetry related to visuospatial episodic long-term memory. A multi-modal rTMS-EEG study. IX Conference of Human Brain Mapping, New York, June 2003.

Di Lazzaro V, Oliviero A, Profice P, Saturno E, Pilato F, Insola A, Mazzone P, Tonali P, Rothwell JC (1998) Comparison of descending volleys evoked by transcranial magnetic and electric stimulation in conscious humans. Electroencephalogr Clin Neurophysiol 109:397-401.

Epstein CM, Sekino M, Yamaguchi K, Kamiya S, Ueno S (2002) Asymmetries of prefrontal cortex in human episodic memory: effects of transcranial magnetic stimulation on learning abstract patterns. Neurosci Lett 320:5-8.

Fitzpatrick SM, Rothman DI (2000) Meeting report: transcranial magnetic stimulation and studies on human cognition. J Cogn Neurosci 12:704-709.

Fletcher PC, Henson RNA (2001) Frontal lobes and human memory. Insights from functional neuroimaging. Brain 124:849-881.

Flitman SS, Grafman J, Wassermann EM, Cooper V, O'Grady J, Pascual-Leone A, Hallett M (1998) Linguistic processing during repetitive transcranial magnetic stimulation. Neurology 50:175-181.

George MS, Nahas Z, Kozel FA, Li X, Denslow S, Yamanaka K, Mishory A, Foust MJ, Bohning DE (2002) Mechanisms and state of art of transcranial magnetic stimulation. J ECT 18:170-181.

Grady CL, Craik IM (2000) Changes in memory processing with age. Curr.Opin Neurobiol 10:224-231.

Grafman J, Pascual-Leone A, Alway D, Nichelli P, Gomez-Tortosa E, Hallett M (1994) Induction of a recall deficit by rapid-rate transcranial magnetic stimulation. NeuroReport 9:1157-1160.

Grafman J, Wassermann, EM (1999) Transcranial magnetic stimulation can modulate learning and memory. Neuropsychologia 37:159-167.

Hallett M (2000) Trascranial magnetic stimulation and the human brain. Nature 406:147-150.

Herwig U, Schönfeldt-Laucona C, Wunderlich AP, von Tiesenhausen C, Thielscher A, Walter H, Spitzer M (2001) The navigation of transcranial magnetic stimulation. Psychiatr Res 108:123-131.

Herwig U, Kolbel K, Wunderlich AP, Thielscher A, von Tiesenhausen C, Spitzer M, Schonfeldt-Lecuona C (2002) Spatial congruence of neuronavigated transcranial magnetic stimulation and functional neuroimaging. Clin Neurophysiol 113:462-468.

Hoffman RE, Cavus I (2002) Slow transcranial magnetic stimulation, long-term depotentiation, and brain hyperexcitability disorders. Am J Psychiatry 159:1093-1102.

Kirchhoff BA, Wagner AD, Maril A, Stern CE (2000) Prefrontal-temporal circuitry for episodic encoding and subsequent memory. J Neurosci 20:6173-80.

Nyberg L, Cabeza R and Tulving E (1996) PET studies of encoding and retrieval: The HERA model. Psychonom. Bull Rev 3:135-148.

Oliveri M, Caltagirone C, Filippi MM, Traversa R, Cicinelli P, Pasqualetti P, Rossini PM (2000) Paired transcranial magnetic stimulation protocols reveal a pattern of inhibition and facilitation in the human parietal cortex. J Physiol 529:461-468.

Paivio A (1986) Mental Representations: A Dual Coding Theory. Oxford University Press, Oxford.

Pascual-Leone A, Walsh V, Rothwell J (2000) Transcranial magnetic stimulation in cognitive neuroscience – virtual lesion, chronometry, and functional connectivity. Curr Opin Neurobiol 10:232-237.

Petrides M (1994) Frontal lobes and working memory: evidence from investigation of the effects of cortical excisions in nonhumans primates. In: Handbook of Neuropsychology (Boller F and Grafmann J, eds), pp59–82, Elsevier, Amsterdam.

Price CJ, Friston KJ (1999) Scanning patients with tasks they can perform. Hum Brain Mapp 8:102-108.

Rossi S, Pasqualetti P, Tecchio F, Sabato A, Rossini PM. (1998) Modulation of corticospinal output to human hand muscles following deprivation of sensory feedback. Neuroimage 8:163-175.

Rossi S, Cappa SF, Babiloni C, Pasqualetti P, Carducci F, Miniussi C, Babiloni F, Rossini PM (2001) Prefrontal cortex in long-term memory: an "interference" approach using magnetic stimulation. Nat Neurosci 4:948-952.

Rossi S, Miniussi C, Pasqualetti P, Babiloni C, Rossini PM, Cappa SF (2003) Role of the prefrontal cortex in visuo-spatial long-term memory along physiological ageing: a rTMS approach. Proceedings of the XII World Congress I.S.B.E.T., Napoli, October 2002, Brain Topogr (in press).

Rossini PM, Pauri F (2000) Neuromagnetic integrated methods tracking human brain mechanisms of sensorimotor areas 'plastic' reorganisation. Brain Res Rev 33:131-154.

Rossini PM, Rossi S (1998) Clinical application of motor evoked potentials. Invited editorial. Electroencephalogr Clin Neurophysiol 106:180-194.

Rossini PM, Barker AT, Berardelli A, Caramia MD, Caruso G, Gracco RQ, Dimitrijevic MR, Hallet M, Katyama Y, Lucking CH, Maertens de Noordhout AL, Marsden CD, Murray NMF, Rothwell JC, Swash M, Tomberg C (1994) Non-invasive electrical and magnetic stimulation of the brain, spinal cord and roots: basic principles and procedures for routine clinical application. Electroencephalogr Clin Neurophysiol 91:79-92.

Sandrini M, Cappa SF, Rossi S, Rossini PM, Miniussi C (2003) The role of prefrontal cortex in a verbal memory task. J Cogn Neurosci 15, 855-861.

Squire LR, Clark RE, Knowlton BJ (2001) Retrograde amnesia. Hippocampus 11:50-55.

Tulving E (2002) Episodic memory: from mind to brain. Annu Rev Psychol 53:1-25.

Tulving E, Kapur S, Craik FIM, Moscovitch M, Houle S (1994) Hemispheric encoding/retrieval asymmetry in episodic memory: positron emission tomography findings. Proc Nat Acad Sci 91:2016-2020.

Walsh V, Cowey A (2000) Transcranial magnetic stimulation in cognitive neuroscience. Nat Rev Neurosci 1:73-79.

Wagner AD, Schacter DL, Rotte M, Koutstaal W, Maril A, Dale AM, Rosen BR, Buckner RL (1998) Building memories: remembering and forgetting of verbal experiences as predicted by brain activity. Science 281:1188-1191.

Wassermann EM (1998) Report on risk and safety of repetitive transcranial magnetic stimulation (rTMS): suggested guidelines from the International Workshop on Risk and Safety of rTMS. June 5-7 1996. Electroencephalogr Clin Neurophysiol 108:1-16.

Wassermann EM, Lisanby SH (2001) Therapeutic application of repetitive transcranial magnetic stimulation: a review. Clin Neurophysiol 112:1367-1377.

Acknowledgements
Authors are grateful to Drs Patrizio Pasqualetti, Marco Sandrini, Katiuscia Sosta, Filippo Carducci, and Fabio Babiloni for their participation in various phases of experimental designs, data acquisition, and analysis.

Chapter 13

FUNCTIONAL NEUROIMAGING AND THE PREFRONTAL CORTEX: ORGANIZATION BY STIMULUS DOMAIN?

Christy Marshuetz† and Joseph E. Bates
Department of Psychology, Yale University, 2 Hillhouse Avenue, Box 208205, New Haven, CT 06520 – 8205, USA
† Correspondence author: christy.marshuetz@yale.edu

Keywords: Prefrontal cortex, spatial memory, object learning, verbal memory, working memory, frontal lobes, human memory, functional MRI, positron emission tomography.

Abstract: Working memory is the set of cognitive operations that maintains and processes information "on-line". It has been characterized both as a mental workspace (Baddeley, 1986) and as a set of operations that allow the efficient allocation of cognitive resources (Carpenter et al., 1990, 1999). Working memory typically is thought to be of limited capacity, between 4-7 items (Miller, 1956; Cowan, 2000), of limited duration, on the order of seconds (Peterson and Peterson, 1959) and as involving a number of separable sub-mechanisms, among these, rehearsal processes, domain-specific storage buffers, and a set of executive processes that are thought to operate on currently active information (Baddeley, 1986; Smith et al., 1996). This chapter focuses on the cognitive operations mediated by the frontal lobes in the service of working memory tasks.

1. INTRODUCTION

One of the most influential cognitive models of working memory is that of Baddeley and colleagues (Baddeley and Hitch, 1974; Baddeley, 1986). The Baddeley model (Fig. 1) includes at least two different storage buffers. The phonological loop stores verbal information in a phonological or articulatory code. Information in this buffer is maintained via sub-vocal (or vocal) rehearsal processes or it is vulnerable to fast decay. The second is a store for visuo-spatial information, and is presumed to have an analogous process for visuo-spatial rehearsal (e.g. Awh et al., 1999). According to the original Baddeley and Hitch (1974) model, there is also a "central executive" that serves to coordinate and control the two storage buffers (Baddeley, 1986). Others (e.g. D'Esposito et al., 1998; Smith et al., 1998; Smith and Jonides, 1999; Smith et al., 2002) have proposed that the central executive can be characterized as a set of "executive processes" which serve to manipulate the information in memory in the service of thought (Jonides, 1995).

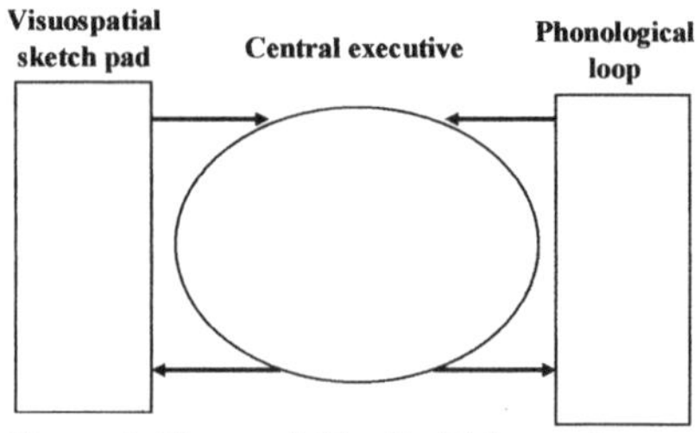

Figure 1: The model by Baddeley and colleagues. After Baddeley (1986).

The principles by which the frontal lobes might be organized are hotly debated, and there are four likely possibilities (Fig. 2), as pointed out by Johnson et al. (2003). The first is that the frontal lobes are "domain-specific": in other words, they are organized by the type of information they process (e.g. Goldman-Rakic, 1987; Wilson et al., 1993; Levy and Goldman-Rakic, 2000). This proposition has great appeal in light of findings that more posterior regions of the brain appear to be segregated by information type (e.g. Ungerleider and Mishkin, 1982). The question of whether or not the data support this view of prefrontal cortex (PFC) organization will constitute the bulk of our review.

The second hypothesis is that the PFC is organized by the type of processing operation that it performs, for example, storage versus executive processing (e.g. Petrides, 1994; Owen et al., 1996, 1998, 1999). The set of possible executive processes is numerous. Among the ones that have been proposed are the initiation and maintenance of goals and context (Cohen and Servan-Schreiber, 1992; Newell, 1992; Meyer and Kieras, 1999), frequency representation and temporal ordering (e.g. Milner et al., 1985), inhibition, selective attention (for a review, see Smith and Jonides, 1999), task switching and the strategic scheduling of task-related processes (Meyer and Kieras, 1999), and monitoring recent actions or stimuli (Petrides, 1994). Many that support the separation-by-process hypothesis have focused on the idea that ventral prefrontal regions mediate storage processes, and dorsal prefrontal regions mediate executive processing (e.g. Owen, 1997; D'Esposito et al., 1998). In this review, our emphasis will not be on complex functions, like alphabetizing letters in memory (e.g. Postle et al., 1999), but rather on the sub-processes involved in performing maintenance tasks, some of which incidentally include processes that might be regarded as "executive".

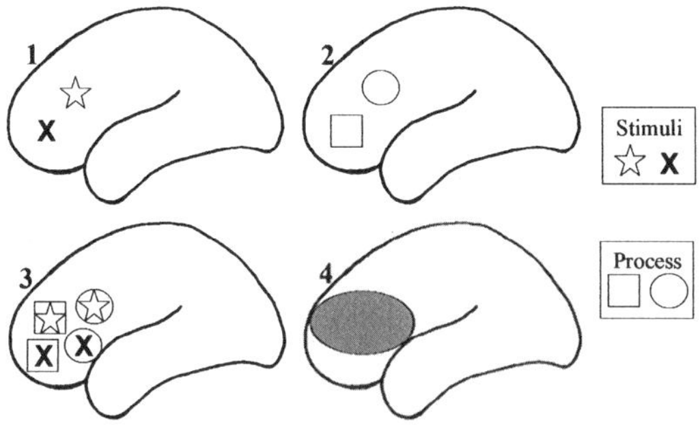

Figure 2: Four ways prefrontal cortex might be organized: 1) Stimulus type only, 2) Cognitive process only, 3) Both stimulus and process, 4) Neither stimulus nor process. After Johnson et al. (2003).

The third hypothesis is based on the logical conclusion that the first two are not mutually exclusive: the PFC might be organized both by process type and by information domain (e.g. Smith and Jonides, 1999; Fletcher and Henson, 2001; Johnson et al., 2003). This hypothesis implies that one might find different prefrontal regions involved in the inhibition of visuo-spatial information and verbal information, for example. Indeed, later we will argue that the processes by which the same cognitive goal is achieved may be fundamentally different depending on the type of information involved.

The final hypothesis is that the frontal lobe's organization is less modular than the first three hypotheses imply, and instead resources can be allocated according to task demands in the service of cognitive control. Therefore, the PFC might actively represent the mappings required to perform a particular task, and not just specific stimuli or actions or types of processes (Miller and Cohen, 2001). Accordingly, the appearance of prefrontal specialization may arise, but as a result of its interactions with other regions, not because sub-regions of PFC are specialized, *per se*. Interactions between regions are what give rise to complex functions, not computations within small regions of neurons themselves. In a sense, this hypothesis is consistent with the third hypothesis, instantiated in a different framework.

We begin this chapter by briefly reviewing the basic neuroanatomy of the frontal lobes, following which we will discuss evidence for the organization of PFC by stimulus domain. We focus on the modality-specific hypothesis for two reasons. The first is that a vast literature examining this claim has accumulated, with a number of recent reviews reaching different conclusions (e.g. D'Esposito et al., 1998; Smith and Jonides, 1999; Postle and D'Esposito, 2000; Fletcher and Henson, 2001). The second is that, whether or not there is support for it, the modality-specific claim has implications for the other hypotheses.

2. ANATOMY OF THE FRONTAL LOBES

Many have argued that what makes humans of greater intelligence than other animals is our highly-developed PFC. The frontal lobes comprise approximately 30% of brain volume in humans (Goldman-Rakic, 1987; Fuster, 1997). They are proportionally largest in humans, smaller in other primates and smaller in other mammals. For example, the PFC accounts for 17% of cortical volume in chimpanzees, and 7% in dogs, and 3.5% in cats (Fuster, 1997). The frontal lobes continue to develop relatively late in life, completing development in adolescence or early adulthood (Yakovlev and Lecours, 1967; Thatcher et al., 1987; Anokhin et al., 1996). They are also the first to begin their decline, roughly in the third decade of life; prefrontal volume loss correlates negatively with age in range of r = -.50 (Raz, 2000).

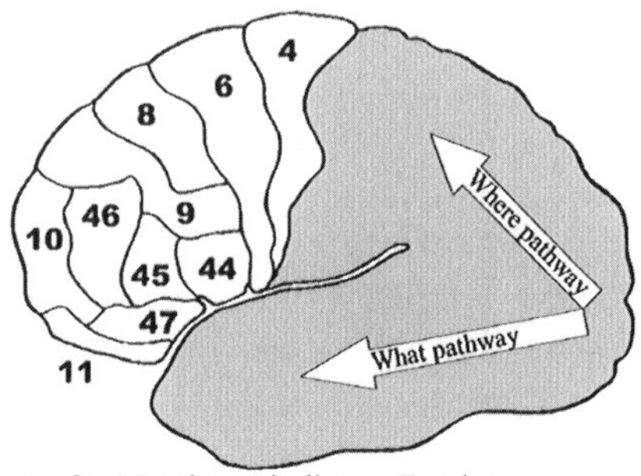

Figure 3: Numbers indicate Brodmann areas in prefrontal cortex, and arrows indicate the dorsal (where) and ventral (what) processing streams.

The PFC (Fig. 3) can be divided into the dorsolateral prefrontal cortex (DLPFC), ventrolateral prefrontal cortex (VLPFC), the orbitofrontal cortex, supplementary motor area (SMA), and premotor cortex. The frontal lobes can be further sub-divided according to Brodmann's areas (Brodmann, 1909). Motor cortex consists of Brodmann's area (BA) 4 and a portion of BA 6. Typically, the DLPFC is regarded as BAs 9, 46 (though some include BA 10 with DLPFC). BA 8 and part of BA 6 lie just behind BA 9 and 46, and are sometimes dubbed SMA. The VLPFC consists of BA 44 and 45 and areas 47; orbitofrontal cortex includes BA 10 and 11. The anterior cingulate cortex (BA 32, some include 24) lies at the midline just above the corpus collosum.

The PFC is well-situated to perform tasks that require the coordination between different brain regions. It is richly interconnected with the parietal lobes, the basal ganglia, hippocampus, and the temporal lobes (Goldman-Rakic, 1987). Furthermore, PFC connectivity hints at regional specialization. Dorsolateral regions (area 46) are heavily interconnected with parietal cortex, whereas more ventral regions (areas 45/12) are interconnected most heavily with the temporal cortex, which is inferior to the parietal cortex (Ungerleider et al., 1998), and different regions of the macaque parietal region are connected with different portions of PFC (Goldman-Rakic, 1987).

3. SUBCOMPONENTS OF WORKING MEMORY

As we mentioned previously, working memory is thought to contain a number of different storage buffers that mediate the retention of different types of information. The major division proposed by Baddeley (1986) is between verbal and visuo-spatial information, and this idea finds support in the behavioral data. For example, requiring subjects to repeat "the the the" interferes more with verbal than spatial memory (Baddeley et al., 1984), whereas requiring subjects to trace a moving stimulus with a finger or stylus interferes more with visuo-spatial memory (Baddeley et al., 1975). Thus, primary task performance suffers more when the secondary task is of the same modality than when it is of the other modality, which is commonly taken as evidence of separable pools of cognitive resources.

In addition to a division between verbal and visuo-spatial information, some have also proposed that the visuo-spatial sketchpad can be further fractionated into at least two separate subsystems, one for object and the other for spatial information (e.g. Goldman-Rakic, 1987; Smith et al., 1995; Courtney et al., 1996; Ungerleider et al., 1998). Although the original Baddeley model does not contain this division, it is not precluded from the model on empirical or theoretical grounds. A division between object and spatial information finds support in the evidence from animal work that suggests that there are at least two different information processing streams, known colloquially as the "what" and "where" systems (Ungerleider and Mishkin, 1982), discussed later in more detail (Fig. 3).

4. EVIDENCE FOR PREFRONTAL ORGANIZATION BY INFORMATION DOMAIN

4.1 Verbal versus Visuo-Spatial Working Memory

Much evidence supports the idea that language is disproportionately represented in the left-hemisphere, although the right hemisphere does appear to have some language capability (Gazzaniga, 1983; Zaidel, 1985; Benson, 1986). For example, clear evidence for left-lateralization of language can be found in the performance of patients who have had surgical recisions (as a therapy for severe epilepsy) of their corpus collosum, the major fiber tract connecting the two hemispheres. Such callosotomy (split-brain) patients often cannot verbally report information presented to their disconnected right hemisphere, whereas other cognitive tasks, like face recognition, are readily performed by the right hemisphere (Gazzaniga and Sperry, 1967; Gazzaniga and Smylie, 1983; Reuter-Lorenz and Baynes, 1992). Second, patients with left-hemisphere damage due to stroke or other types of cerebral damage experience a greater degree of language impairment than those with right-hemisphere damage (Benson, 1986). Finally, two of the most critical areas in language production and comprehension, Broca's and Wernicke's areas, reside in the left hemisphere whereas the corresponding right-hemisphere structures are smaller (Falzi et al., 1982).

Like verbal processing, spatial processing likely occurs to some degree in both hemispheres, but unlike verbal coding, it is thought to rely disproportionately on the right hemisphere. Split-brain patients can perform visuo-spatial pattern tasks better with their left hands (controlled by the right hemisphere) than with their right (Gazzaniga, 1970; Gazzaninga and LeDoux, 1978). Furthermore, intact subjects also exhibit a left-hemisphere dominance for verbal information and a right-hemisphere dominance for visuo-spatial information (e.g. Kimura, 1973). In light of this verbal-left, spatial-right distinction in hemispheric specialization, researchers have speculated that these divisions might be also hold for working memory.

Reuter-Lorenz and Miller (1998) have examined the extent to which split-brain patient V.P. shows laterality effects on spatial and verbal working memory tasks. V.P. was required to attend to a central fixation point and was presented with items to-be-remembered either on the left or right side of the visual field. Thus, information was only presented to one hemisphere at a time. Reuter-Lorenz and Miller found a clear double-dissociation between verbal and spatial working memory, with V.P. exhibiting superior performance for verbal material presented to the left hemisphere and superior performance for spatial material presented to her right hemisphere (Fig. 4). This finding is strong evidence of a left-right laterality difference for verbal and spatial information in working memory, at least in patient V.P. Although the left-right distinction seems to hold for verbal and spatial cognition, the split-brain evidence does not speak to the issue of whether the left-right division is honored by the frontal lobes, or whether the differences observed by Reuter-Lorenz and Miller results from material-specific storage buffers in posterior cortical regions. Thus, we turn to data from neuroimaging experiments to provide evidence for more specific information about regional involvement.

One area of PFC that is normally activated by verbal working memory tasks is Broca's area (e.g. Paulesu et al., 1993; Awh et al., 1996; Schumacher et al., 1996; Smith et al., 1996). The involvement of Broca's area and left premotor cortex in verbal articulatory and rehearsal tasks has been confirmed by a number of studies (Paulesu et al., 1993; Awh et al., 1996). These findings imply that there is some PFC specialization by stimulus modality, since Broca's area is rarely observed in, for example, spatial- or object-memory tasks.

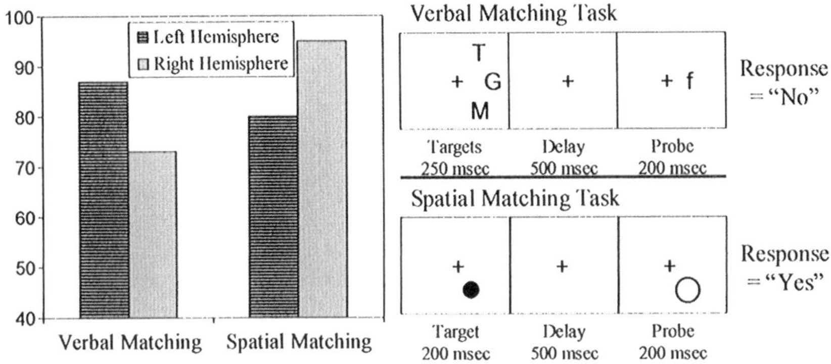

Figure 4: Patient V.P.'s performance (left) on a spatial and verbal working memory tasks (right).

Other PFC regions have been found to be involved in working memory, including SMA, DLPFC and other portions of ventrolateral PFC, and the evidence for the specialization of these regions is less clear. Activation in DLPFC has been presumed by many to be the result of executive processes or "manipulation" of information in memory (e.g. Smith et al., 1998; D'Esposito et al., 1999; Postle et al., 1999).

A number of neuroimaging studies have contrasted PFC activity in verbal and spatial working memory. Smith et al. (1996) used positron emission tomography (PET) to directly compare verbal and spatial working memory in a variant of the *n*-back task (Fig. 5A), called the 3-back task. The *n*-back is a task in which subjects see a series of items and must respond positively to any item that matches the one seen n trials ago. In these 3-back experiments, stimuli in both conditions were letters that changed location on the screen. Thus, stimuli were identical in both the spatial- and verbal-memory tasks, but subjects attended to different aspects of the stimuli, responding to 3-back letter matches in the verbal condition, and 3-back location matches in the spatial condition.

Smith and colleagues reported greater activity in the left hemisphere in the verbal relative to the spatial task, and more right-hemisphere activity in the spatial task than in the verbal task. First, Broca's area was significantly active in the verbal-memory task, but not in the spatial-memory task. Beyond that, the left and right hemisphere PFC activations were more weakly dissociated. In the verbal-memory task, two other DLPFC sites were activated on the left (BA 9/46), and only one on the right. In the spatial-

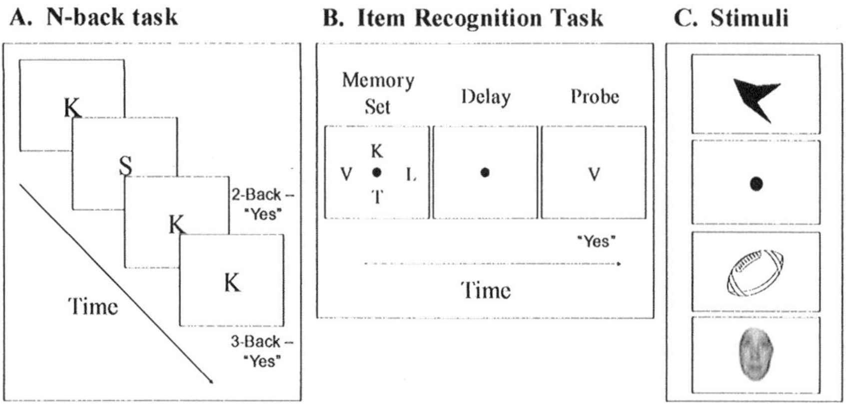

Figure 5: A. The 2- and 3-back tasks with verbal stimuli. B. The item-recognition task. C. Types of stimuli used in the n-back and item-recognition experiments.

memory task versus the spatial control, two activation foci were reported on the left (SMA and BA 46/10).

In the right hemisphere, there were also two prefrontal areas of activation, one in SMA in BA 9/46. Note also that the z-score values for the verbal task were all higher in the left than in the right hemisphere, and vice-versa for the spatial task. Thus, both tasks activated both hemispheres, but there were three left-hemisphere areas of activation in the verbal task as compared with one in the right hemisphere. In the spatial task, there were two areas of activation in each hemisphere.

Nystrom and colleagues (Nystrom et al., 2000) employed a similar task in a functional magnetic resonance imaging (fMRI) experiment and examined the neural activation not just for main effects of stimulus type, but also for interactions of load and stimulus type. They found a main effect of stimulus-type in three prefrontal regions: right and left mid-frontal or SMA (BA 6/8), right VLPFC (BA 45/47), and in each of those, activity was greater in the spatial than the memory condition. They also observed two other right-hemisphere frontal regions in which there was an interaction of load and stimulus type, and did not observe any activity in or near Broca's area either in the verbal or spatial tasks. Thus, their results do not replicate the findings of Smith et al. (1996), who found evidence for hemispheric specialization. However, Broca's area was not active in the verbal condition, suggesting the possibility of low statistical power.

At least two other experiments employing the *n*-back task have reported similar neural activations between different stimulus types, lending support to the findings of Nystrom et al. (2000). For example, D'Esposito et al. (1998) compared verbal and spatial versions of the 2-back task; but this time did not use letters in the spatial condition, rather they used a set of twelve locations aligned along a circle and indicated by a small square. They reported largely bilateral prefrontal activity in both tasks. Another experiment by Hautzel et al. (2002) also reported no significant differences in PFC when they used abstract shapes, nameable objects, letters, and spatial locations as stimuli. The lack of detectable differences may have been due to the analysis method (a fixed-effects analysis that underestimated the between-subjects variance) employed by D'Esposito et al. (1998), although there is no such explanation for the lack of statistical differences in the experiment by Hautzel, et al. (2002).

In light of the studies described above, it is fair to conclude that evidence from the *n*-back tasks is mixed. However, one must use caution in interpreting null results. The D'Esposito et al. (1998) study employed twelve spatial locations on an imaginary circle and may have encouraged subjects to code the locations verbally ("twelve o'clock") rather than spatially. The potential for this sort of confound can also be found in the task designs used by Smith et al. (1996) and Nystrom et al. (2000): The same verbal stimuli were used in both the spatial and verbal tasks; the tasks differed only by the stimulus dimension to which subjects were instructed to attend. This type of design has advantages; the tasks are exactly the same except for the instructions. However, even though not required to do so, subjects may automatically identify the letters, engaging verbal articulatory processes, as occurs in the Stroop task (Stroop, 1935).

Other experiments have contrasted verbal and spatial working memory using the much simpler item-recognition task (Fig. 5B). Item-recognition tasks require that subjects maintain a small set of items over a short delay and respond to a recognition probe. Following the probe, they are given a new set of to-be-remembered items. Thus, they are simpler than the *n*-back tasks which, in addition to memory storage, require subjects to keep track of the order of the stimuli and update the contents of working memory on every trial.

Smith and colleagues (Smith et al., 1996) performed a second experiment, contrasting spatial and verbal versions of the item-recognition task. In the verbal task, the stimuli were letters, and in the spatial-memory task, the stimuli were dots at specific to-be-remembered spatial locations. They found activity that was almost exclusively lateralized to the left hemisphere in the verbal task, including activity in left BAs 6 and 44. In the spatial item-recognition task, the activity was lateralized to the right hemisphere, including BAs 6 and 47.

Another PET experiment employing an item-recognition task was performed by Reuter-Lorenz et al. (2000). In order to explicitly examine PFC for laterality effects, they defined regions-of-interest from the literature on verbal and spatial working memory, and then applied each region and its opposite-hemisphere homologue to activation in PFC. They found a cross-over interaction: activity was significantly greater in the left-hemisphere in the verbal task and significantly greater in the right hemi-sphere in the spatial task. Interestingly, consistent with the results from the 2-back task reported in Smith et al. (1996), lateralization was greater in the verbal task than in the spatial task.

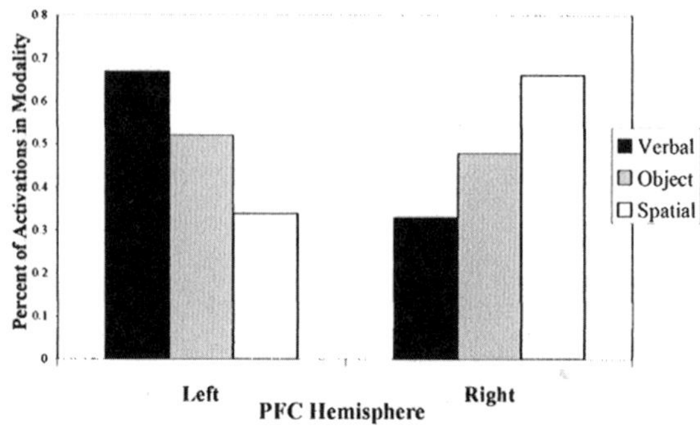

Figure 6: Proportions of PFC activations in the left and right hemispheres in verbal and spatial working memory tasks. Studies included are indicated by an * in the references.

One drawback to the PET experiments just discussed is that both were between-subjects comparisons, and there is a possibility that the subjects in the two experiments were simply different from one another. An fMRI experiment free of this concern was preformed by Prabhakaran et al. (2000). They had subjects perform two variants of the item-recognition task, one verbal and one spatial. Subjects had to remember four items, either letters or spatial locations (indicated with brackets). Prabhakaran and colleagues reported only a single focus of activation in the PFC in the verbal task; an area of left inferior PFC in the vicinity of Broca's area. In the spatial task, the only significant prefrontal regions were in the more superior portions of the right hemisphere. Thus, in a task that required just maintenance and rehearsal, the verbal task was lateralized to the left hemisphere and did not activate DLPFC, and the spatial task was lateralized to the right hemisphere.

The studies described above present evidence for both sides of the debate: some studies have found relative PFC lateralization for verbal and spatial working memory, whereas others have failed to find hemispheric differences. In order to get a clearer picture of the overall trends in the literature, we have constructed a graphical representation of the results across 23 neuroimaging reports, some of which included multiple experiments. We included available studies that employed storage-plus-rehearsal tasks and the *n*-back task for which stereotaxic coordinates

(Talairach and Tournoux, 1988) were available[1]. All prefrontal coordinates from all studies were included, except a small number in anterior cingulate in which the activation fell directly on the midline (i.e. an X-axis coordinate value of 0). Other anterior cingulate areas were included because such activations also often extend into BA 6[2]. We then performed a t-test on the X-coordinate values for each area of activity, categorized by stimulus modality. What we found confirmed the left-right verbal-spatial division in PFC: the mean X-coordinate value for verbal tasks was -14, whereas for the spatial tasks, the average X-coordinate value was 11.8. This difference was significant: $t(165)=-4.54$, $p<0.001$. Furthermore, objects (discussed in more detail below) occupied a middle ground, with more bilateral distribution of activity (mean X=-5.3). A graphical depiction of the results can be found in Figure 6.

What can we conclude from about the lateralization of verbal and spatial working memory across all of these studies? Although there is some bilaterality in both the verbal and spatial tasks, activity in the verbal tasks seems to favor the left hemisphere and spatial tasks seem to rely more on right-hemisphere neural regions. This finding is in good agreement with both the behavioral and patient evidence and furthermore, is consistent with the split between verbal and visuo-spatial information in the model of Baddeley and Hitch (1974). One question is how to interpret hemispheric bilaterality to the extent that it exists. Do these bilateral activations reflect noise in the data? Do they represent functional recruitment of opposite-hemisphere homologues as has been suggested in the aging literature (e.g. Reuter-Lorenz et al., 2000)? Are they reflective of variable strategies across studies? These are all important questions for further research.

---

[1] According to the Talariach and Tournoux (1988) coordinate system, the X-axis starts at 0, with negative values indicating left-hemisphere brain locations. The Y-axis indicates how far forward (positive coordinates) or how far back a region is (negative coordinates). The Z-axis indicates how high (positive values) or low (negative values). Some researchers have indicated left-hemisphere activations with positive X-axis coordinate values instead of negative ones. We examined each paper carefully to determine whether positive X values indicated left or right-hemisphere and corrected the coordinates accordingly for the sake of our analysis.

[2] We assumed that near midline activations (e.g. X=2) would fall equally randomly in the left and right hemispheres and would not bias the analysis. All areas with a negative X-value were counted as left hemisphere activations, and all with positive X-values were counted as right-hemisphere activations. Only three activation foci were excluded due to an X-value of 0.

4.2 Object versus Spatial Working Memory

Visual information is projected from the retina of the eye to primary visual cortex (V1 or BA 17) in the occipital lobes, and becomes divided as it leaves V1 and travels forward through the brain. Spatial location information travels via a dorsal route, finding its way to parietal cortex (Mishkin et al., 1983). Object, or "what", information takes a lower ventral route, into temporal cortex (Fig. 3), where much information about objects and their uses is thought to reside (e.g. Warrington and McCarthy, 1983; Warrington and Shallice, 1984; Chao et al., 1999).

The what/where dorsal-ventral distinction has been found in primate and human studies, and evidence suggests that it may carry forth into the frontal lobes. First, there is anatomical evidence that the ventrolateral frontal areas are richly interconnected with inferotemporal cortex (Webster et al., 1994), whereas the mid-dorsolateral frontal regions receive input from posterior parietal cortex (Cavada and Goldman-Rakic, 1989). Furthermore, some have reported individual cells within PFC that respond differentially to object and spatial information. For example, Wilson et al. (1993) found ventral PFC cells that code for object information during a delay, and cells lying in dorsolateral PFC that code for spatial information. Finally, lesion studies of the PFC support the dorsal-spatial and ventral-object distinction in memory (Petrides, 1994; Fuster, 1997). Thus, Goldman-Rakic and colleagues (e.g. Goldman-Rakic, 1987; Levy and Goldman-Rakic, 2000) and others have argued that PFC is organized according to the type of information, rather than specific operations performed on those representations.

Evidence for this information-type hypothesis has also been found in the human neuroimaging literature on working memory. For example, Smith et al. (1995) used PET to examine the differences between spatial and object memory. In their first experiment, they compared data from a spatial 3-dot item-recognition task from another study, with data from an object item-recognition task in which they used abstract shape drawings. As described earlier, in the spatial item-recognition task, Smith and colleagues observed activity in right BA 47 and right premotor cortex. In the object-memory task, they observed activation of left inferior PFC (BA 44) and anterior cingulate cortex (BA 32). Thus, the activation that they observed was lateralized, with left-hemisphere activity in the object-memory task and right-hemisphere activity in the spatial-memory task. However, although some of the spatial-memory activity was quite superior and some object-related activity quite inferior, there was also inferior PFC activation in the spatial task; which does not support the idea of a strict dorsal-spatial ventral-object division.

Smith and colleagues (Smith et al., 1995) reported a second experiment and this time, employed the same subjects and stimuli in both tasks. In both the spatial and object tasks, subjects saw two abstract shapes and had to remember them over a three-second delay. After the delay, a single shape appeared, and subjects had to make a recognition response to either the object or the object's location. As in their first experiment, activation in the object-memory task was lateralized to the left hemisphere, and activity in posterior regions roughly honored the what/where distinction. However, in this case, no area of activation in the PFC was observed in the object-memory task. In the spatial task, activity was observed in premotor cortex (BA 6), and dorsolateral PFC (BA 46), both on the right and in anterior cingulate (BA 32). Thus, their results supported a division of object and spatial memory in posterior regions, and the spatial activation in PFC was relatively dorsal, but inconsistent with their other experiment; they did not find ventral activity in PFC in the object task.

Courtney et al. (1998) conducted an experiment similar to those reported in Smith et al. (1995), using event-related fMRI. Subjects saw a set of faces in a number of locations on the screen, and the subjects had to respond either to the locations or to the identity of the faces. In the spatial task, the activity that they observed was in bilateral superior frontal gyrus (in or near BA 6/8). In the face working memory task, the activation that they observed was primarily in the left inferior frontal gyrus (in the approximate vicinity of BA 10/47). Although they did observe some overlap between the information domains, the face-related sites that were superior and bilateral were a small subset of those activated in the spatial-memory sites and vice-versa.

Another within-subjects PET study, by Courtney et al. (1996), also employed an item-recognition task. In their experiment, a face could appear in one of 24 locations marked on the screen. Subjects saw three faces sequentially, and after a delay, a recognition probe appeared that was either in the same location or was the same face as one of the ones they had just seen. As in other studies, posterior activation sites honored the dorsal-ventral division. In their contrast of the face-memory and sensory-motor control condition, they observed activity in several right frontal-lobe sites, one on the border of dorsal and ventral PFC (BA 45/46), one at the midline (anterior cingulate), and a third near the border between inferior and orbito-frontal cortex (BA 11/47). Furthermore, they observed activity in the left PFC in the vicinity of Broca's area (BA 44). In the location-versus-control comparison, they observed no significant PFC activation. When they directly compared the face and spatial conditions, they found activity in a number of prefrontal sites, with the face-memory sites being more inferior (right BA 9/45/46 and BA 11), and the spatial-memory sites being more superior (left and right BA 6/8). Thus, their findings for the spatial task were in partial agreement with those of Smith et al. (1995). Common to both experiments was activity in superior regions of PFC; however the "object" memory activity was more superior and lateralized to the right hemisphere.

Nystrom et al. (2000), discussed earlier, also included a task contrasting spatial and object working memory. They employed a 2-back task in which subjects saw a set of abstract shapes. In the spatial-memory task, they were required to remember the locations of the objects; in the object-memory task, they were required to remember the identity of the objects. During the inter-trial interval, words appeared on the screen, and subjects were asked to read them aloud; this manipulation was designed to prevent the subjects from verbally encoding the shapes or locations. The results provide some support for both the idea that object memory is left-lateralized and more inferior, and spatial-memory is more right-lateralized and superior. They found main effects of spatial memory in right BA 6/8 (middle frontal gyrus) and effects of object memory in the left inferior frontal gyrus, BA 44/45 and BA 9/8/44. There were also a set of complex interactions of condition and memory load, especially in BA 8/9. Thus, their data provide some support for the dorsal-ventral where/what distinction in PFC as well as some support for the lateralization of spatial-memory to the right hemisphere and object to the left, although the conclusions must be tempered by the complex set interactions in a number of prefrontal regions.

Further support for the left-right object-spatial division can be found in a study by McCarthy and colleagues (McCarthy et al., 1996). They used fMRI to contrast two demanding spatial and object working memory tasks. Items, an abstract object or location indicated by a square, appeared every 1500 ms. Subjects' task was to respond "yes" to any item that matched any of the ones they had previously seen during that block of trials. In the shape task, the activity in the middle and inferior frontal regions was greater in the left hemisphere than in the right. Activity was less evident in the superior frontal gyrus, but was relatively right-hemisphere lateralized. For locations, superior, middle, and inferior frontal activity was right-lateralized, and inferior frontal gyrus was relatively less active than in the shape task.

Belger and colleagues (Belger et al., 1998) contrasted spatial and object working memory using a sequential-presentation item-recognition task. In their task, subjects saw a series of three items, either locations indicated by small squares, or abstract shapes. Consistent with the studies described above, they found a greater number of activated voxels in the superior frontal gyrus (focus in BA 32/10) in the right hemisphere than the left in the spatial working memory task. In the object working-memory task, they observed greater activity on the left overall, with a bigger left-right difference in inferior frontal gyrus (BA 46/10 and BA 11/47) than in the middle frontal gyrus, which had a smaller left-right difference, again, hinting at both a dorsal-ventral difference between spatial and object memory as well as a left-right difference.

One concern is that the objects used in experiments contrasting object versus spatial working memory are often quite different from one another, which may lead to inconsistencies across experiments. Sala et al. (2003) explicitly varied the type of object used in their "what" versus "where" short-term memory experiments. In their first experiment, they contrasted spatial and object memory for faces and houses using a variant of the item-recognition task. They found that object memory activated large regions of the inferior frontal gyrus, medial frontal gyrus, and insula on the left. Furthermore, large regions at the midline in the area of the cingulate and supplementary motor area (SMA) were also activated. When they examined regions that were more active in the spatial-memory delay than in the object-memory delay, they found more bilateral and superior activation, with activity on the right and left in the superior frontal sulcus. In the second experiment, they used only houses as stimuli. Again, what they found was bilateral superior frontal sulcus activity in the spatial task (with more activity on the right than left) and greater activity in more ventral areas of PFC on the left in the object-memory task. Thus, the results are in agreement with the dorsal-ventral hypothesis, but only moderately supportive of a left-right hemispheric specialization for objects and spatial memory in PFC. They also examined the difference between face and house-identity related activity in a similar paradigm as in the first two experiments, but this time, did not include a spatial-location memory condition. They found that there was greater left-hemisphere activity in the inferior frontal gyrus in the face identity task than the house identity task, and greater middle frontal gyrus activity on the right in the house-identity task than the face identity task, and greater superior frontal sulcus activation bilaterally. They concluded that the dorsal, "where" pathway is involved in processing the house stimuli more than the face stimuli because spatial relationships are inherently part of the object. Furthermore, houses may be perceived as landmarks that play a role in navigation. They argued that the use of some types of objects, like houses, may engage dorsal pathway mechanisms more than other objects, like faces, and that this difference may explain some of the inconsistency in the neuroimaging literature.

Thus, there is partial convergence between the studies: there are hints of left lateralization in the object-memory tasks, and when present, object-memory activity appears to be more inferior than in spatial memory tasks. Spatial-memory activity, when observed, seems to be more superior than that in the object-memory tasks and also may be relatively lateralized to the right hemisphere.

Some studies have failed to find support for strict frontal-lobe divisions by information type. For example, Postle et al. (1999) used a task that required subjects to remember an object then a location or vice-versa, all within trial. Thus, effects could be examined in the same region over time. What they found was that there was no area in PFC in which there was clear evidence for the what/where distinction: rather, in the regions they identified, activity seemed to be sensitive to both conditions. Similar results have been reported by a number of others (Owen et al., 1998; Postle et al., 2000; Hautzel et al., 2002).

In order to answer the question of whether the existing data taken together support the left-right or dorsal-ventral distinction, or both, we constructed a graphical representation similar to the one we presented earlier. We included studies that reported Talairach coordinates from analyses representing differences between either object-versus-spatial comparisons or object and spatial conditions versus minimal memory control conditions. All Talairach z-axis values of 30mm or below were assigned to the "ventral" category and all sites of activation with a z-axis value of 31mm or above were entered into to the "dorsal" category (thus, these are relative terms; for example, some of the "dorsal" activities were not in DLPFC, but rather were in SMA). What we found was partially consistent with the dorsal-ventral object-spatial

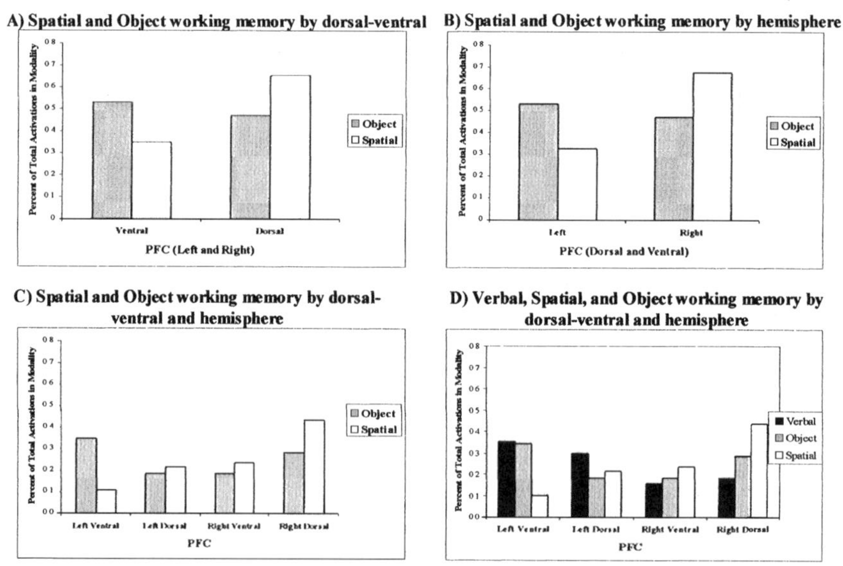

Figure 7: A. Activations in spatial and object working memory tasks in dorsal and ventral PFC. B. Activations in spatial and object working memory tasks in the left and right hemisphere. C. Spatial and object activations split both by dorsal-ventral and hemisphere. D. Figure includes activations from verbal working memory tasks.

division of PFC. The mean Z-value for spatial memory was significantly more dorsal (mean Z=34.7) than for object memory (mean Z=27.6), $t(95)=-2.07$, $p<0.05$. Thus, relative to each other, spatial-memory is more dorsal and object-memory more ventral, but strong conclusions must be tempered by the even dorsal-ventral split in the object-memory experiments. The results can be seen in Figure 7A.

Another question was whether the object-related activations tended to be more left-lateralized and spatial more right-lateralized. As in our graph of verbal and spatial activations, negative X-axis values were categorized as "left" and positive X-axis values were categorized as "right". We found that object memory was relatively more left lateralized (mean X=-5.3) than spatial memory (mean X=11.8), $t(95)=-2.75$, $p<0.01$) (Fig. 7B). Finally, when divided by both dorsal-ventral and left-right (Fig. 7C), the greatest proportion of spatial activations occur in the right dorsal region, and the greatest proportion of object-memory activations occur in the left ventral region.

Taken together, the studies on verbal, spatial, and object working memory seem to support left-right laterality differences, as well as a division between dorsal and ventral regions. We have seen here that dorsal regions, especially in the right hemisphere, appear to be especially involved in spatial memory. A disproportionate number of activations in the object-memory task fall in the left ventral PFC, though of the three types of stimuli we have considered, the results for objects is the least clear.

5. ORGANIZATION BY BOTH INFORMATION TYPE AND PROCESS?

We have suggested that PFC is organized loosely by stimulus domain. As we have pointed out in our introduction, the question of whether PFC is organized by stimulus type or information processing operation is often pitted as if they are mutually exclusive possibilities, with one serving as the null hypothesis for the other. As others have pointed out, we need not reject the organization of PFC by processing function simply because we have evidence that PFC is organized by information processing type (e.g. Smith and Jonides, 1999; Fletcher and Henson, 2001). In order to test the possibility that PFC is organized by both information domain and process, Johnson and colleagues (2003) explicitly compared two different processing operations, a minimal-recall task ("refresh") and a minimal-recognition task ("note"), with two different information-types, words and nameable objects. In their first experiment, subjects saw items, followed by a very brief delay (550 ms). Then, subjects were either shown the same item, a new item, or a black dot. When they saw a dot, they were instructed to recall ("refresh") the item they had just seen.

Johnson et al. (2003) reported reliable activity in left BA 9/10 when the items were words, whereas the recall-related activity in the objects condition was more posterior and inferior, in left BA 9/44. In their second experiment, subjects saw either words or nameable objects. After a delay, subjects made a decision (without overt button-press response) about whether the item was old or new (they refer to this process as "noting", Johnson, 1992). This time, Johnson et al. found right-hemisphere activation in both the object and word tasks. For the object-memory task, activity occurred in right BA 9/44 and was greatest when the item was new and when the subject simply had to passively view the item without making a yes/no decision. In the words task, activation was slightly more anterior (right BA 9) and was greatest when the item was new and the subject had to note that it was old or new. Thus, with these very minimal processing demands and different stimulus types, dissociations were observed between both processing operation and stimulus type

Interestingly, this set of experiments highlights the point that processes are often fundamentally different from one stimulus type to another. For example, the act of calling to mind a word is likely to involve articulatory mechanisms, whereas the analogous act with an object is likely to engage visual imagery mechanisms. The same holds for verbal and spatial rehearsal mechanisms. Although selective attention has been discussed as a potential mechanism for spatial rehearsal (Awh and Jonides, 2001), the relationship between that and verbal rehearsal is *analogous* but not *equivalent*. Attentional and articulatory rehearsal may accomplish the same thing, but do so through different means. One might say that when it comes to many mental operations, function follows form and to the extent that different "forms" are represented in different parts of cortex, different functions must be as well.

6. CONCLUSIONS

The evidence found in the neuroimaging literature seems to support a division of PFC that is consistent with a loose organization based on stimulus modality. This evidence is not entirely consistent, but the general trend across studies seems to favor three principles. The first is that verbal working memory relies on left-hemisphere mechanisms, whereas spatial working memory relies on right-hemisphere mechanisms. The second is that object working memory is less lateralized than either spatial working memory or verbal working memory. This may be because objects occupy a representational "middle-ground", or because subjects simply use a mixture of strategies to maintain them in memory. The third principle is that working memory for objects is mediated by more left ventral PFC regions and spatial information is mediated by more dorsal regions, especially on the right. These findings imply that the "what" and "where" processing division extends into PFC.

One important question is why our conclusions are different from those found in other reviews of this topic (e.g. D'Esposito et al., 1998; Postle, 2000; Fletcher and Henson, 2001), but consistent with others (e.g. Smith and Jonides, 1999). One contributing factor is that the patterns of results shift over time as new studies become available, and here we have included a number of studies not included in the earlier reviews (e.g. Johnson et al., 2003; Sala et al., 2003). The second reason is that the object-working memory results are particularly inconsistent, possibly due to differences in the types of objects that have been used. The third is that some earlier meta-analyses have relied on qualitative description and visualization as a means of determining whether or not prefrontal divisions exist, and complicated patterns are difficult to detect by such means. The final factor is that we have used Talaraich coordinates in our analyses of dorsal and ventral PFC activation, and small portions of BA 9 and 46 in humans extend into areas that are quite ventral. Meta-analyses relying on Brodmann's areas without taking into consideration how superior or inferior the activation lies have classified some very ventral activations as "dorsal" because BA 9 and 46 are often regarded as "dorsolateral" PFC. Our approach takes into account how superior or inferior an activation is without regard for Brodmann's area.

To the extent that different processes are needed to perform the same function (e.g. subvocal rehearsal and selective attention as mechanisms for rehearsal in the service of memory), there is also a division of the PFC by processing operation. It is not yet certain exactly what computations give rise to these divisions, and it is possible that such divisions are an emergent property of PFC interactions with other regions of the brain; an idea consistent with the views of Miller and Cohen (2001). Although we have suggested that functions follow form, there may be other executive processes that are modality-neutral (like inhibiting an already-chosen motor response, or maintaining task goals). Thus, although we have not ruled out the possibility that the frontal cortex is divided by processing type, we have found support for modality-specific activation in the frontal lobes in working memory.

REFERENCES

Anokhin AP, Birbaumer N, Lutzenberger W, Nikolaev A, Vogel F (1996) Age increases brain complexity. Electrocencephalogr Clin Neurophysiol 99:63–68.

Awh E, Jonides J (2001) Overlapping mechanisms of attention and spatial working memory. Trends Cogn Sci 5:119–126.

Awh E, Jonides J, Smith EE, Schumacher EH, Koeppe RA, Katz S (1996) Dissociation of storage and rehearsal in verbal working memory: evidence from positron emission tomography. Psychol Sci 7:25 – 31.

Awh E, Jonides J, Smith EE, Buxton RB, Frank LR, Love T, Wong EC, Gmeindl L (1999) Rehearsal in spatial working memory: evidence from neuroimaging. Psychol Sci 10:433–437.

Baddeley AD (1986) Working Memory. OUP, Oxford.

Baddeley AD, Hitch G (1974) Working memory. In: Recent Advances in Learning and Motivation, Vol 8 (Bower GA, ed), pp47-89. New York: Academic Press.

Baddeley AD, Grant S, Wight E, Thomson N (1975) Imagery and visual working memory. In: Attention and Performance V (Rabbitt PMA and Dornic S, eds), pp205–217, Academic Press, London.

Baddeley AD, Lewis VJ, Villar G (1984) Exploring the articulatory loop. Q J Exp Psychol 36:233–252.

* Belger A, Puce A, Krystal JH, Gore JC, Goldman-Rakic P, McCarthy G (1998) Dissociation of mnemonic and perceptual processes during spatial and nonspatial working memory using fMRI. Hum Brain Mapp 6:14–32.

Benson DF (1986) Aphasia and lateralization of language. Cortex 22:71–86.

* Braver TS, Cohen JD, Nystrom LE, Jonides J, Smith EE, Noll DC (1997) A parametric study of prefronal cortex involvement in human working memory. Neuroimage 5: 49–62.

Brodmann K (1909) Vergleichende Localisationslehre der Grosshirnrinde in ihren Prinzipien dargestellt auf Grund des Zellenbaues. Barth, Leipzig.

Carpenter PA, Just MA, Shell P (1990) What one intelligence test measures: a theoretical account of the processing in the Ravens Progressive Mattrices Test. Psychol Rev 97:404–431.

Carpenter PA, Just MA, Keller TA, Eddy W Thulborn K (1999) Graded functional activation in the visuospatial system with the amount of task demand. J Cogn Neurosci 11:9–24.

Cavada C, Goldman-Rakic PS (1989) Posterior parietal cortex in rhesus-monkey: II. evidence for segregated corticocortical networks linking sensory and limbic areas with the frontal lobe. J Comp Neurol 287:422-445.

Chao LL, Haxby JV, Martin A (1999) Attribute-based neural substrates in temporal cortex for perceiving and knowing about objects. Nat Neurosci 2:913–919.

Cohen JD, Servan-Schreiber D (1992) Cortex, context and dopamine: a connectionist approach to behavior and biology in schizophrenia. Psychol Rev 99:45–77.

* Courtney SM, Ungerleider LG, Keil K, Haxby JV (1996) Object and spatial visual working memory activate separate neural systems in human cortex. Cereb Cortex 6:39–49.

* Courtney SM, Petit L, Maisog JM, Ungerleider LG, Haxby JV (1998) An area specialized for spatial working memory in human frontal cortex. Science 279:1347–1351.

Cowan N (2000) The magical number 4 in short-term memory: A reconsideration of mental storage capacity Behav Brain Sci 24:87–185.

* D'Esposito M, Aguirre GK, Zarahn E, Ballard D, Shin RK, Lease SJ (1998) Functional MRI studies of spatial and nonspatial working memory. Cogn Brain Res 7:1–13.

D'Esposito M, Postle BR, Ballard D, Lease J (1999) Maintenance versus manipulation of information held in working memory: an fMRI study. Brain Cogn 41:66–86.

Falzi G, Perronne P, Vignolo LA (1982) Right-left asymmetry in anterior speech region. Arch Neurol 39:239–240.

Fletcher PC, Henson RNA (2001) Frontal lobes and human memory: insights from functional neuroimaging. Brain 124:849–881.

Fuster JM (1997) The Prefrontal Cortex: Anatomy, Physiology, and Neuropsychology of the Frontal Lobe. Lippincott-Raven Publishers, New York.

Gazzaniga MS (1970) The Bisected Brain. Appleton-Century-Crofts, New York.

Gazzaniga MS (1983) Right hemisphere language following brain bisection: a 20-year perspective. Am Psychol 38:547–549.

Gazzaniga MS, LeDoux JE (1978) The Integrated Mind. New York: Plenum Press.

Gazzaniga MS, Smylie CS (1983) Facial recognition and brain asymmetries: clues to underlying mechanisms Ann Neurol 13:536–540.

Gazzaniga MS, Sperry RW (1967) Language after section of the cerebral commissures. Brain 90:131–148.

Goldman-Rakic PS (1987) Circuitry of primate prefrontal cortex and regulation of behavior by representation memory. In: Handbook of Physiology, vol 5, The Nervous System (Plum F and Mountcastle VB, eds), pp373–417, American Physiological Society, Bethesda MD.

Hautzel H, Mottaghy FM, Schmidt D, Zemb M, Shah NJ, Muller-Gartner HW, Krause BJ (2002) Topographical segregation and convergence of verbal, object, shape, and spatial working memory in humans. Neurosci Lett 323:156–160.

Johnson MK (1992) MEM: mechanisms of recollection. J Cogn Neurosci 4: 268–280.

* Johnson MK, Raye CL, Mitchell KJ, Greene EJ, Anderson AW (2003) fMRI evidence for organization of prefrontal cortex by both type of process and type of information. Cereb Cortex 13:265–273.

Jonides J (1995) Working memory and thinking. In: An Invitation to Cognitive Science: Thinking, vol 3 (Smith EE and Osherson DN, eds). pp215-233, MIT Press, Cambridge MA.

* Jonides J, Smith EE, Koeppe RA, Awh E, Minoshima S, Mintun MA (1993) Spatial working memory in humans as revealed by PET. Nature 363:623–625.

* Jonides J, Schumacher EH, Smith EE, Lauber E, Awh E, Minoshima S, Koppe RA (1997) Verbal working memory load affects regional brain activation as measured by PET. J Cogn Neurosci 9:462–475.

* Jonides J, Smith EE, Marshuetz C, Koeppe RA, Reuter-Lorenz PA (1998) Inhibition of verbal working memory revealed by brain activation. Proc Natl Acad Sci USA 95:8410–8413.

Kimura D (1973) Asymmetry of human brain. Sci Am 228:70-78

Levy R, Goldman-Rakic PS (2000) Segregation of working memory functions within the dorsolateral prefrontal cortex. Exp Brain Res 133:23–32.

* Marshuetz C, Smith EE, Jonides J, DeGutis J, Chenevert TL (2000) Order information in working memory: fMRI evidence for parietal and prefrontal mechanisms. J Cogn Neurosci 12(S2):130–144.

* McCarthy G, Puce A, Constable RT, Krystal JH, Gore JC, Goldman-Rakic P (1996) Activation of human prefrontal cortex during spatial and nonspatial working memory tasks measured by functional MRI. Cereb Cortex 6:600–611.

Meyer DA, Kieras DE (1999) Precis to a practical unified theory of cognition and action: some lessons from EPIC computational models of human multiple-task performance. In: Attention and Performance XVII. Cognitive Regulation of Performance: Interaction of Theory and Application (Gopher D and Koriat A, eds), pp17–88, MIT Press, Cambridge MA.

Mishkin M, Ungerleider LG, Macko KA (1983) Object vision and spatial vision: two cortical pathways. Trends Neurosci 6:414–417.

Miller GA (1956) The magic number seven, plus or minus two: some limits on our capacity for processing information. Psychol Rev 63:81–97.

Miller EK, Cohen JD (2001) An integrative theory of prefrontal cortex function. Annu Rev Neurosci 24:167–202.

Milner B, Petrides M, Smith ML (1985) Frontal lobes and the temporal organization of memory. Hum Neurobiol 4:137–142.

Newell A (1992) Unified theories of cognition and the role of SOAR. In: SOAR: A Cognitive Architecture in Perspective (Michon JA and Akyurek A, eds), pp25–79, Kluwer, Amsterdam.

* Nystrom LE, Braver TS, Sabb FW, Delgado MR, Noll DC, Cohen JD (2000) Working memory for letters, shapes, and locations: fMRI evidence against stimulus-based regional organization in human prefrontal cortex. Neuroimage 11:424–446.

Owen AM (1997) The functional organization of working memory processes within the human lateral frontal cortex: the contribution of functional neuroimaging. Eur J Neurosci 9:1329–1339.

Owen AM, Evans AC, Petrides M (1996) Evidence for the two-stage model of spatial working memory processing within the lateral frontal cortex: a positron emission tomography study. Cereb Cortex 6:31–38.

* Owen AM, Stern CE, Look RB, Tracey I, Rosen BR, Petrides M (1998) Functional organization of spatial and nonspatial working memory processing within the human lateral frontal cortex. Proc Natl Acad Sci USA 95:7721–7726.

* Owen AM, Herrod NJ, Menon DK, Clark JC, Downey SPMJ, Carpenter A, Minhas PS, Turkheimer FE, Williams EJ, Robbins TW, Sahakian BJ, Petrides M, Pickard DJ (1999) Redefining the functional organization of working memory processes within human lateral prefrontal cortex. Eur J Neurosci 11:567–574.

* Paulesu E, Frith CD, Frackowiak SJ (1993) The neural correlates of the verbal component of working memory. Nature 362:342–344.

Peterson LR, Peterson M (1959) Short-term retention of individual items. J Exp Psychol 58:193–198.

Petrides M (1994) Frontal lobes and working memory: evidence from investigations of the effects of cortical excisions in nonhuman primates. In: Handbook of Neuropsychology, vol 9 (Boller F and Grafman J, eds), pp59–82, Elsevier, Amsterdam.

* Petrides M, Alivisatos B, Evans AC, Meyer E (1993) Functional activation of the human frontal cortex during the performance of verbal working memory tasks. Proc Natl Acad Sci USA 90:878–882.

* Pollmann S, von Cramon DY (2000) Object working memory and visuospatial processing: functional neuroanatomy analyzed by event-related fMRI. Exp Brain Res 133:12–22.

Postle BR, D'Esposito M (2000) Evaluating models of the topographical organization of working memory function in frontal cortex with event-related fMRI. Psychobiology 28:132-145.
Postle BR, Berger JS, D'Esposito M (1999) Functional neuroanatomical double dissociaton of mnemonic and executive control processes contributing to working memory performance. Proc Natl Acad Sci USA 96:12959–12964.
Postle BR, Stern CE, Rosen BR, Corkin S (2000) An fMRI investigation of cortical contributions to spatial and nonspatial visual working memory. Neuroimage 11:409–432.
* Prabhakaran V, Narayanan K, Zhao Z, Gabrieli JDE (2000) Integration of diverse information in working memory within the frontal lobe. Nature 3:85–90.
Raz N (2000) Aging of the brain and its impact on cognitive performance: integration of structural and functional findings. In: Handbook of Aging, 2nd Ed (Craik FIM, Salthouse TA, eds), pp1–90. New Jersey: Lawrence Erlbaum Associates.
Reuter-Lorenz PA, Baynes K (1992) Modes of lexical access in the callosotomized brain. J Cogn Neurosci 4:155–164.
Reuter-Lorenz PA, Miller AC (1998) The cognitive neuroscience of human laterality: lessons from the bisected brain. Curr Dir Psychol Sci 7:15–20.
* Reuter-Lorenz PA, Jonides J, Smith EE, Hartley A, Miller A, Marshuetz C, Koeppe RA (2000) Age differences in the frontal lateralization of verbal and spatial working memory revealed by PET. J Cogn Neurosci 12:174–187.
* Rypma B, Prabhakaran V, Desmond JE, Glover GH, Gabrieli JD (1999) Load-dependent roles of frontal brain regions in the maintenance of working memory. Neuroimage 9:216–226.
* Sala JB, Rama P, Courtney SM (2003) Functional topography of a distributed neural system for spatial and nonspatial information maintenance in working memory. Neuropsychologia 41:341–356.
* Schumacher EH, Lauber E, Awh E, Jonides J, Smith EE, Koeppe RA (1996) PET evidence for an amodal verbal working-memory system. Neuroimage 3:79–88.
Smith EE, Jonides J (1999) Storage and executive processes in the frontal lobes. Science 283:1657–1661.
* Smith EE, Jonides J, Koeppe RA, Awh E, Schumacher EH, Minoshima S (1995) Spatial versus object working memory: PET investigations. J Cogn Neurosci 7:337–356.
* Smith EE, Jonides J, Koeppe RA (1996) Dissociating verbal and spatial working memory using PET. Cereb Cortex 6:11–20.

Smith EE, Jonides J, Marshuetz C, Koeppe RA (1998) Components of verbal working memory: evidence from neuroimaging. Proc Natl Acad Sci USA 95:876–882.

Smith EE, Marshuetz C, Geva A (2002) Working memory: findings from neuroimaging and patient studies. In: Handbook of Neuropsychology, Vol. 7: The Frontal Lobes (Boller F, Grafman, J, eds), pp55-72. Amsterdam: Elsevier.

Stroop JR (1935) Studies of interference in serial verbal reactions. J Exp Psychol 18:643–662.

Talairach J, Tournoux P (1988) Co-planar Stereotaxic Atlas of the Human Brain. Thieme, New York.

Thatcher RW, Walker RA, Giudice S (1987) Human cerebral hemispheres develop at different rates and ages. Science 236:1110–1113.

Ungerleider LG, Mishkin M (1982) Two cortical visual systems. In: Analysis of Visual Behavior (Ingle DJ, Goodale MA and Mansfield RJW, eds), pp548-586, MIT Press, Cambridge MA.

Ungerleider LG, Courtney SM, Haxby JV (1998) A neural system for human visual working memory. Proc Natl Acad Sci USA 95:883–890.

Warrington E, McCarthy R (1983) Category specific access dysphasia. Brain 106:859–878.

Warrington E, Shallice T (1984) Category specific semantic impairments. Brain 107:829–854.

Webster MJ, Bachevalier J, Ungerleider LG (1994) Connections of inferior temporal areas TEO and TE with parietal and frontal cortex in macaque monkeys. Cereb Cortex 5:470–483.

Wilson FA, O'Scalaidhe SP, Goldman-Rakic PS (1993) Dissociation of object and spatial processing domains in primate prefrontal cortex. Science 260:1955–1958.

Yakovlev PV, Lecours AR (1967) The myelogenetic cycles of regional maturation of the brain. In: Regional Development of the Brain in Early Life (Minkowski A, ed), pp3-70, Davis, Philadelphia.

Zaidel E (1985) Language in the right hemisphere. In: The Dual Brain: Hemispheric Specialization in Humans (Benson DF and Zaidel E, eds), pp205-231, Guilford Press, New York.

* indicates inclusion in statistical analyses.

Acknowledgements

The authors wish to acknowledge the helpful comments of Marcia K. Johnson and Karen Mitchell on an earlier draft of this chapter. This work was in part supported by publication grant to CM from the American Association of University Women.

Index

E

F

K

L

M

N

O

P